ECONOMIC ANALYSIS OF TELECOMMUNICATIONS

Theory and Applications

ECONOMIC ANALYSIS OF TELECOMMUNICATIONS

Theory and Applications

edited by

Léon COURVILLE
L'Institut d'Economie Appliquée
Ecole des Hautes Etudes Commerciales
Montreal, Canada

Alain DE FONTENAY
Department of Communications
Government of Canada
Ottawa, Canada

Rodney DOBELL
School of Public Administration
University of Victoria
Victoria, Canada

1983

NORTH-HOLLAND
AMSTERDAM · NEW YORK · OXFORD

ISBN: 0 444 86674 4

Published by:
ELSEVIER SCIENCE PUBLISHERS B.V.
P.O. BOX 1991
1000 BZ AMSTERDAM
THE NETHERLANDS

Sole distributors for the U.S.A. and Canada:
ELSEVIER SCIENCE PUBLISHING COMPANY, INC.
52 VANDERBILT AVENUE
NEW YORK, N.Y. 10017

Library of Congress Cataloging in Publication Data
Main entry under title:

Economic analysis of telecommunications.

"Papers [selected from those] presented at a Conference on Economic Analysis of the Telecommunications Industry in Canada held at Ecole des Hautes Etudes Commerciales in Montreal, March 4, 5, 6, 1981"--Introd.
1. Telecommunication--Canada--Congresses.
2. Telecommunication--Economic aspects--Congresses.
I. Courville, Léon. II. De Fontenay, Alain.
III. Dobell, Rodney. IV. Conference on Economic Analysis of the Telecommunications Industry in Canada (1981 : Ecole des hautes études commerciales, Montréal, Québec)
HE7814.E26 1983 384'.0971 83-8070
ISBN 0-444-86674-4

PRINTED IN THE NETHERLANDS

INTRODUCTION

The papers contained in this book were selected from amongst a much larger set of papers presented at a Conference on Economic Analysis of the Telecommunications Industry in Canada held at Ecole des Hautes Etudes Commerciales in Montreal March 4, 5, 6, 1981. That conference was organized jointly by professors Leon Courville of Ecole des Hautes Etudes Commerciales and Professor Rodney Dobell of the University of Victoria, under the sponsorship of the Department of Communications, Government of Canada, in collaboration with Dr. Alain de Fontenay of that department.

The purpose of the conference itself was to take stock of past economic research into characteristic features of the telecommunications sector in Canada, to draw out the results of primary interest, and to link these to major questions of corporate management or public policy. The goal was to provide a careful appraisal of results to date, and a link to usable policy conclusions. By examining the validity of contemporary analytical work, it was hoped, participants would be able to identify key topics for further research.

In editing the present volume it was necessary to act ruthlessly in selecting and condensing a relatively small amount of material from the 1600 pages of unedited proceedings. This selection was accomplished, so far as possible, in such a way as to draw out the main themes or lessons for management from the vast body of existing statistical and econometric work on telecommunications. The resulting material has been organized in three main parts: production analysis, demand analysis, and welfare considerations and regulation. Within each part, the line of argument flows from descriptive statistical background, through more theoretical analysis, to implications for managers or regulators.

Because of stringent pressures on space, introductory editorial observations are held to the minimum necessary to introduce the papers in each section and set them into some perspective.

It is perhaps appropriate that the conference organizers (and present editors) record here their gratitude to all the contributors from government, industry, research institutes, and universities who worked very hard both in preparation and in discussion of papers and commentary. In particular it is significant to note that major companies in the industry were extraordinarily open and helpful in making material available for inclusion in papers presented to the conference, as well as in financial contributions by Teleglobe, the Government of Quebec, Bell Canada, and Quebec Telephone toward overall sponsorship of the conference.

It is appropriate as well to acknowledge the individual help of Francine Lefebvre and Kaaren Cummings, without whose efforts the conference could not have been held, nor this volume produced.

As noted above, the book is divided into three major parts. The first begins with an up-to-date survey of recent results in the analysis of production conditions for telephone services, with particular attention to conflicting estimates of economies of scale or scope and estimates of rates of technological progress. Applications to management decisions are developed, drawing on work undertaken within Electricite de France and A.T. & T., as well as Bell Canada and other Canadian telephone companies. These center on information systems incorporating estimates of total factor productivity routinely reported on a regular basis as part of the normal financial information system.

The second part is devoted to analysis of demand conditions. Beginning with a survey of recent results in the modelling of demand for telephone services, the analysis carries on to new estimates of demand for long-distance services and a global model of residential demand. Particular attention then is directed to interpretation of the results from experiments in local measured services, the implications of two-part tariff structures, and the value of a measured service option open to the customer at the end of each billing period.

Part III develops some welfare implications of pricing decisions, emphasizing theoretical and empirical aspects of cross-subsidy problems, along with new work on the investment and financing decisions of a regulated firm.

PART I - PRODUCTION ANALYSIS

In the first section of Part I, devoted to an analysis of production conditions in telecommunications, three papers explore the statistical evidence on economies of scale or scope (variety) and on rates of technological change. They suggest that recent work with flexible functional forms and dual cost functions in place of production functions leads to sharper (and possibly larger) estimates of the extent of scale economies, while at the same time raising questions whether profit maximization or cost minimization models are more appropriate, and whether work can usefully proceed on the basis of aggregate ouptut measures.

Melvyn Fuss observes that estimates of rates of technological change and scale elasticities will always be confounded in time-series data for the period of post-war expansion up to 1980, and suggests that one could account for the observed residual growth of revenues in excess of the value of inputs by adopting either a high estimate of scale elasticity coupled with a very low estimate of technological change, or an assumption of constant returns to scale coupled with an estimate of technological change at a rate equal to the highest rates experienced in Canadian manufacturing industry. He concludes that scale elasticity estimates substantially above unity imply rates of technological change which seem to be unreasonably low when compared with estimates drawn from Canadian manufacturing.

Lau Christensen and his collaborators, on the other hand, base their estimates on a single-output cost function (using data for the U.S. Bell system) and report robust evidence of large economies of scale, with most estimates of scale elasticities falling between 1.4 and 1.6.

The results of Ferenc Kiss and his collaborators, beginning with a 3-output, 3-input generalized translog (GTL) cost function, confirm the date problems involved in the disaggregation of output (revenue) data in the telecommunications sector. They do, however, report a consistent picture from work with a two-output model, results which again suggest substantial overall economies of scale.

A first step toward a translation of these general econometric results into a form more directly applicable in management and regulatory decisions is taken with the estimation of rates of increase of total factor productivity, taken as one overall measure of economic performance for a corporation. The paper of Ferenc Kiss, "Productivity Gains in Bell Canada", introduces some of the issues of productivity measurement faced by telephone compagnies, and derives an estimate of about 3.5 percent per year productivity gain in Bell Canada over the period 1958-1980. Analysis of the pattern of productivity increase from year to year suggests an association of productivity increase with the introduction of particular technical changes, but overall, the paper reports, only about 25 percent of Bell Canada's productivity gains was generated directly by technological change: about three-quarters was due to economies of scale realized as a result of growing demand.

The paper of Denny, de Fontenay and Werner adopts a comparison across firms, and finds estimates comparable to those of Kiss for productivity growth in Bell Canada but derives larger rates of productivity increase for Alberta General Telephone Company and B.C. Telephone Company. These somewhat tentative results suggest a number of topics for fruitful work on interfirm comparisons.

In the third section of Part I, the translation of these total factor productivity calculations into corresponding dollar impacts and hence into a tool of direct use to management is completed. The paper of J.N. Reimeringer of Electricite de France reports on the pioneering work of that agency in the development of economic performance measures and their incorporation into a conventional management decision framework. In particular, the use of TFP targets as a device for the devolution of management authority will be of interest to regulatory bodies. Similar applications, including the integration of the factor productivity calculations with standard financial statements of income and expenditure, are described by Chaudry and Burnside of A.T. & T., with detailed illustrations simulating the consequences of alternative productivity trends.

Attempting to relax the restrictive framework of the ex post accounting identity on which the NIPA analysis of Chaudry and Burnside is based, Denny, de Fontenay and Werner introduce a target rate of return of investment, and thereby transform the factor productivity calculation from a decomposition of an accounting identity to a target setting device for planning purposes.

Thus, Part I provides a survey -- with applications primarily to Canadian Telephone Companies -- of the current state of play in the analysis of production conditions, productivity increases, technological change, and economies of scale. The impression left is one of substantial overall economies of scale, realized in large part through the introduction of technological change in response to rising demand, and leading to high observed rates of productivity increase. These econometric features may be open to some dispute because the sophistication of the models and statistical specifications employed press far beyond the limits of available data. And certainly it appears clear that estimates of scale elasticity are highly sensitive to changes in the specification of production conditions.

But in any case the development of performance indicators based upon a decomposition of accounting data into price and quantity components leading to implied productivity statistics or targets, and a calculation of the dollar value of physical productivity gains, can only be of growing interest to management and regulatory bodies in a world increasingly concerned with monitoring and improving productivity performance. Whether one sees this device as a way of enabling senior management to deal with major issues while delegating operating responsibilities, or a way of specifying what contributions enterprises should make to their parent company, or a way for the government to regulate, directly or through agencies, the activities of state enterprises and private corporations, target setting and monitoring on the basis of total factor productivity measures is clearly one application of academic production analysis which appears to have demonstrated its utility in practice.

PART II - DEMAND ANALYSIS

Part II is devoted to the analysis of demand for telephone services, taking into account both the demand for access to the communications network and demand for utilization of particular services. A survey of the major results of statistical and econometric work to date is offered by Lester Taylor and followed up by a brief account by Marshall Lee and Alain de Fontenay of further empirical investigations in Canada.

Taylor's survey suggests that just as there are critical uncertainties (noted above) surrounding the presence of economies of scale in production of telecommunications services, and hence the character of the industry as a natural monopoly, so there are critical uncertainties surrounding estimates of the price and income elasticities of demand for those services. For long haul toll calls Taylor interprets the evidence to suggest that demand may be price elastic, and is certainly income elastic. Nevertheless, the need for more empirical work is clear. Much more important, in his view, is the need for vastly expanded research into the characteristics of demand for access, local use, and terminal equipment (as discussed in section III below). So far as the evidence to date is concerned, Taylor concludes the demand for access to the telephone network is probably high income and price elastic, while local calling is price inelastic but income elastic.

The Lee - de Fontenay paper proposes a more elaborate model, with income and price elastitities varying with distance, to describe B.C.-Alberta interprovincial calling. The results, while not definitive, provide some indication that -- in contrast to some other evidence before Canadian regulatory bodies -- long distance calling is price elastic. The importance, for policy purposes, of work to resolve this empirical controversy is obvious.

Moving on into new theoretical territory, a paper originally contributed in french examine the threshold income giving rise to a demand for access to the telecommunication network. The paper of Curien and Vilmin uses a classical microeconomic model to determine the threshold income at which a utility maximizing individual would pay a specified access price, and then introduces an income distribution to permit aggregation across all individuals and thus derivation of an overall access demand and traffic demand. An econometric model based on this theoretical structure is developed and its results compared against historical experience before being used for a forecast of demand in France to 1985.

The third section of this portion of the book devoted to demand analysis offers a direct response to Taylor's plea for more work on demand for access and local services. In the first of two papers reporting on some of the fascinating recent experimental work in the field of demand analysis, Wilkinson studies the impact of a shift to local measured service -- usage-sensitive pricing -- on aggregate measures of demand such as monthly calls and minutes per mainstation. Data were drawn from local measured service experiments conducted in three exchanges within the GTE system. On the basis of deviations from trends established prior to the introduction of measured service, Wilkinson infers that significant repression of residence local calling, but not of business local calling, results from the introduction of measured service.

In the second paper based directly on experimental data, Wong proposes a procedure for estimating the shift in the entire subscriber distribution of monthly local telephone usage in response to a change to usage sensitive pricing. He finds some support for his estimates from empirical observations in a Denver, Colorado flat-to-measured service conversion in 1971 (Metropac). He also finds encouraging consistency with the results reported by Wilkinson. Further development of analysis based on such experimental and quasi-experimental data appears likely to be an important priority in the future.

Following this analysis of experimental data, it is natural to ask under what circumstances individual consumers would pay more or less following a shift to measured service. To answer this question, it is necessary to be more precise as to whether the consummer may elect to be billed at either a flat or measured rate and whether this choice must be made in advance or can be postponed until after the monthly telephone usage is known. Dansby provides an elegant discussion of these questions in his examination of the value of a measured service option. In particular, he demonstrates that an ex post billing option (in which consumers pay a premium to have their bill computed each billing period according to the two part tariff which yields the smaller expenditure) need not, when the firm's profit are taken into account along with consumers' surplus, provide an unambiguous increase in aggregate welfare. For some demand distributions, use of this ex post option will reduce profits and aggregate welfare.

In two further papers the design of rate structures for local telephone calls is explored. Mitchell starts from the technology and cost characteristics of the local network, and emphasizes the desirability of some form of peak load pricing within the context of some workable rate structure. Brander and Spencer, on the other hand, begin from the theory of non-uniform pricing in the case of a public utility capable of discriminating among different types of consumers, but conclude that the efficiency gains from adopting Ramsey-optimal pricing structures are likely to be rather small (and the transfers or cross-subsidies rather large): they voice the suspicion that a simple usage-sensitive pricing structure incorporating peak load prices close to marginal costs (with subsidy free licence fees to make up any deficits) may be the best approach to pricing in light of differing demand conditions.

PART III - WELFARE CONSIDERATIONS AND REGULATION

The first section of Part III follows directly from the discussion of local telephone pricing with which Part II closed. Rheaume explores the relationship between Ramsey prices and subsidy-free prices, and develops a theoretical model to generate welfare-optimal prices that are subsidy-free. In place of the willingness to pay principle which lies behind Ramsey pricing, Rheaume embraces a principle of anonymous equity equivalent to subsidy-free pricing, and elaborates a welfare model which achieves anonymous equity efficiently. thus deriving the welfare-optimal subsidy-free prices of his title.

For operational purposes, tests to identify the presence of cross-subsidization may take many forms. Autin and Leblanc employ a large simulation model of the TCTS telecommunications system in order to evaluate empirically the significance of cross-subsidy problems according to four selected tests. Subject to a number of qualifications both with respect to model structure and empirical content, they find no evidence of cross-subsidy in the tests applied at the level of the national system.

The second section of Part III moves on to examine investment and financing decisions for a regulated firms. This section includes three papers which are concerned with areas of interest in both finance and economics. Indeed, it has long been recognized that the study of regulation isolated financial decisions from production decisions. These papers examine the interrelation between the two. The paper of Bernstein investigates the behavior of the firm regulated by a general constraint on the sources and uses of funds. The paper links somewhat financial and output decisions as the capital structure of the regulated firm and its investment decisions are determined within the context of the models developed in the paper. The paper has a set of interesting conclusions: the most important is that the regulated will underutilize internal funds as a result of the constraint placed on the flow of funds. The second paper of the section is that of Berkowitz and Cosgrove. They set out to integrate financial and "real" aspects of the regulated firm. Their modelling allows for a simultaneous determination of the level of capacibbty, the means of financing investment and the price of output of the regulated firm. They postulated a regulatory framework whereby rate of return regulation is enforced but calculated on the basis of marginal costs as opposed to embedded costs. They show that regulation leads to production inefficiency as well as financial inefficiency. The regulated firm will underutilize debt as a mean of financing. This result is at variant with that of Bernstein. Part of this divergence is due to differences in regulatory rules postulated by the respective papers. As both papers look at alternative regulatory processes, their different conclusions is a source of debate as to the appropriate type of regulation.

The third paper also formally addresses the issue of the interrelation of production and financial decisions. The theory of arbitrage pricing enables professor Perrakis to derive and characterize the value of the regulated firm: contrary to most work done in finance, he investigates backward-looking regulation and finds that, in the context of value maximization, the AJ results do not hold and that there is a positive leverage effect. Professor Perrakis extends his one period model to a multi-period model; he notes that backward-looking regulation has disturbing implications. His models show that cost-padding behavior is optimal; also, regulators can affect whether or not capital gains or losses are transferred to stockholders and the magnitude of the transfer depending on how regulation react to observed changes in realized rates of return.

Thus this tour of recent work, through research on production conditions and management applications of productivity analysis (for either corporate or regulatory decisions), demand analysis and applications to decisions on pricing structures, welfare analysis and appraisal of regulatory processes, comes to a close with recognition that regulatory bodies are unlikely to have sufficient information on production, cost, and demand conditions, but perhaps must seek much more information on public perceptions and political preferences. Given the lack of convergence of results in the already highly-sophisticated research on economic aspects of telecommunications, this broadening of the scope of analysis can leave no one optimistic that the end of the road for research into the telecommunications industry or the applications to telecommunications regulation is near at hand.

Montreal, February 1983.

Leon Courville
Rod Dobell
Alain de Fontenay

CONTENTS

PART 2 : DEMAND ANALYSIS

Modelling and Telecommunications Demand

New Theoretical Developments

Local Measured Service

PART 3 : WELFARE CONSIDERATIONS AND REGULATION

Pricing Implications

Investment Decisions Under Regulation

PART 1
PRODUCTION ANALYSIS

Production Conditions and the "Residual"

Economic Analysis of Telecommunications:
Theory and Applications
L. Courville, A. de Fontenay and R. Dobell (eds.)

A SURVEY OF RECENT RESULTS IN THE ANALYSIS OF PRODUCTION CONDITIONS IN TELECOMMUNICATIONS

M.A. Fuss

University of Toronto

1. Introduction

Recent advances in the econometric literature (utilizing the duality between cost and production) have made possible representation of the telecommunications production process by a structure of technology sufficiently general so as to capture its important features. The purpose of this paper is to survey the empirical results emanating from the application of these relatively new techniques to telecommunications.

In general there are three characteristics of production which are of interest: (1) factor substitution possibilities, (2) output expansion (scale) effects, and (3) the rate and bias of technological progress. The second characteristic has traditionally been given the most attention in telecommunications since it is closely connected to the natural monopoly question. However, estimation of the rate of technical change can also be an input into the natural monopoly issue in a dynamic context. Finally, biases in technical change and factor substitution characteristics have implications for capital accumulation and employment opportunities.

The studies analysed in this survey utilize second order flexible functional forms. The major advantage of these forms is their ability to represent multi-input, multi-output production processes characterized by variable elasticities of substitution and transformation, non-homothetic output expansion effects, and biased technical change. Earlier attempts to estimate the telecommunications production process employed functions which implied unitary, or at least constant elasticities of substitution and transformation, homogeneous expansion effects, and Hicks-neutral technical change (Dobell et al (1972), Mantell (1974), Vinod (1976), Waverman (1976)). The empirical results contained in papers reviewed in this survey demonstrate that the above restrictions are inappropriate for telecommunications. Formal tests carried out by Fuss and Waverman (1977), Denny et al. (1979) and Nadiri and Shankerman (1979) confirm these less rigorous impressions.

2. Choice of Behavioural Model and Functional Form

2.1 Choice of the Behavioural Model

The empirical literature to be surveyed in this paper places the cost function at the centre of the estimation procedure.[1] The choice of the cost function rather than the production function as the basic building block is due to a number of advantages possessed by the cost function. Most important public policy issues in telecommunications require a knowledge of the cost structure, and the cost function is the most direct way of obtaining this needed information.[2] Since the observation unit is usually the individual firm, factor prices are likely to be more exogenous than factor inputs, reducing the possibility of simultaneous equations bias.[3] It is easier to specify a functional firm which represents a sufficiently general technology using a cost function, particularly when output disaggregation is necessary. Finally, application of Shephard's Lemma provides a direct, simple way of generating a system of factor demand functions. Estimation of the demand system along with the cost function increases the number of observations without increasing the number of parameters. Generation of a system of factor demand functions from the production function is difficult unless constant returns to scale is imposed a priori.

The main disadvantage associated with estimation of the cost function is the need to assume cost minimizing behaviour. Since most telecommunications firms are monopolists in at least one of their service categories, competitive pressures cannot be relied on to force cost-minimizing input choices. In addition, investor-owned telecommunication firms are subject to rate of return regulation which may induce Averch-Johnson effects.[4]

The production of telecommunications services is a capital-intensive process, characterized by the use of physically long-lived capital-equipment much of which is "putty-clay" in nature. Hence a dynamic intertemporal cost-minimizing model with increasing marginal costs of adjustment is appropriate. Such a model has not been estimated in telecommunications. The question arises as to which of two polar cases provides the best approximation to this model: (1) a long-run constant marginal costs of adjustment model (the unrestricted cost function), or (2) a model which does not attempt to explain the time path of capital services and treats capital as exogenous (the restricted cost function). All the studies surveyed in this paper estimate the unrestricted cost function model. Denny et al. (1979) and Christensen et al. (1980) also consider the restricted cost function model. The former set of authors present a detailed conceptual comparison of the two models and conclude that for Canadian telecommunications the unrestricted cost function is the more appropriate approximation.

For investor-owned telecommunications firms, it is reasonable to assume present value (profit) maximizing behaviour, subject to regulatory constraint. Fuss and Waverman (1977) developed an econometric model in which the telecommunications firm chooses the profit maximizing levels of toll services but is constrained by the regulatory autorities to charge a price for basic local service below the profit-maximizing price. This restricted profit maximizing assumption results in an additional estimating equation for each toll service considered. The Fuss-Waverman model has been used by Denny et al (1979), Breslaw and Smith (1980), and Waverman (1980). The last-named authors showed that the model could be obtained from a dynamic specification in which the objective is to maximize the intertemporal utility function of the investor-owners of the firm.

2.2 Choice of the Functional Form

All but one of the studies surveyed employ the translog second order approximation to an arbitrary cost function. The translog approximation to the cost function $C(Z_1...Z_n)$ takes the form

$$(1)\quad \log C(Z_1...Z_n) = \alpha_0 + \sum_{i=1}^{n} \alpha_i \log Z_i + \tfrac{1}{2} \sum_{i=1}^{n} \alpha_{ii}(\log Z_i)^2$$

$$+ \sum_{i \neq j} \sum \alpha_{ij} \log Z_i \log Z_j$$

where the Z_i are the exogenous variables, usually factor prices, outputs and technology shift variables. In the case of the restricted cost function, one of the exogenous variables is capital. Nadiri and Shankerman (1979), Breslaw and Smith (1980), Denny and Fuss (1980), and Christensen et al. (1980) all use the translog model in the form given by (1). Fuss and Waverman (1977) restrict technical change to be capital-augmenting, while Denny et al. (1979) assume technical change is output-augmenting.

Fuss and Waverman (1980) estimated a generalization of the translog model which applies a Box-Cox transformation to the output levels. In this hybrid translog form, the output components of $\ln Z_1 ... \ln Z_n$ are replaced by

$$(2)\quad Q^*_j = \frac{Q_j^\theta - 1}{\theta}$$

where Q_j is output j and θ is a parameter to be estimated. Since $\lim_{\theta \to 0} \frac{Q_j^\theta - 1}{\theta} = \ln Q_j$, the hybrid translog cost function can be used to investigate the effects on the estimated cost structure of departures from the translog maintained hypothesis.

3. Estimation Procedures

The translog cost model is usually estimated as a systems multivariate regression model consisting of the cost function and (n-1) factor share equations, using the Zellner iterative estimation procedure. This approach was adopted by Nadiri and Shankerman (1979) and Christensen et al. (1980). Denny and Fuss (1980) employed the two-stage estimation procedure suggested by Fuss (1977) for the case of a large number of inputs.

The inclusion of profit-maximizing bahaviour adds "revenue share" equations to the system and renders outputs endogenous. In addition, estimates of service demand elasticities are necessary. Simultaneous equations estimation techniques were used by Fuss and Waverman (1977,

Table 1

A Comparison of Basic Features of Studies of Telecommunication Production Technology

Features	Fuss-Waverman (1977)	Denny et al. (1979)	Nadiri-Shankerman (1979)	Breslaw-Smith (1980)
Data Set	Bell Canada, 1952-75	Bell Canada, 1952-76	A.T.&T., 1947-76	Bell Canada, 1952-78
Outputs	Q_1 = message toll Q_2 = private line + WATS + TWX + miscellaneous Q_3 = local	Q_1 = message toll Q_2 = private line + WATS + TWX Q_3 = local + miscellaneous	Single aggregate output	Q_1 = local Q_2 = message toll + WATS
Inputs	Aggregate capital, labour, materials	Aggregate capital, labour, materials	Aggregate capital, labour, mateirals, research and development	Aggregate capital, labour, materials
Functional Form	Translog Cost Function	Translog Cost Function	Translog Cost Function	Translog Cost Function
Technical Change Indicators	Capital augmenting time trend	Output augmenting access to direct distance dialing facilities and conversion to modern switching facilities - 2 indicators	Time trend	Index of switching and accessibility to the system
Behavioural Assumptions	Constrained profit maximization	Constrained profit maximization	Cost minimization	Constrained profit maximization
Method of Estimation	Iterative 3SLS - cost and demand systems estimated separately	Iterative 3SLS - cost and demand systems estimated separately	Iterative Zellner	Full Information Maximum Likelihood - cost and demand systems estimated separately

Table 1 (continued)

Features	Denny-Fuss (1980)	Christensen et al. (1980)	Fuss-Waverman (1980)
Data Set	Bell Canada, 1952-72	A.T.&T. 1947-77	Bell Canada, 1952-77
Outputs	Single aggregate output	Single aggregate output	Q_1 = message toll + WATS Q_2 = private line + TWX Q_3 = local + miscellaneous
Inputs	Aggregate capital, materials 4 occupational labour groups	Aggregate capital, labour, materials	Aggregate capital, labour materials
Functional Form	Two-stage translog cost function	Translog cost function	Hybrid Translog cost function
Technical Change Indicators	Access to direct distance dialing facilities	Distributed lag function of R+D expenditures by A.T.&T.	Output augmenting access to direct distance dialing facilities and conversion to modern switching facilities (2 indicators)
Behavioural Assumptions	Two-stage cost minimization	Cost minimization	Constrained profit maximization
Method of Estimation	Iterative Zellner	Two-stage iterative Zellner	Iterative 3SLS - cost and demand systems estimated simultaneously

1980), Denny et al. (1979), and Breslaw and Smith (1980) to overcome potential simultaneous equations bias. In addition, Breslaw and Smith (1980) and Fuss and Waverman (1980) estimated the factor demand and service demand systems together, incorporating the across-equations constraints implied by the presence of service demand elasticities in both systems. The result is a fully efficient estimation procedure for the Fuss-Waverman model.

Table 1 presents a summary comparison of the basic features, discussed above, of the studies of telecommunications production technology surveyed in this paper. While these studies differ in a number of details they all take as their starting point the duality theory between cost and production as embodied in the cost function.

4. Evidence on Factor Substitution

4.1 Measurement of Factor Substitution

The two most common measures of factor substitution effects are the constant output Allan-Uzawa (A-U) partial elasticity of substitution and the constant output cross-partial elasticity of demand. The A-U elasticity can be calculated from the cost function C as

(3) $$\sigma_{ij} = \frac{CC_{ij}}{C_i C_j}$$

where C_i, C_j and C_{ij} are partial derivatives of the cost function. The cross-partial elasticity of demand for factor i with respect to a change in the price of factor j can be calculated as

(4) $$\varepsilon_{ij} = \frac{\partial \log X_i}{\partial \log p_j} = S_j \cdot \sigma_{ij}$$

where S_j is the cost share of factor j .

I will use the constant output cross-partial elasticities to compare estimates of factor substitution. If $\varepsilon_{ij} > 0$ factors are substitutes, if $\varepsilon_{ij} < 0$, they are complements and if $\varepsilon_{ij} = 0$ they are independent. For the translog and hybrid translog cost functions, ε_{ij} can be calculated as

(5) $$\varepsilon_{ij} = \frac{1}{S_i} \left[\gamma_{ij} + S_i S_j\right]$$

$$\text{where } \gamma_{ij} = \frac{\partial^2 \log C}{\partial \log p_i \, \partial \log p_j}$$

In addition, the own price elasticity of demand (output constant) can be calculated as

$$(6) \qquad \varepsilon_{ii} = \frac{1}{S_i} \left[\gamma_{ii} - S_i + S_i^2\right]$$

$$\text{where} \quad \gamma_{ii} = \frac{\partial^2 \log C}{\partial \log p_i^2}$$

Table 2 contains a summary of the substitution characteristics of aggregate inputs as estimated by the various studies under review. Table 3 presents the own price elasticities of demand, estimated either at the mean of the data set or for a year at the midpoint of the data. Capital and labour and labour and materials are estimated as substitutes in production. Conflicting results have been obtained for capital and materials. However studies reporting both complementary and substitutability characteristics agree that the capital-materials input mix is the least responsive to relative price changes. In all cases the cross-price effects are statistically significant indicating that the choice of inputs by the telecommunications firm is responsive to changes in factor prices. Estimates of own price elasticities vary considerably across studies. Nevertheless there is general agreement that demands for aggregate inputs are inelastic and that the capital input is the least responsive to price changes. This latter fact is hardly surprising since capital equipment in this industry have long physical lifetimes and second hand markets are essentially non-existent.

The only evidence on factor substitution among disaggregated inputs is contained in a study by Denny and Fuss (1980). They disaggregated the labour category into operators, plant craftsmen, clerical workers and white collar (executive and supervisory) personnel. All factors were found to be substitutes except operators and capital and clerical workers and plant craftsmen. Demands were inelastic except for operators and white collar personnel. The operators/capital complementarity is particularly interesting because a large number of studies have found unskilled labour and capital to be substitutes. However this result can be explained by the fact that technical change has led to a substitution of capital for operators which dominates the price complementarity (see section 6). This substitution relationship would have been attributed to a factor price effect if Hicks neutral technical change had been imposed, as was the case with previous studies.

5. Evidence on Scale Effects

Perhaps the most important production characteristic for policy purposes is the behaviour of costs as outputs vary; since this behaviour can establish whether or not a telecommunications firm such as Bell Canada or A.T.& T. is a natural monopoly over some range of its service offerings. Baumol (1977) has refined the definition of a natural monopolist and shown that the basic requirement is that the cost function be "sub-additive". A

Table 2

Factor Substitution Characteristics - Aggregate Inputs

Factors	Fuss-Waverman (1977)	Denny et al. (1979)	Nadiri-Shankerman (1979)	Breslaw-Smith (1980)	Fuss-Waverman (1980)
Capital-Labour	substitutes	substitutes	substitutes	substitutes	substitutes
Capital-Materials	substitutes	complements	substitutes	substitutes	complements
Labour-Materials	substitutes	substitutes	substitutes	substitutes	substitutes

Table 3

Own Price Elasticities of Demand - Aggretage Inputs*

Factors	Fuss-Waverman (1977)	Denny et al. (1979)	Nadiri-Shankerman (1979)	Breslaw-Smith (1980)	Fuss-Waverman (1980)
Capital	-.671	.019 (.026)	-.26 (.04)	-.369	0**
Labour	-.989	-.397 (.044)	-.55 (.06)	-.773	-.437 (.033)
Materials	-1.02	-.541 (.087)	-1.12 (.13)	-.577	-.371 (.060)

* Standard errors in parentheses.

** Constrained estimate. Unconstrained estimation yielded a positive price elasticity which was statistically insignificantly different from zero.

firm's cost function is sub-additive if it can produce any configuration of outputs at a lower cost than that attained by multi-firm production. Baumol shows that a firm may exhibit diseconomies of large-scale production and still be a natural monopolist under the sub-additivity definition, or, conversely if it produces more than one product may exhibit increasing returns to scale and still not be a natural monopolist. Hence the preoccupation with economies of scale to the exclusion of other output characteristics of the cost structure for the multi-product firm is misplaced.

The additional concept that needs to be considered is that of "economies of scope". A production technology exhibits economies of scope when for any configuration of multiple outputs, these outputs can be produced at less cost by a firm which operates a multi-product technology than if the same outputs were produced by a number of firms each operating a single product technology. While the necessary conditions for sub-additivity have yet to be established, Baumol demonstrates that the simultaneous existence of economies of scale and economies of scope are sufficient to insure sub-additivity.[6] Panzar and Willig (1977) have shown that a natural monopolist (defined in terms of a sub-additive cost function) may not be sustainable in the face of competitive entry in one of the multi-product monopolist's markets. A monopolist's pricing strategy is said to be sustainable if it can find a set of stationary product and quantity prices which does not attract rivals into the industry. Baumol, Bailey and Willig (1977) demonstrate that a natural monopolist (again defined as a multi-product firm with a sub-additive cost function) is sustainable if it chooses the Ramsey-optimal rate structure. The Ramsey-optimal rate structure is equivalent to the inverse elasticity rule for quasi-optimal pricing when the cross-price elasticities of demand for the multi-product firm's outputs are zero.

While sub-additivity is the basic cost concept of interest, it is very difficult to test per se. However the sufficient conditions for sub-additivity - economies of scale and economies of scope - are more amenable to the formulation of testable hypotheses. In this section we will survey the recent empirical results concerning economies of scale and scope. We begin with some necessary definitions of scale and scope in terms of characteristics of the cost function.

5.1 Tests of Overall Economies of Scale and Overall Economies of Scope

The starting point for testing the natural monopoly hypothesis is the construction of a test for overall (aggregate) economies of scale. Overall economies of scale exist if an increase in all outputs of λ% leads to a cost increase of less than λ% . As shown by Panzar and Willig (1979) and Fuss and Waverman (1977), local overall economies of scale are measured by the scale elasticity

$$S = \frac{1}{\sum_{j=1}^{n} \varepsilon_{CQ_j}} \qquad (7)$$

where ε_{CQ_j} is the cost-output elasticity of the j-th output. If $S > 1$, economies of scale prevail locally, if $S < 1$ diseconomies of scale prevail and if $S = 1$, constant returns to scale prevail. Of course in the aggregate output specification, $N = 1$.

A global test of economies of scope can be formulated in the following way. Suppose an N output production process can be represented by the joint cost function

$$(8) \quad C = C(Q_1, Q_2, \ldots Q_N)$$

where factor prices and any other arguments of the cost function have been suppressed for simplicity. Overall economies of scope can be determined by comparing the cost of producing each output separately (the "stand alone" cost) with the actual joint cost. The relevant expression is

$$(9) \quad SC = \sum_{j=1}^{N} C_j(Q_j) - C(Q, Q_2, \ldots Q_N)$$

If $SC > 0$, economies of scope exist; if $SC < 0$, diseconomies of scope exist and independent production is cost-minimizing. If $SC = 0$, joint production neither yields cost savings nor causes cost increases.

It should be noted that to compute a global necessary and sufficient test of overall economies of scope requires that one be able to compute stand-alone costs. In telecommunications this would require observations on independent production of outputs such as message toll, competitive and local services. Clearly, we do not have the required set of observations and hence a global test of overall economies of scope is not possible.

A local sufficient test of overall economies of scale is possible. Panzar and Willig (1977) have shown that

$$(10) \quad \frac{\partial^2 C}{\partial Q_i \, \partial Q_j} < 0 \qquad i,j = 1,\ldots N; \quad i \neq j$$

is sufficient for the existence of overall economies of scope. However, as noted by Fuss and Waverman (1977) and Baumol, Fischer and Nadiri (1978), the local nature of this test makes it a very weak one. We conclude that there exists no satisfactory test of overall economies of scope due to data limitations.

5.2 Product-Specific Economies of Scope and Economies of Scale

One particular public policy issue of considerable importance is the question of whether competition in the provision of certain services should be encouraged. Researchers can shed light on this issue by attempting to estimate the extent of product specific economies of scope and economies of scale in the provision of private line services. One requirement for computing product-specific economies of scope is that one observe a production process in which a zero amount of the product under consideration is produced. For private line services in Canada this requirement is approximately met, since Bell Canada produced a very small output of this service in the early 1950's, which is part of the data sample. Unfortunately, a second requirement for computing private line-specific economies of scope is that one observe independent production of this output, so that stand-alone costs can be estimated. The hybrid translog cost function, unlike the ordinary translog function permits the estimation of stand-alone costs. However this estimation requires extrapolation of the cost function well outside the observed data points for toll and local services, and thus considerable caution must be exercised in interpreting the results.

The test for product-specific economies of scope is as follows. Suppose private line service is the j-th service output. Product-specific economies of scope with respect to private line service exist if

$$(11)\quad C(Q_1, Q_2, \ldots Q_{j-1}, 0, Q_{j+1}, \ldots Q_N) + C(0, \ldots 0, Q_j, 0, \ldots 0) - C(Q_1, Q_2, \ldots Q_N) > 0$$

Panzar and Willig (1979) have defined the degree of product specific economies of scope as

$$(12)\quad SC_j = \frac{C(Q_1,Q_2,\ldots Q_{j-1},0,Q_{j+1},\ldots Q_N)+C(0,\ldots 0,Q_j,0,\ldots 0)-C(Q_1,\ldots Q_N)}{C(Q_1,\ldots Q_N)}$$

If $SC_j > 0$, SC_j measures the proportionate increase in cost from separating private line services from the production of other services. If $SC_j < 0$, it measures the proportionate cost decrease from independent production of private line services.

Panzar and Willig (1979) have also proposed a measure of product specific economies of scale. They define the degree of product j specific economies of scale as

$$(13)\quad S_j = \frac{IC_j}{Q_j \frac{\partial C}{\partial Q_j}}$$

where $IC_j = C(Q_1,Q_2,\ldots Q_N) - C(Q_1,\ldots Q_{j-1},0,Q_{j+1},\ldots Q_N)$ is the incremental cost of producing product j. It can be shown that (13) can be written in the form

$$(14)\quad S_j = \frac{IC_j}{C} \bigg/ \varepsilon_{CQ_j}$$

If $S_j > 1$, there exists product j specific economies of scale (locally). If $S_j < 1$, there exists diseconomies of scale and if $S_j = 1$, there exists constant returns to scale.

One final test of product-specific returns to scale is of interest. Suppose private line services are produced by two firms in the amounts $Q^1{}_2$ and $Q^2{}_2$, so that industry output is $Q_2 = Q^1{}_2 + Q^2{}_2$. We are interested in whether the takeover of firm 2's output by firm 1 would allow firm 1 to produce the additional output under increasing returns to scale (declining average incremental cost). Fuss and Waverman (1980) have shown that the degree of returns to scale associated with this takeover can be computed as

(15)
$$\tilde{S}_2 = \left[\frac{IC_2}{C} \Big/ \varepsilon_{CQ_2} \right] \cdot \left[1 - \frac{Q_2^1}{Q_2} \right]$$

where IC_2, C, and ε_{QC} are all evaluated at Q_2. If $\tilde{S}_2 > 1$, then the additional production is subject to increasing returns to scale. If $\tilde{S}_2 <$ 1 decreasing returns to scale prevail and if $\tilde{S}_2 = 1$ the additional production is subject to constant returns to scale. This final test has an obvious application to the Bell Canada-CNCP Interconnection case since it can be used to test whether the efficient market structure (in a static sense) is for Bell to become the monopoly supplier of private line services.

5.3 Testing for the Presence of Economies of Scale and Scope

For flexible functional forms measures of economies of scale and scope are functions of the data as well as the estimated parameters. Hence testing of hypotheses requires one to use the approximation methodology suggested by Denny and Fuss (1977). The usual procedure is to test a hypotheses at the mean by transforming (scaling) the data so that all variables equal unity at the mean observation. for the translog and hybrid translog models, the resulting test statistic is usually a function of parameter estimates alone, since $\log Z_i$ and $\frac{Q_j^\theta - 1}{\theta}$ are both zero when Z_i, $Q_j = 1$.

We will illustrate the procedure for the translog and hybrid translog specifications used by Denny et al (1979) and Fuss and Waverman (1980). The translog specification used by Denny et al (1979) is

(16)
$$\alpha 0 + \sum_i \alpha_i \log p_i + \sum_k \beta_k \log Q^*_k$$

$$+ \tfrac{1}{2} \sum_i \gamma_{ii} (\log p_i)^2 + \sum_i \sum_{\substack{j \\ i \neq j}} \gamma_{ij} \log p_i \log p_i$$

$$+ \tfrac{1}{2} \sum_{k} \delta_{kk} (\log Q^*_k)^2 + \sum_{k \neq \ell} \sum \delta_{k\ell} \log Q^*_k \log Q^*_\ell$$

$$+ \sum_{i} \sum_{k} \rho_{ik} \log p_i \log Q^*_k$$

where P_i is an input price, Q^*_K is a technical change augmented output (see section 6 for details), i,j are indexed over inputs and k, ℓ are indexed over outputs. The aggregate scale elasticity is given by

(17)

$$S = \frac{1}{\sum_{\ell} \varepsilon_{CQ_\ell}} = \left[\sum_{\ell} \quad (\beta_\ell + \sum_{k} \delta_{k\ell} \log Q^*_k + \sum_{i} \rho_{i\ell} \log p_i) \right]^{-1}$$

which reduces to $S = \left[\sum_{\ell} \beta_\ell \right]^{-1}$ at the transformed means.

The hybrid translog cost function used by Fuss and Waverman (1980)[7] is specified as

(18)

$$\log C = \alpha_0 + \sum_{i} \alpha_i \log p_i + \sum_{k} \beta_k \left[\frac{Q^{*\theta}_k - 1}{\theta} \right]$$

$$+ \tfrac{1}{2} \sum_{i} \gamma_{ii} (\log p_i)^2 + \sum_{\substack{i \\ i \neq j}} \sum_{j} \gamma_{ij} \log p_i \log p_j$$

$$+ \tfrac{1}{2} \sum_{k} \delta_{kk} \left[\frac{Q^{*\theta}_k - 1}{\theta} \right]^2 + \sum_{\substack{k \\ k = \ell}} \sum_{\ell} \delta_{k\ell} \left[\frac{Q^{*\theta}_k - 1}{\theta} \right] \left[\frac{Q^{*\theta}_\ell - 1}{\theta} \right]$$

$$+ \sum_{i} \sum_{k} \rho_{ik} \log p_i \left[\frac{Q^{*\theta}_k - 1}{\theta} \right]$$

The aggregate scale elasticity is given by

(19)

$$S = \left[\sum_{\ell} \varepsilon CQ_{\ell}\right]^{-1} = \left[\sum_{\ell} Q_{\ell}^{*\theta} \cdot \left\{\beta_{\ell} + \sum_{k} \delta_{\ell k} \left[\frac{Q_{k}^{*\theta} - 1}{\theta}\right] + \sum_{i} \rho_{i\ell} \log p_{i}\right\}\right]^{-1}$$

which also reduces to $S = \left[\sum_{\ell} \beta_{\ell}\right]^{-1}$ at the transformed means. In general estimates of β_{ℓ} will differ for the two functional forms (16) and (18) thus providing different estimates of aggregate returns to scale.

Local overall economies of scope for the translog model can be tested at the transformed means by computing the test statistics for cost complementarities (Fuss and Waverman (1977))

(20)

$$\frac{\partial^{2} C}{\partial Q_{k}\, \partial Q_{\ell}} = \beta_{k}\, \beta_{\ell} + \delta_{k\ell} \qquad k \neq \ell$$

The identical formula can be used to test for cost complementarities in the case of the hybrid translog function.

The translog cost function is undefined whenever one of the outputs is zero. As we have seen above, necessary and sufficient tests of economies of scope and tests of product-specific economies of scale require a cost function which is defined at zero levels of output. Fuss and Waverman (1980) circumvented that problem through their use of the hybrid translog function. They showed that for the hybrid function the degree of product

j specific economies of scale could be computed at the transformed means as

(21)

$$S_{j} = \frac{\exp[\alpha_{0}] - \exp\left[\alpha_{0} - \frac{\beta_{j}}{\theta} + \frac{\delta_{jj}}{2\theta^{2}}\right]}{\alpha_{j} \cdot \exp[\alpha_{0}]}$$

Similarly $\tilde{S}_{j}$ can be computed as

(22)

$$\tilde{S}_{j} = \frac{\exp[\alpha_{0}] - \exp\left[\alpha_{0} - \frac{\beta_{j}}{\theta} + \frac{\delta_{jj}}{2\theta^{2}}\right]}{\alpha_{j} \cdot \exp[\alpha_{0}]}$$

Finally, Fuss and Waverman (1980) showed that the degree of product specific economies of scope with respect to output 2 (private line services) for a 3 output hybrid translog cost function, calculated at the transformed mean, could be obtained as

(23)

$$SC_2 = \frac{\exp\left[\alpha_0 - \frac{\beta_2}{\theta} + \frac{\delta_{22}}{2\theta}\right] + \exp\left[\alpha_0 - \frac{1}{\theta}(\beta_1+\beta_3) + \frac{1}{2\theta^2}(\delta_{11}+\delta_{33}+2\delta_{13})\right] - \exp[\alpha_0]}{\exp[\alpha_0]}$$

5.4 Evidence on Aggregate Economies of Scale

Table 4 presents a summary of the estimates of aggregate economies of scale, calculated at the mean of the sample (Denny et al (1979), Nadiri and Shankerman (1979), Christensen (1980), Fuss and Waverman (1980)) or a midpoint observation (Fuss and Waverman (1977), Breslaw and Smith (1980)) along with approximate 95% confidence intervals where available.

The estimates for Bell Canada based on the translog cost function appear to indicate that the aggregate scale elasticity is in the neighborhood of 1.4. If this were true a 1% increase in all outputs would result in only a 0.7% increase in (long-run) total costs, a very substantial efficiency effect. The estimates of the scale elasticity for A.T.& T. based on the translog function are even higher. On the other hand, Fuss and Waverman's (1980) estimate based on the hybrid translog model is substantially below the other Canadian estimates and the U.S. estimates. Hence it is important to investigate the relationship between their estimate and the previous ones.

Table 1 demonstrates that the Fuss-Waverman (1980) structure differs from the previous Canadian studies in a number of ways related to data sets, output variable definition, technical change specifications, behavioural assumptions and estimation procedures. Yet from Table 4 it appears that the mean scale elasticity estimate for Bell Canada is invariant to these differences, as long as the translog cost function is used. Only when Fuss and Waverman switch to the hybrid translog specification does the scale elasticity change substantially. Essentially the decomposition of efficiency gains between scale effects and technological change effects is highly sensitive to variation in functional forms, even among second order flexible forms. This fact creates a real dilemma for policy decision-makers who wish to base their decisions, at least in part, on those empirical estimates of scale economies which have been provided by the most current research.

Since the hybrid translog function contains the ordinary translog function as a nested special case traditional methods of statistical inference are available for discriminating among them. Recall that the hybrid tanslog function approaches the ordinary translog function as θ approaches 0. However at $\theta = 0$, the likelihood function becomes degenerate and hence this value cannot be imposed in estimating the hybrid function. Nevertheless the translog function can be approximated as closely as desired by choosing θ close to 0. Fuss and Waverman chose $\theta = 0.01$. At

Table 4

Estimates of Aggregate Scale Economies for Telecommunications Productions

	Fuss-Waverman (1977)	Fuss-Waverman (1977) revised*	Denny et al. (1979)	Nadiri-Shankerman (1979)	Breslaw-Smith (1980)
Point Estimate	1.02	1.45	1.47	2.12	1.29
95% Confidence Region	(1.15, 0.89)		(1.59, 1.37)	(1.75, 2.69)	

	Denny-Fuss (1980)	Christensen et al. (1980)**	Fuss-Waverman (1980) translog	Fuss-Waverman (1980) hybrid translog
Point Estimate	1.46	1.73	1.43	0.94
95% Confidence Region	(2.15, 1.10)	(1.94, 1.56)	(1.63, 1.26)	(1.09, 0.83)

* As reported in Denny et al. (1979) using revised Bell Canada data.

** My best guess as to the preferred estimate - corresponds to Table 6, specification (10) based on Bell R&D Expenditures.

that point, $\frac{Q_j^{*\theta} - 1}{\theta}$ is virtually identical to log Q^*_j. The hybrid translog function with $\theta = 0.01$ and the ordinary translog function yield essentially identical empirical estimates. A likelihood ratio test of the null hypothesis $\theta = 0.01$ yielded the test statistic 11.92. The Chi-squared critical value is 3.84 (6.64) at the 5% (1%) significance level. At any reasonable significance level the null hypothesis was rejected, which implies rejection of the ordinary translog model and its associated estimates of substantial overall scale economies. Fuss and Waverman go on to conclude on the basis of these results that any aggregate economies of scale which exist are modest at best. This is the correct conclusion on the basis of formal statistical inference. However I think that the important lesson to be learned is that we still do not know the extent of aggregate scale economies in telecommunications despite the enormous amount of research effort devoted to that topic. The main value of Fuss and Waverman's research on this issue is to point out the danger of accepting for policy purposes at this time the evidence generated by the ordinary translog cost function estimates - that telecommunications production is subject to substantial increasing returns to scale in the aggregate. Guilkey and Lovell (1980) noted in their Monte Carlo study a tendency of the translog function to overestimate returns to scale. Perhaps this phenomenon is at work here. In any case it is important to determine whether the Canadian and U.S. studies which use a single aggregate output suffer from the same lack of robustness to functional form specification as do the three output Canadian studies.

5.5 Evidence on Economies of Scope and Product-Specific Economies of Scale

In order to provide evidence on economies of scope and product-specific economies of scale one must estimate a multi-output technology. No U.S. studies have appeared as yet which disaggregate output, hence all evidence to date comes from Canadian studies. Fuss and Waverman (1977) using the local test (see equation (10)) find no statistically significant economies of scope. They do find insignificant cost complementarities between local and toll services and between toll and competitive services. Breslaw and Smith (1980) also using (10), found cost complementarities between local and toll services which were "unimportant relative to marginal cost". Fuss and Waverman (1980), using the global test outlined earlier found no statistically significant economies of scope between private line services and the other services. The evidence to date would appear to indicate that cost savings in telecommunications due to economies of scope are, at best, minor relative to aggregate costs.

The only study to investigate product-specific economies of scale was Fuss and Waverman (1980). Using equation (21) they estimated that Bell Canada produced private line services subject to increasing returns to scale. However, from the estimation of equation (22) they determined that these returns to scale would be exhausted if Bell became the monopoly supplier of private line services. They concluded that there was no statistically significant static efficiency-related evidence that competition should not be encouraged in the provision of private line services.

6. Evidence on Technical Change

A number of authors being surveyed have noted that the most difficult problem in estimating cost functions for telecommunications is the specification of technical change. In order to specify technical change one looks for an indicator of shifts in the cost function, i.e., a reason why costs might decline for a given set of factor prices and outputs. The most common technical change indicator used in econometric studies is the passage of time. This method has been used in telecommunications by Fuss and Waverman (1977) and Nadiri and Shankerman (1979). While time is simple to compute, it is itself a rather explicit indication of ignorance regarding the process of technical change. The research and development (R&D) expenditure pattern is another technical change indicator often used in econometric studies. Telecommunications studies which have used R&D effort include Nadiri and Shankerman (1979) and Christensen et al (1980). The main problem with this indicator is that it is a measure of input into the innovative process rather than output from the process. Outputs from the innovative process which have become embodied in telecommunications production include direct distance dialling facilities and improved (modern) switching facilities. These indicators of innovative activity have been used by Denny et al (1979), Denny and Fuss (1980), Breslaw and Smith (1980), Fuss and Waverman (1980) and Christensen et al (1980). They also played an important role in the decomposition analysis of Denny, Fuss and Waverman (1979) where total factor productivity growth was decomposed into scale effects and effects due to technical change-inducing innovative activity. The innovations indicators used have been: (1) the percentage of telephones with access to direct distance dialling facilities (Denny et al, Denny, Fuss and Waverman, Denny and Fuss, Fuss and Waverman), (2) the percentage of long distance calls directly dialed (Christensen et al), (3) the percentage of telephones connected to central offices with modern switching facilities (Denny et al, Denny, Fuss and Waverman, Fuss and Waverman, Christensen et al) and (4), an index combination of (1) and (3) (Breslaw and Smith (1980)). These measures come closest to the spirit of technological change indicators but they suffer from the disadvantage that only a small number of major innovations are covered. Small-scale continuous technical change is not represented, nor is there any indicator which might represent improvements in outside plant (transmission facilities).

The technical change indicators have been incorporated into the specification of the cost function in a number of ways. First, the measure of technical change can be treated as just another variable in the second order expansion (Nadiri and Shankerman (1979), Denny and Fuss (1980), Breslaw and Smith (1980)). Second technical change can be specified as augmenting: capital augmenting (Fuss and Waverman (1977)), all factors augmenting (Christensen et al (1980)) and output augmenting (Denny et al (1979), Denny, Fuss and Waverman (1979), Fuss and Waverman (1980)). At this point there does not appear to be evidence that any one method of incorporating technical change is the superior one.

In cost function models the rate of technical change is measured by the proportionate downward shift of the cost function over time.[8] For Bell Canada during the period 1952-76 this rate has been estimated at 0.8% (Denny et al (1979)). There exists no comparable estimate for A.T.& T. The direct technical change effect estimate given by Nadiri and Shankerman (1979) of 1.2% appears to be comparable but this appearance is misleading. Nadiri and Shankerman did not permit the trend toward a higher scale elasticity over time contained in their estimated structure to affect their decomposition of total factor productivity growth into technical change effects and scale effects. Hence their estimated rate of technical change is upward biased, and potentially seriously so. There is no evidence bearing on the question of whether technical change in telecommunications has been slower in Canada than in the United States.

The estimated rate of technical change can be used as an aid in the evaluation of scale elasticity estimates. Denny et al (1979) have shown that average cost can be decomposed into factor price effects, scale effects, and technical change effects by the formula

$$\dot{C/Q} = \sum_i S_i \dot{p}_i + (S^{-1} - 1)\dot{Q} + \dot{B} \tag{24}$$

where C is cost; Q is output in the single output case and cost elasticity weighted aggregate output in the multi-output case; S is the overall scale elasticity; and B is technical change. The dot represents a rate of change. The rate of change of cost efficiency is given by $\dot{C/Q} - \Sigma S_i \dot{p}_i$ which can easily be calculated from time series data. The difficult problem is to decompose this cost efficiency into scale and technical change effects. The greater the scale elasticity S, the lower the rate of technical change $\dot{B}$. For Denny et al. (1979), $\dot{B}$ = 0.008 (see above) and S = 1.47 (at the mean). May and Denny (1979) have calculated $\dot{B}$ at about 0.010 for total manufacturing during the post World War II period. One would not expect technical change in telecommunications industry to be 20% below that for total manufacturing; hence S = 1.43 is likely to be an overestimate, a fact which is consistent with the analysis of Section 5.

Evidence relating to the bias in technical change among aggregate factors appears to be consistent in Canada and the United States. Denny et al (1979) and Nadiri and Shankerman (1979) found technical change to be capital using and labour saving. Denny et al found technical change to be materials saving whereas Nadiri and Shankerman found it to be materials neutral. With respect to specific technical change indicators, Denny et al found that the diffusion of direct distance dialing facilities through the Bell Canada telecommunications network was capital using and materials and labour saving. In contrast they estimated that the conversion to modern switching facilities resulted in savings with respect to all three factors of production.

Denny and Fuss (1980) also found technical change as represented by access to direct distance dialing facilities to be capital using and labour and materials saving. Among the occupational categories, the severity of the labour-saving impact was felt in inverse relation to the skill level associated with the occupation. Technical change had its strongest effect on the operators category, the category which one could expect to be the most directly influenced by the direct-distance dialing innovation.

7. A Final Overview

Recent empirical studies of telecommunications production have utilized the theory of duality between cost and production and the availability of flexible functional forms to allow for the possibility of general substitution effects, non-homothetic scale effects, and non-neutral technical change effects. An important result of this research activity has been a substantial advance in the level of methodology applied in the telecommunications area. Much has been learned about substitution possibilities and the nature of technical change. Output expansion effects bearing on the natural monopoly question remain controversial. Perhaps this is inevitable, given the highly trended output and technical change indicator data in the typical time series data set which is available in telecommunications. This trending makes the separation of efficiency gains into those due to scale economies and those due to technical change very difficult. Because of this difficulty any researcher trying to establish the extent of scale economies in telecommunications from time series data should report as much detail on technical change estimates as on scale estimates. It is only relative to the reasonableness of technical change estimates that the appropriateness of scale estimates can be evaluated. This fact has not often been appreciated by those doing research in this area, including the present author.

Reliable estimates of economies of scale and economies of scope will probably have to await the development of a pooled time series-cross section data base. Such a data base might be formed from the operating companies of the U.S. Bell System or several Canadian telephone companies. In this latter regard, I note with approval the recent efforts of a number of Canadian telephone companies, in cooperation with the Department of Communications, to begin the construction of the needed data.

FOOTNOTES:

[1] The cost function was first applied to telecommunications by Waverman (1976) who utilized a Cobb-Douglas cost function. Fuss and Waverman (1977) applied the multiple output cost function to telecommunications and were the first to exploit the empirical implications of duality theory in telecommunications applications.

[2] Virtually all regulatory problems in telecommunications are linked to characteristics of the cost structure. Prominent examples are the economies of scale - economies of scope natural monopoly debate featured in the recent CNCP-Bell interconnection case, and the discussion concerning cross-subsidization of basic local service found in recent Bell Canada rate increase application hearings.

[3] Telecommunications firms compete in local and national markets for labour and material inputs. Firms other than Bell Canada purchase equipment in international markets. Bell Canada purchases virtually all its equipment requirements from its subsidiary, Northern Telecom., and may engage in artificial transfer pricing. However, detailed regulatory scrutiny of Bell's equipment purchases probably results in purchase prices which reflect international competitive conditions.

[4] Fuss and Waverman (1977) developed a cost function model which incorporates the rate of return constraint. Fuss and Waverman (1980) attempted to estimate this model for Bell Canada but were unsuccessful. This lack of success suggests that the Averch-Johnson effect may be unimportant in Canadian telecommunications.

[5] Denny and Fuss (1980) employ a two-stage translog specification for which the calculation of the price elasticities are more complex than equations (5) and (6). The reader is referred to their article for details.

[6] Baumol calls these output characteristics "decreasing ray average cost" and "transray convexity" respectively.

[7] This function applies a Box-Cox transformation to outputs. The form used here was first proposed by Caves, Christensen and Tretheway (1980). The use of the Box-Cox transformation has a long history in econometrics. For a recent example of its use in the context of a single output cost function, see Berndt and Khaled (1979).

[8] For the link between the downward shift of the cost function and the upward shift of the production function as measures of technical change see Denny, Fuss and Waverman (1979).

REFERENCES:

[1] Baumol, W.J., On the Proper Cost Tests for Natural Monopoly in a Multiproduct Industry, American Economic Review (December 1977) 809-822.

[2] Baumol, W.J., Bailey, E.E. and Willig, R.D., Weak Invisible Hand Theorems on the Sustainability of Prices in a Multiproduct Monopoly, American Economic Review (June 1977) 350-365.

[3] Baumol, W.J., Fischer, D. and Nadiri, M.I., Forms for Empirical Cost Functions to Evaluate Efficiency of Industry Structure, Paper No. 30, Centre for the Study of Business Regulation, Graduate School of Business Administration, Duke University, Durham, N.C., 1978.

[4] Berndt, E. and Khaled, M., Parametric Productivity Measurement and Choice Among Flexible Functional Forms, Journal of Political Economy 87 (1979) 1220-1245.

[5] Breslaw, J. and Smith, J.B., Efficiency, Equity and Regulation: An Econometric Model of Bell Canada, Report to the Department of Communications, Final Report (March 1980).

[6] Caves, D.W., Christensen, L.R. and Tretheway, M.W., Flexible Cost Functions for Multiproduct Firms, Review of Economics and Statistics (August 1980) 477-481.

[7] Christensen, L.R., Cummings, D. and Schoech, P.E., Econometric Estimation of Scale Economies in Telecommunications, paper presented at the conference "Telecommunications in Canada: Economic Analysis of the Industry" (1980).

[8] Denny, M. and Fuss, M., The Use of Approximation Analysis to Test for Separability and the Existence of Consistent Aggregates, American Economic Review (June 1977) 404-418.

[9] Denny, M. and Fuss, M., The Effects of Factor Prices and

Technological change on the Occupational Demand for Labour: Evidence from Canadian Telecommunications, Institute for Policy Analysis Working Paper No. 8014, University of Toronto (July 1980).

[10] Denny, M., Everson, C., Fuss, M. and Waverman, L., Estimating the Effects of Diffusion of Technological Innovations in Telecommunications: The Production Structure of Bell Canada, paper presented at the Seventh Annual Telecommunications Policy Research Conference April 29 - May 1, 1979, published in the Canada Journal of Economics (February 1981) 24-43.

[11] Denny, M., Fuss, M. and Waverman, L., The Measurement and Interpretation of Total Factor Productivity in Regulated Industries, with an Application to Canadian Telecommunications, paper presented at the National Science Foundation Conference on Productivity Measurement in Regulated Industries, Madison, Wisconsin, April 30 - May 1, 1979; published as Chapter 8 in T. Cowing and R. Stevenson (eds.) Productivity Measurement in Regulated Industries, (New York: Academic Press, 1981) 179-218.

[12] Dobell, A.R., Taylor, L.D., Waverman, L., Liu, T. and Copeland, M.D.G., Telephone Communications in Canada: Demand, Production and Investment Decisions, The Bell Journal of Economics and Management Science 5 (1972) 175-219.

[13] Fuss, M.A., The Demand for Energy in Canadian Manufacturing: An Example of the Estimation of Production Structures with Many Inputs, Journal of Econometrics (January 1977) 89-116.

[14] Fuss, M. and Waverman, L., Regulation and the Multiproduct Firm: The Case of Telecommunications in Canada, paper presented at the NBER Conference on Public Regulation, Washington, December, forthcoming as Chapter 6 in G. Fromm (ed.) Studies in Public Regulation (Cambridge: M.I.T. Press, 1981). This paper also appeared as, Multi-product, Multi-input Cost Functions for a Regulated Utility: The Case of Telecommunications in Canada, Institute for Policy Analysis, Working Paper No. 7810 (1977).

[15] Fuss M. and Waverman, L., The Regulation of Telecommunications in Canada, Draft of the Final Report to the Economic Council of Canada Regulation Reference (June 1980) Final Draft, (February 1981).

[16] Guilkey, D.K. and Lovell, C.A.K., On the Flexibility of the Translog Approximation, International Economic Review (February 1980) 137-148.

[17] Mantell, L.H., An Econometric Study of Returns to Scale in the Bell System, Staff Research Paper, Office of Telecommunications Policy, Executive Office of the President, Washington, D.C. (February 1974).

[18] May, J.D. and Denny, M., Post-War Productivity in Canadian Manufacturing, Canadian Journal of Economics, XII (1979) 29-41.

[19] Nadiri, M.I. and Shankerman, M., The Structure of Production, Technological Change and the Rate of Growth of Total Factor Productivity in the Bell System, paper presented at the N.S.F.

Conference on Productivity in Regulated Industries, University of Wisconsin, April 30 - May 1, 1979; published in T. Cowing and R. Stevenson (eds.) Productivity Measurement in Regulated Industries (New York: Academic Press, 1981).

[20] Panzar, J.C. and Willig, R.D., Free Entry and the Sustainability of Natural Monopoly, The Bell Journal of Economics, vol. 8, no. 1 (1977) 1-22.

[21] Panzar, J.C. and Willig. R.D., Economies of Scope, Product Specific Economies of Scale, and the Multiproduct Competitive Firm, Bell Laboratories Economics Discussion Paper No. 152 (August 1979).

[22] Vinod, H.D. Application of New Ridge Regression Methods to a Study of Bell System Scale Economies, Journal of the American Statistical Association (December 1976).

[23] Waverman, L. The Regulation of Intercity Telecommunications, Chapter 7 in A. Phillips (ed.) Promoting Competition in Regulated Markets, Brookings Institution, Washington, D.C. (1976).

Economic Analysis of Telecommunications:
Theory and Applications
L. Courville, A. de Fontenay and R. Dobell (eds.)

ECONOMETRIC ESTIMATION OF SCALE ECONOMIES IN TELECOMMUNICATIONS

L.R. Christensen, D. Cummings, and P.E. Schoech*

University of Wisconsin-Madison

1. Introduction

In most countries the provision of telecommunications is considered to be a natural monopoly. This view also prevailed in the U. S. for several decades. In the early 1970s, however, the Federal Communications Commission began to permit competitive entry in U.S. telecommunications. This decision is consistent with the view that scale economies are not substantial in telecommunications, and therefore that little if any efficiency would be sacrificed by allowing entry. In an industry as large as telecommunications a small decline in efficiency would represent a substantial welfare loss to consumers. Thus, the question of whether or not there are significant scale economies in telecommunications is vital for the formulation of appropriate public policy.

The two principal approaches to the study of scale economies are engineering cost studies and econometric estimation of the structure of cost and production. The engineering approache employs detailed specifications of technology while the econometric approach provides a broad view of the relationship among the major aggregate economic variables. Thus, the engineering approach is more suitable for studying scale economies in specific services, and the econometric approach is more suitable for assessing the importance of scale economies in the overall provision of telecommunication services. Evidence from aggregate econometric analysis does not in itself provide sufficient information on which to base policy for specific services. On the other hand, evidence on the degree of scale economies for the entire system can provide a useful check on the reasonableness of estimates from the body of engineering analyses.[1] Thus the engineering and econometric approaches are best viewed as domplementary rather than competing methodologies for assessing the importance of scale economics.

In the 1970s there were several econometric studies of scale economies in the U.S. Bell System using the production function approach. Mantell (1974) and Vinod (1976) are examples of such studies, which typically estimated Bell System scale economies to be 1.2 or less. Since the mid-1970s econometric studies of the structure of production have shifted from direct estimation of the production function to estimation of the neoclassical cost function. It is generally agreed that the cost function provides a more attractive stochastic specification for regulated industries, and also provides a more direct approach to the estimation of scale economies.

Recently there have been two cost function studies of North American telecomunications. Denny, Fuss, Everson and Waverman (1979) have analyzed the structure of production for Bell Canada and Nadiri and Schankerman (1980) have investigated the structure of production for the U.S. Bell System. Both studies found scale economies that were much larger than those found in most of the earlier production function studies. Denny et al. reported a mean estimate of scale economies of 1.46 with an estimate of 2.23 for 1976 -- the most recent year in the sample.[2] Nadiri and Schankerman presented two alternative models with mean scale economies of 1.75 and 2.12[3].

The assessment of scale economies in telecommunications was not the primary objective of either the Denny et al. or Nadiri-Schankerman studies. Their estimates of scale economies deserve serious scrutiny, however, since they are far above estimates from production function studies. The purpose of this paper is to investigate the importance of scale economies in telecommunications using data from the U.S. Bell System. Like Denny et al. and Nadiri and Schankerman, we exploit the neoclassical cost function; but our study goes beyond the previous studies by making use of a much more detailed data set and by employing a wide range of alternative specifications to assess the robustness of our findings.

We estimate numerous alternative specifications of the translog total cost function. Our primary representation of the level of technology is based on a distributed lag function of R & D expenditures by AT&T. For this representation of the level of technology, estimated scale economies cluster around 1.5. We also estimate scale economies with four alternative representations of the level of technology, all except one of which result in the same or higher estimates of scale economies. However, when we allow for autocorrelated disturbances, the version which initially had lower scale estimates also results in scale economies greater than 1.5.

We use two approaches to investigate whether our estimates of scale economies are capturing changes in cost due to differrential utilization of quasi-fised inputs. First, we estimate the translog variable cost function with three different quasi-fised inputs. All three specifications indicate scale economies are in the neighbourhood of 1.5. Second, we partition the sample into observations reflecting relatively high and relatively low utilization of quasi-fixed factors of production. Total cost function estimates for the two subsamples result in estimated scale economies that bracket those of the full sample. Furthermore, the hypothesis of equal scale economies in the two subsamples cannot be rejected. We conclude that our estimates of scale economies are not biased by changes in utilization of inputs over the postwar period.

Our principal finding is that scale economies for telecommunications fall somewhere between those reported by the older production function studies and the recent cost function studies. Taking into account the 95% confidence bounds associated with our estimates of scale economies, our research indicates that Bell System scale economies are in the range from 1.3 to 1.7. This finding is consistent with the view that the proliferation of suppliers of telecommunications would result in a large sacrifice of efficiency due to foregone scale economies.

Comparing estimates of scale economies from engineering and econometric studies involves numerous complications. In particular, our estimates are most appropriately interpreted as an average of scale economies over the multitude of services provided by the Bell System, whereas engineering estimates generally relate to specific services. Nonetheless, it is interesting to note that Meyer et. al. (1979) have interpreted the results of engineering studies of long-distance telecommunications as indicating scale economies that fall in the range from 1.1 to 1.5, a range that overlaps substantially with our estimates.

2. The Translog Cost Function

The translog functional form was proposed by Christensen, Jorgenson, and Lau (1971, 1973), and was first used to represent a cost function allowing for the presence of scale economies by Christensen and Greene (1976). The translog form has been used in the telecommunications cost function studies of Denny et al. (1979) and Nadiri and Schankerman (1980) and in numerous other empirical applications. We write the translog cost function in the following form:

$$(1)\quad \ln C = \alpha_0 + \alpha_Y \ln Y + \sum_i \beta_i \ln P_i + \omega_A \ln A + \frac{1}{2}\delta_{YY}(\ln Y)^2 + \frac{1}{2}\sum_i\sum_j \gamma_{ij}\ln P_i \ln P_j + \frac{1}{2}\omega_{AA}(\ln A)^2 + \sum \rho_{Yi}\ln Y \ln P_i + \sum \phi_{iA}\ln A \ln P_i + \phi_{YA}\ln A \ln Y$$

where $\gamma_{ij} = \gamma_{ji}$, C is total cost, Y is the level of output, the P_i are the prices of the inputs, and A represents the level of technology.

Any cost function must be homogeneous of degree one in input prices, which implies the following restrictions on the parameters of (1):

$$(2)\quad \sum_i \beta_i = 1, \sum_i \rho_{Yi} = 0, \sum_i \phi_{iA} = 0, \sum_i \gamma_{ij} = 0,\ \forall j.$$

Shephard's Lemma (1953) allows us to equate the cost shares (M_i) of the inputs to the logarithmic derivatives of the cost function with respect to the input prices:

$$(3)\quad M_i = \beta_i + \rho_{Yi}\ln Y + \sum_j \gamma_{ij}\ln P_j + \phi_{iA}\ln A.$$

We follow standard practice in specifying classical disturbances for (1) and (3). The parameter of the cost function can thus be obtained by treating (1) and (3) as a multivariate regression and using a modification of Zellner's (1962) technique for estimation.[4]

Successful estimation of a cost function as general as (1) with time series data is rare. The number of parameters to be estimated is too large for the limited variation found in time series date.[5] Thus, rather than begin with (1) in its general form, as a point of departure we specify a restrictive version of (1). In particular we specify (1) with only the first order terms of each argument included. This specification implies a homogeneous structure of production, i.e., permits non-constant returns to scale, but does not permit variation in the degree of scale economies. In addition, this specification restricts all elasticities of substitution to be equal to unity. After estimating this relatively simple form of the cost function, we estimate numerous more general versions. This procedure permits us to investigate the sensitivity of the results to changes in specification; furthermore, it indicates the point at which the model becomes too general for successful estimation.

The degree of scale economies can be computed from any cost function as the inverse of the elasticity of total cost with respect to output: $SCE = (\partial \ln C/\partial \ln Y)^{-1}$. For the translog cost function (1) this yields:

$$(4) \quad SCE = (\alpha_Y + \delta_{YY} \ln Y + \phi_{YA} \ln A + \sum_i \rho_{Yi} \ln P_i)^{-1},$$

in which case SCE is a function of the levels of output, technology, and input prices. For the cases in which the parameters on the second order term in output and the interaction terms between output and the other arguments are zero, the degree of scale economies is constant at α_Y^{-1}. For our empirical work all variables have been normalized to unity in 1961, thus, α_Y^{-1} provides the estimate of SCE in 1961. Since the 1961 values of the variables are approximately equal to their sample means, α_Y^{-1} also provides a good approximation to SCE evaluated at the sample mean.

3. Data

The most difficult problem in the estimation of cost functions for telecommunications is how to represent the level of technology. Several representations have been suggested in the literature. Vinod (1976) has proposed using a distributed lag of R & D expenditures to represent the level of technology. He has constructed two variables based on this approach; the first uses R & D expenditures by AT&T, which we hereafter refer to as Bell R & D. The second Vinod index uses R & D expenditures by AT&T and Western Electric, which we hereafter refer to as Bell and Western R & D. Denny et al. (1979) claim that the introduction of direct-distance dialing facilities and the changeover to modern switching equipment have been the two most important innovations in telecommunications in the postwar period. These innovations can be represented by the percentage of long distance calls directly dialed and the percentage of telephones connected to central offices with modern switching facilities.

It appears to us that the Bell R & D variable has the most justification as a representation of the level of technology for telecomunications. Thus, we adopt it as our primary specification. However, since a reasonable case can be made for the other measures as well, we consider them as alternative specifications. In addition to these four specifications of technology, which are specific to the Bell System, we also use an exponential time trend. We include a time trend since it is the variable that is used most widely in econometric studies to represent the level of technology.

Aside from the alternative representations of the level of technology, the data required to estimate the cost function have been discussed in Christensen, Cummings, and Schoech (1980). Thus, we provide here only a brief overview of the methods used to derive these data. The basic approach to the data has been to collect information at a very detailed level and then use the Tornqvist (or translog) index number procedure to aggregate up to the variable required for the cost function.[6] The Tornqvist index is superlative in the sense of Diewert (1976), and thus does not entail restrictive assumptions about the structure of production at the detailed level.

The output variable for the Bell System is based on its five principal sources of revenue: local, intrastate, and interstate services, directory advertising, and miscellaneous. These revenue categories are deflated by appropriate price indexes and then combined into a Tornqvist index of aggregate output.

In our cost function estimation we distinguish the three principal input categories of labor, capital, and intermediate or purchased materials. Our approach to materials is the same as for output. We were able to obtain data for six major categories of materials: electricity, accounting machines, advertising, stationery and postage, services from Bell Telephone Laboratories, and "all other". These expenditure categories are deflated by appropriate price indexes and then aggregated. The steps used to obtain indexes of labor and capital input were more extensive.

Our index of labor input for the Bell System is based on hours worked by Bell System employees distinguished by occupation, experience, and age. In all, we used one hundred different categories of hours worked, which were combined into a Tornqvist index of labor input using relative wages as weights.

Our index of capital input is based on detailed data for twenty different types of owned tangible assets. For each of the twenty categories we obtained a time series of investment expenditures, which we then deflated by specific price indexes. The resulting real investment figures were used in conjunction with capital stock benchmarks and rates of replacement to obtain capital stock series via the perpetual inventory method. The benchmarks and replacement rates were based on surveys of Bell System Capital Stock for 1958, 1965, and 1970. These capital stocks, their asset prices, and rates of replacement were used along with the Bell System's cost of capital and tax information to compute capital service price weights. These weights were constructed following the methodology originally proposed by Christensen and Jorgenson (1969) and modified for regulated firms by Caves, Christensen, and Swanson (1980a). We computed capital input for the Bell System as a Tornqvist index of the twenty types of owned capital, and one category of rented capital, using service price weights.[7]

The price indexes for capital, labor, and materials were obtained by dividing the annual expenditures for each category by the quantity indexes described above. Total cost is taken to be the sum of annual costs for capital, labor, and materials.

4. Estimates of Scale Economies

As discussed in Section 2, our point of departure is the translog cost function with only the first order terms of each argument allowed to appear with non-zero coefficients. For the level of technology we use

Vinod's variable that is based on Bell R & D expenditures. The parameter estimates for this specification are presented in the first column of Table 1. Since this cost function is homogeneous in output, scale economies are equal to α_Y^{-1} for every data point. We find α_Y = .669, with a standard error of .019. Taking the inverse of $_Y$ and its 95% confidence interval yields 1.50 as the estimate of SCE with 1.41 and 1.59 as lower and upper bounds.

We now generalize the basic specification allowing for non-zero coefficients on second order terms in the arguments of the cost function. The second and third columns of Table 1 provide the parameter estimates with, respectively, second order terms in the level of technology and input prices. The term in technology is not significant, but some of the price terms are. The addition of these terms has little impact on estimated scale economies. With second order technology and price terms SCE remains at 1.50 with lower 95% confidence bounds (hereafter simply "lower bounds") of 1.42 and 1.40, respectively.

The addition of the second order term in output (column four in Table 1) introduces $\delta_{YY}\ln Y$ into the SCE formula, in addition to α_Y. Thus, SCE becomes a function of the level of output, and the translog cost function becomes homothetic rather than homogeneous. The parameter α_{YY} is not significant, but the estimated SCE rises slightly; in 1961 SCE is 1.53 and in 1977 it is 1.59. The lower bounds are 1.43 and 1.42 respectively.

The fifth through seventh columns of Table 1 present the parameter estimates resulting from entering the second order terms in pairs. The eighth column of Table 1 results from entering all these types of second order terms simultaneously. There is no change in SCE from entering second order terms in technology and prices. Nor is there much change when second order terms in output and prices are entered. However, when second order terms in output and technology are entered together (with or without the second order terms in prices), the results change markedly. SCE are found to be approximately 1.7 in 1961 and 2.7 in 1977. These estimates suggest that specifications with second order terms in both output and technology may be too general for successful estimation.

In the ninth column of Table 1 we present the general translog cost function, subject to the condition that output enters only through a linear term. Technology and price terms are allowed to enter linearly, quadratically, and with interactions between technology and prices. SCE is somewhat reduced from the previous specifications; the estimate is 1.40 with a lower vound of 1.33. However, contrary to the previous specifications the neoclassical curvature conditions are not satisfied for all the data points. The cost function is not concave for eight years in the middle of the sample. We were able to overcome this problem by restricting slightly the form in which the technology index enters the cost function, as described in the next paragraph.

Changes in technology have often been modelled as augmenting individual inputs. The dual formulation of this approach is the specification that changes in technology diminish the prices of the inputs in the cost function. Each price, P_i, in the cost function is replaced by $P_i\ A^{\lambda i}$. Thus, we have the level of technology, A, entering the cost function via the three parameter λ_K, λ_L, and λ_M. This is one less than the four independent parameters associated with A in the ninth column of Table 1:

$\omega_A, \omega_{AA}, \phi_{KA}, \phi_{LA}, \phi_{MA}$ (with the restriction that $\phi_{KA} + \phi_{LA} + \phi_{MA} = 1.0$, to preserve linear homogeneity of the cost function in the input prices). The principal disadvantage of the factor augmenting specification is that the cost function becomes nonlinear in the parameters, and therefore is more difficult to estimate.

In the tenth column of Table 1 we present the factor augmented version of the homogeneous translog cost function. Contrary to the previous specification, the estimated cost function is concave at all sample points. SCE is estimated to be 1.41 with a lower bound of 1.34.

We have found that our cost function estimates eith second order terms in output tend to differ greatly from the other specifications. Nonetheless, for the sake of completeness we present two very general specifications in the eleventh and twelfth columns of Table 1. The latter is the general translog form and the former restricts the interactions between output and all other arguments to be zero. Both versions indicate SCE in 1961 to be between 1.5 and 1.6 with lower bounds between 1.4 and 1.5. SCE in 1977 is indicated to be very large, but the bounds are quite wide. Neither of these estimated cost functions is concave over the full sample.

The SCE estimates from the specifications in Table 1 are summarized in Table 2. The estimates for 1961 and 1977 are presented, along with their lower bounds.

Table 2 provides evidence that SCE is 1.3 or higher. The lowest point estimate is 1.40 with a 95% confidence interval bounded below by 1.33. The most general forms of the cost function indicate much higher SCE, but the estimates are not stable. SCE varies little across many of the specifications. The exceptions are cases in which output is allowed to enter through second order terms and cases in which technology is allowed to enter in a very general way. The former exceptions result in very high SCE, and the later exceptions result in SCE that is somewhat lower than the remaining specifications.

We proceed to investigate how the estimates of SCE in Table 2 are affected by using four alternative specifications of the level of technology. We use specification (1) to represent the bulk of the SCE estimates and specification (10) to represent the lowest estimate from a cost function that is concave. The parameter estimates for these regressions are presented in Table 3. Both of these specifications are homogeneous in output and thus have SCE that is the same for all sample points. The SCE estimates for these two specifications are presented in Table 4 for all five indexes of technology. Table 3 provides some support for our choice of the Bell R & D index as the primary representation of the level of technology. Its coefficient has a higher t-ratio than any of the other indexes of technology.

The results in Table 4 indicate that representing the level of technology by the percentage of long distance calls directly dialed does not change the estimates of SCE. Two of the remaining representations of technology -- access to modern switching facilities and time -- lead to substantially higher estimates of SCE. The forme leads to point estimates of 1.75 and 1.67, both with lower bounds of approximately 1.5. The latter leads to point estimates of 2.14 and 2.05, but the confidence bounds are very wide, with a lower bound of approximately 1.3. The final alternative representation of technology is based on R & D expenditures by Bell and Western Electric. This specification provides somewhat lower estimates of SCE and somewhat wider confidence bounds than the first two sets of estimates in Table 4. We conclude from the estimates in Tables 3 and 4 that time does not provide a useful representation of the level of technology in telecommunications, and we do not consider it further.

It is possible to estimate translog cost functions containing more than one index of technology. This approach has been followed by Nadiri and Schankerman (1980), who included both time and an R & D index. We have estimated several such models, but we do not present the estimates. These models generally resulted in higher estimates of SCE, but usually one or both of the technology indexes was not significant. Time is generally not significant when entered with another index of technology. Modern Switching also continues to be insignificant when it is entered with Direct Distance Dialing. Entering Bell and Western R & D indexes separately results in higher SCE, but the Western R & D index is not significant.

For the four satisfactory representations of technology in Table 4 the estimates of SCE range from 1.30 to 1.75, with lower bounds that range from 1.17 to 1.53. These results are based on an estimation method that allows for contemporaneous correlation among the disturbances of the estimating equations, but not for any serial correlation of the disturbances. We investigate the robustness of our results by permitting serial correlations in the manner discussed by Berndt and Savin (1975). We permit two distinct non-zero correlation coefficients -- one for the cost function and one for the share equations.[8] We repeat the regressions for specifications (1) and (10) for the four satisfactory representations of technology. The parameter estimates are presented in Table 5, and the implied SCE are presented in Table 6.

For specification (1), SCE for three of the representations of technology are higher in Table 6 than in Table 4, and SCE for the other one is slightly lower. Furthermore, the lower bounds are all higher than the corresponding ones in Table 4, and are in the narrow range from 1.47 to 1.54. For specification (10) we were able to obtain convergence in our non-linear estimation for only two of the four representations of technology. The resulting estimates of SCE are substantially higher than their counterparts in which serial correlation of the disturbances was not recognized. They are also higher than for specification (1) either with or without the recognition of serially correlated disturbances. Since they allow for serial correlation, the estimates in Table 6 are preferred to those in Table 4.

5. Estimates of Scale Economies Allowing for Changes in the Utilization of Capital and Labor

It might be claimed that the estimates of scale economies reported in the previous section reflect variations in the utilization of factors of production in addition to the exploitation of existing scale economies. In this section we use two approaches to explore such a possibility. First, we replace the assumption of full static equilibrium with an assumption of partial equilibrium. Second, we maintain the assumption of full static equilibrium but estimate separate cost functions for periods corresponding to peaks and troughs of the business cycle.

Brown and Christensen (1980) and Caves, Christensen, and Swanson (1980b) have discussed estimation of the translog variable cost function implied by partial static equilibrium. Rather than minimization of total cost conditional on the levels of output, the behavioral assumption becomes minimization of variable cost conditional on the level of output and the level of any inputs that are quasi-fixed. Specification (10) of the variable cost function can be written:

$$\ln CV = \alpha_0 + \alpha_Y \ln Y + \alpha_Z \ln Z^* + \sum_i \beta_i \ln P_i^* + 1/2\, \delta_{YY} (\ln Y)^2$$

$$+ 1/2\, \delta_{ZZ} (\ln Z^*)^2 + \delta_{YZ} \ln Y \ln Z^*$$

$$+ 1/2 \sum_i \sum_i \gamma_{ij} \ln P_i^* \ln P_j^*$$

$$+ \sum \eta_{Yi} \ln Y \ln P_i^* + \sum_i \eta_{Zi} \ln Z^* \ln P_i^*$$

where $P_i^* = P_i A^{\lambda i}$, $Z^* = Z A\lambda_i$, and CV is the cost of the variable inputs. This form specializes to specification (1) if the λ_is, δ_{ij}s, γ_{ij}s, and $_{ij}$s are restricted to be zero and a first order term in technology is added.

We have estimated the translog variable cost function for three alternative specifications in which a portion of the Bell capital stock or labor force is treated as quasi-fixed: first, all tangible assets in which the lag between order and installation exceeds one year, including buildings, central office equipment (COE) and large private branch exchanges (LPBX); second, all employees with five or more years of experience; and third, all management employees, regardless of experience, and all non-management employees with five or more years of experience. All of these variants have been estimated using the technology variable based on the Bell R & D expenditures -- for specifications (1) and (10). In addition specification (1) has been run with allowance for serially correlated disturbances. We were not able to achieve convergence for specification (10) with serial correlation permitted.

The parameter estimates for the nine variable cost functions are presented in Table 7. Caves, Christensen, and Swanson (1980b) have shown that SCE can be computed directly from the parameters of the variable cost function as:

$$SCE = (1 - (\partial \ln CV/\partial \ln Z) / (\partial \ln CV/\partial \ln Y)$$

SCE computed from this formula are presented in Table 8. The SCE are very similar to those in Tables 4 and 8, but the confidence bounds are somewhat wider. Estimated SCE range from 1.42 to 1.60, and their lower bounds range from 1.32 to 1.41.

Our second approach to allowing for variation in the utilization of factors of production is to divide our sample into two subsamples, the first of which represents relatively high utilization, and the second of which represents relatively low utilization. We have used a combination of the U.S. business cycle and cycles in Bell System output in partitioning the sample. We include the following years in the subset reflecting relatively high utilization: 1948, 1950, 1951, 1955, 1956, 1959, 1964, 1965, 1966, 1968, 1969, 1972, 1973, 1976, and 1977.

We have estimated the translog cost function specifications (1) and (10) with our primary representation of the level of technology for the two subperiods. The number of contiguous observations in each subsample is so small (eight and seven) that allowing for serial correlation is not appropriate. The parameter estimates are presented in Table 9 and the implied SCE in Table 10. The SCE are very similar to the corresponding estimates in Table 4. In fact the estimates in Table 4 (1.50 and 1.41) both fall between the estimates for the two subsamples. The bounds in Table 10 are somewhat wider, however, reflecting the fact that each subsample is much smaller than the full sample. Since the estimate of SCE for both specification (1) and (10) are based on a single parameter, it is straightforward to determine whether the differences between the estimates for years of high and low utilization are statistically significant. The t-ratios for the tests of equality are .7 and 1.4, and thus we cannot reject the hypothesis of equal SCE from the subsamples.

Neither the translog variable cost function estimates, nor the splitsample estimates of this section provides any estimates of SCE that are substantially different from those of the previous section. We conclude that there is no evidence of upward bias in an estimated SCE due to a failure to control for capacity utilization.

Although the principal motivation for estimating the variable cost function was a concern over the effects of differential capacity utilization, the similarity of the results from the total and variable cost functions also permit us to infer that our results are not biased due to the Averch-Johnson (AJ) (1962) effect. If, as a result of rate of return regulation, a firm does not attempt to minimize total cost, estimates from a total cost function might be invalid. The AJ model specifies that a firm will use more than the optimal amount of capital. Whether or not the model is realistic (a matter of great controversy), the firm will attempt to minimize variable cost conditional on the level of capital. Therefore, the variable cost model will be valid even if there is an AJ effect. The fact that we obtain very similar estimates of scale economies with the total and variable cost models provides evidence that any AJ effect which might exist for AT&T is not important enough to invalidate estimates of scale economies from the total cost model.

FOOTNOTES:

[*] A preliminary version of this paper was presented at the meetings of the Econometric Society held in Denver, Colorado, September, 1980. The authors wish to thank Douglas Caves, Thomas Cowing, and Zvi Griliches for helpful comments.

[1] For example, engineering estimates of large and pervasive scale economies for electric power generation have never been substantiated by econometric analyses. See Christensen and Green (1970) and Weiss (1975) for discussion.

[2] This estimate is based on the cost-output elasticities reported by Denny, Fuss, and Waverman (1979), which were attributed to Denny, Fuss, Everson, and Waverman (1979).

[3] Not enough information was provided to compute scale economies for individual years.

[4] The covariance matrix of the multivariate regression is singular. We overcome this problem by deleting one of the share equations at the second stage of the Zellner procedure. This provides estimates that are invariant with respect to which equation is deleted and are asymptotically equivalent to maximum likelihood estimates.

[5] Both Denny et al. (1979) and Nadiri-Schankerman reported difficulties in using the general translog specification with telecommunication data.

[6] The index can be written:

$$\ln (X_1/X_0) = \Sigma\bar{w}_i \ln (X_{1i}/X_{0i})$$

where $\bar{w}_i$ is the arithmetic average of the expenditure weights in periods 0 and 1.

This index is one of many discussed by Fisher (1922). It has been recommended for applications by Tornqvist (1936) and subsequently by Theil (1965) and Kleok (1966). It has been used extensively by Christensen and Jorgenson (1973) and others. Diewert (1976) has shown that this index is exact for a homogeneous translog function.

[7] Bell System rented capital consists almost entirely of buildings.

[8] We were not able to attain convergence with more general specifications of the disturbance structure.

REFERENCES:

[1] Averch, H. and Johnson, L.L., Behavior of the Firm Under Regulatory Constraint, American Economic Review, 52 (December 1962) 1052-69.

[2] Berndt, E.R. and Savin, N.E. Estimation and Hypothesis Testing in Singular Equation Systems with Autoregressive Disturbances Econometrica, 43 (September-November 1975) 937-957.

[3] Brown, R.S. and Christensen, L.R. "Estimating Substitution Possibilities in a Model of Partial Static Equilibrium," Discussion Paper No. 8007, Social Systems Research Institute, University of Wisconsin-Madison, 1980.

[4] Caves, D.W., Christensen, L.R. and Swanson, J.A., Producitvity in U.S. Railroads, 1951-1974, Bell Journal of Economics (Spring 1980a).

[5] Caves, D.W., Christensen, L.R. and Swanson, J.A., Productivity Growth, Scale Economies, and Capacity Utilization in U. S. Railroads, 1955-1974, Discussion Paper No. 8002, Social Systems Research Institute, University of Wisconsin-Madison (1980 (b)).

[6] Christensen, L.R., Cummings, D. and Schoech, P.E., Productivity in the Bell System, 1947-1977, paper presented at the Eighth Annual Telecommunications Policy Research Conference (April 27-30, 1980).

[7] Christensen, L.R., and Greene, W.H., Economies of Scale in U.S. Electric Power Generation, Journal of Political Economy 84 (August 1976) 655-676.

[8] Christensen, L.R. and Jorgensen, D.W., Measurement of U.S. Real Capital Input, 1929-1967, Review of Income and Wealth, Series 15, 4 (December 1969) 293-320.

[9] Christensen, L.R. and Jorgenson, D.W., Measuring the Performance of the Private Sector of the U.S. Economy, 1929-1969, in M. Moss, ed., Measuring Economic and Social Performance, New York, National Bureau of Economic Research, (1973) 233-338.

[10] Denny, M., Fuss, M., Everson, C. and Waverman, L., Estimating the Effects of Diffusion of Technological Innovations in Telecommunications: The Production Structure of Bell Canada, Paper presented at the Seventh Annual Telecommunications Policy Research Conference (April 29-May 1, 1979).

[11] Denny, M., Fuss, M. and Waverman, The Measurement and Interpretation of Total Factor Productivity in Regulated Industries, with an Application to Canadian Telecommunications, Working Paper No. 7911, Institute for Policy Analysis, University of Toronto (1979).

[12] Diewert, W.E., Exact and Superlative Index Numbers, Journal of Econometrics 4 (1976) 115-145.

[13] Fisher, I., The Making of Index Numbers (Houghton Mifflin, Boston, 1922).

[14] Kloek, T., Indexcijfers: enige methodologisch aspecten (The Hague, Pasmans, 1966).

[15] Mantell, L.H., An Econometric Study of Returns to Scale in the Bell System, Staff Research Paper, Office of Telecommunications Policy, Executive Office of the President, Washington, D.C. (February 1974).

[16] Meyer, J.R., Wilson, R.W., Baughcum, M.A., Burton, E. and Caouette, L., The Economics of Competition in the Telecommunications Industry (Charles River Associates Incorporated, Boston, Mass., August 1979).

[17] M.I. Nadiri and Mark A. Schankerman, The Structure of Production, Technological Change and the Rate of Growth of Total Factor Productivity in the Bell System, forthcoming in Productivity Measurement in Regulated Industries T. Cowing and R. Stevenson, eds. (Academic Press, 1980).

[18] R.W. Shepherd, Cost and Production Functoins (Princeton U. Press, Princeton, N.J. 1953).

[19] H. Theil, The Information Approach to Demand Analysis, Econometrica, 33, No. 1 (January 1965) 67-87.

[20] L. Tornqvist, The Bank of Finland's Consumption Price Index, Bank of Finland Monthly Bulletin 10 (1936) 1-8.

[21] H.D. Vinod, Application of New Ridge Regression Methods to a Study of Bell System Scale Economies, Journal of the American Statistical Association (December 1976).

[22] L.W. Weiss, Antitrust in the Electric Power Industry, in Promoting Competition in Regulated Markets, A. Phillips, ed. (D.C.: Brookings Institution, Washingtion, 1975).

[23] A. Zellner, An Efficient Method of Estimating Seemingly Unrelated Regressions and Tests for Aggregation Bias, Journal of the American Statistical Association, 58 (December 1962) 977-992.

Table 1

Parameter Estimates for Twelve Variations of the Translog Cost Function With the Level of Technology Variable (A) Based on Bell R & D Expenditures

(Standard errors in parentheses)

Parameter	(1) First Order Terms	(2) Second Order Technology	(3) Second Order Prices	(4) Second Order Output
α_0	9.060 (.003)	9.060 (.004)	9.060 (.025)	9.062 (.003)
α_Y	.669 (.019)	.665 (.021)	.667 (.017)	.653 (.023)
β_K	.486 (.007)	.484 (.009)	.493 (.008)	.480 (.009)
β_L	.396 (.008)	.398 (.009)	.401 (.009)	.401 (.009)
β_M	.118 (.002)	.118 (.002)	.107 (.003)	.118 (.002)
ω_A	-.102 (.021)	-.098 (.023)	-.098 (.018)	-.086 (.025)
ω_{AA}	---	-.003 (.020)	---	---
γ_{KK}	---	---	.047 (.019)	---
γ_{LL}	---	---	-.032 (020)	---
γ_{MM}	---	---	.035 (.025)	---
γ_{KL}	---	---	.010 (.018)	---
γ_{KM}	---	---	-.057 (.015)	---
γ_{LM}	---	---	.022 (.012)	---
δ_{YY}	---	---	---	-.021 (.016)

Table 1 (continued)

Parameter	(5) Second Order Technology, Prices	(6) Second Order Prices, Output	(7) Second Order Technology, Output	(8) Second Order Technology, Prices, Output
α_0	9.060 (.003)	9.062 (.003)	9.053 (.004)	9.052 (.003)
α_Y	.665 (.018)	.653 (.021)	.589 (.028)	.584 (.026)
β_K	.492 (.008)	.491 (.006)	.478 (.009)	.491 (.007)
β_L	.401 (.009)	.402 (.009)	.404 (.009)	.403 (.009)
β_M	.107 (.003)	.107 (.003)	.118 (.002)	.106 (.003)
ω_A	-.098 (.021)	-.083 (.023)	-.049 (.025)	-.043 (.023)
ω_{AA}	.005 (.019)	---	.203 (.063)	.212 (.055)
γ_{KK}	.048 (.020)	.054 (.020)	---	.059 (.020)
γ_{LL}	-.031 (.022)	-.023 (.022)	---	-.018 (.022)
γ_{MM}	.035 (.025)	.035 (.025)	---	.039 (.025)
γ_{KL}	.009 (.020)	.002 (.020)	---	-.001 (.020)
γ_{KM}	-.057 (.015)	-.056 (.015)	---	-.058 (.016)
γ_{LM}	.022 (.012)	.021 (.012)	---	.019 (.013)
δ_{YY}	---	-.017 (.015)	-.177 (.051)	-.181 (.045)

Table 1 (continued)

Parameter	(9) Second Order Interactions Technology, Prices	(10) Second Order Prices, Factor Augmenting Technology	(11) General Except Output Interactions	(12) General
α_0	9.060 (.003)	9.064 (.003)	9.051 (.004)	9.050 (.004)
α_Y	.715 (.019)	.708 (.018)	.635 (.028)	.641 (.034)
β_K	.518 (.004)	.516 (.004)	.518 (.004)	.527 (.004)
β_L	.370 (.003)	.371 (.003)	.370 (.004)	.363 (.003)
β_M	.112 (.002)	.113 (.002)	.112 (.002)	.110 (.001)
ω_A	-.138 (.021)	---	-.078 (.025)	-.084 (.034)
ω_{AA}	.113 (.019)	---	.287 (.063)	.303 (.322)
γ_{KK}	.277 (.019)	.266 (.018)	.278 (.019)	.328 (.015)
γ_{LL}	.212 (.020)	.192 (.017)	.213 (.020)	.241 (.018)
γ_{MM}	-.017 (.020)	-.034 (.019)	-.017 (.021)	.051 (.023)
γ_{KL}	-.253 (.016)	-.246 (.015)	-.254 (.016)	-.291 (.014)
γ_{KM}	-.024 (.008)	-.020 (.008)	-.024 (.008)	-.037 (.007)
γ_{LM}	.041 (.017)	.054 (.016)	.041 (.017)	-.014 (.018)
δ_{YY}	---	---	-.143 (.050)	-.103 (.223)
ϕ_{KA}	.151 (.009)	---	.151 (.010)	.025 (.026)
ϕ_{LA}	-.152 (.009)	---	-.152 (.009)	-.029 (.024)
ϕ_{MA}	.001 (.006)	---	.000 (.006)	.004 (.008)
λ_K	---	.161 (.028)	---	---
λ_L	---	-.376 (.057)	---	---
λ_M	---	-.589 (.253)	---	---
ρ_{YK}	---	---	---	-.110 (.022)
ρ_{YL}	---	---	---	-.122 (.021)
ρ_{YM}	---	---	---	.012 (.010)
ϕ_{YA}	---	---	---	-.027 (.262)

Table 2

Scale Economies and Lower Bounds (95%) Implied by Estimates in Table 1 for 1961 and 1977

Specification	1961 Scale	1961 Lower Bound	1977 Scale	1977 Lower Bound
(1)	1.50	1.41	1.50	1.41
(2)	1.50	1.42	1.50	1.42
(3)	1.50	1.40	1.50	1.40
(4)	1.53	1.43	1.59	1.42
(5)	1.50	1.42	1.50	1.42
(6)	1.53	1.44	1.58	1.42
(7)	1.70	1.55	2.65	1.82
(8)	1.71	1.57	2.71	1.92
(9)	1.40	1.33	1.40	1.33
(10)	1.41	1.34	1.41	1.34
(11)	1.58	1.45	2.16	1.59
(12)	1.56	1.41	2.65	1.76

Table 3

Parameter Estimates for Specifications (1) and (10) Using Five Alternative Representations of the Level of Technology

Specification (1): First Order Terms

Parameter	(a) Bell R & D	(b) Direct Distance Dialing	(c) Modern Switching	(d) Time	(e) Bell & Western R & D
α_0	9.060 (.003)	9.048 (.003)	9.051 (.005)	9.054 (.005)	9.061 (.004)
γ_Y	.669 (.019)	.669 (.026)	.573 (.041)	.468 (.144)	.706 (.041)
β_K	.486 (.007)	.485 (.008)	.487 (.008)	.487 (.008)	.486 (.008)
β_L	.396 (.008)	.397 (.008)	.395 (.008)	.395 (.008)	.396 (.008)
β_M	.118 (.002)	.118 (.002)	.118 (.002)	.118 (.002)	.118 (.002)
ω_A	-.102 (.021)	-.197 (.053)	.008 (.150)	.008 (.144)	-.120 (.038)

Table 3 (continued)

Specification (10): Second Order Prices, Factor Augmenting Technology

Parameter	(a) Bell R & D	(b) Direct Distance Dialing	(c) Modern Switching	(d) Time	(e) Bell & Western R & D
α_0	9.064 (.003)	9.051 (.003)	9.062 (.005)	9.063 (.005)	9.068 (.004)
α_Y	.708 (.018)	.710 (.026)	.597 (.032)	.487 (.140)	.771 (.041)
β_K	.516 (.004)	.535 (.003)	.535 (.003)	.529 (.004)	.518 (.004)
β_L	.371 (.003)	.353 (.002)	.354 (.003)	.360 (.003)	.370 (.003)
β_M	.113 (.002)	.112 (.001)	.111 (.001)	.112 (.001)	.112 (.001)
γ_{KK}	.266 (.018)	.231 (.011)	.198 (.009)	.239 (.009)	.221 (.016)
γ_{LL}	.192 (.017)	.208 (.009)	.169 (.010)	.180 (.010)	.196 (.014)
γ_{MM}	-.034 (.019)	-.017 (.010)	-.003 (.002)	-.004 (.003)	-.015 (.013)
γ_{KL}	-.246 (.015)	-.248 (.009)	-.201 (.009)	-.212 (.010)	-.236 (.015)
γ_{KM}	-.020 (.008)	-.023 (.005)	-.029 (.004)	-.027 (.004)	-.025 (.005)
γ_{LM}	.054 (.016)	.399 (.009)	.032 (.003)	.032 (.004)	.040 (.011)
λ_K	.161 (.028)	.376 (.091)	1.486 (.196)	.036 (.009)	.114 (.068)
λ_L	-.376 (.057)	-.569 (.185)	.815 (.498)	.016 (.012)	-.295 (.138)
λ_M	-.589 (.253)	-1.956 (.884)	-9.401 (2.013)	-.145 (.057)	-.969 (.557)

Table 4

Scale Economies and Lower Bounds for Specifications (1) and (10) Using Five Alternative Representations of the Level of Technology

Technology Variable Based on :	Specifications (1) SCE	(1) Lower Bound	(10) SCE	(10) Lower Bound
Bell R & D Expenditures	1.50	1.41	1.41	1.34
Direct-Distance Dialing	1.49	1.39	1.41	1.31
Access to Modern Switching Facilities	1.75	1.53	1.67	1.51
Time	2.14	1.32	2.05	1.30
Bell & Western R & D Expenditures	1.42	1.27	1.30	1.17

Table 5

Parameter Estimates for Specifications (1) and (10) Using Four Alternative Representations of the Level of Technology and Allowing for Serially Correlated Disturbances

Specification (1): First Order Terms

Parameter	Bell R & D	Direct Distance Dialing	Modern Switching	Bell & Western R & D
α_0	9.074 (.006)	9.073 (.007)	9.078 (.007)	9.078 (.007)
α_Y	.634 (.024)	.617 (.030)	.578 (.036)	.561 (.057)
β_K	.464 (.014)	.465 (.014)	.464 (.015)	.465 (.016)
β_L	.393 (.011)	.393 (.012)	.394 (.013)	.393 (.013)
β_M	.142 (.010)	.142 (.011)	.142 (.1011)	.142 (.011)
ω_A	-.069 (.025)	-.096 (.059)	-.035 (.129)	.008 (.053)
ρ_C	.181 (.048)	.227 (.047)	.255 (.047)	.256 (.050)
ρ_S	.914 (.028)	.915 (.028)	.916 (.028)	.915 (.028)

Table 5 (continued)

Specification (10): Second Order Prices, Factor Augmenting Technology

Parameter	(a) Bell R & D	(b) Direct Distance Dialing	(c) Modern Switching	(d) Bell & Western R & D
α_0	9.080 (.007)			9.081 (.007)
α_Y	.577 (.031)	Convergence not achieved		.524 (.057)
β_K	.514 (.023)			.520 (.023)
β_L	.354 (.021)			.349 (.021)
β_M	.132 (.010)			.131 (.011)
γ_{KK}	-.075 (.069)			.094 (.093)
γ_{LL}	-.019 (.080)			-.009 (.110)
γ_{MM}	-.170 (.119)			-.202 (.152)
γ_{KL}	-.095 (.056)			-.143 (.073)
γ_{KM}	.056 (.043)			.050 (.048)
γ_{LM}	.114 (.098)			.153 (.126)
λ_K	-.204 (.262)			.001 (.161)
λ_L	.222 (.327)			.069 (.175)
λ_M	.084 (.226)			.108 (.141)
ρ_C	.254 (.052)			.274 (.050)
ρ_S	.868 (.039)			.865 (.040)

Table 6

Scale Economies and Lower Bounds for Specifications (1) and (10) Using Four Alternative Representations of the Level of Technology and Allowing for Serially Correlated Disturbances

Technology Variable Based on:	Specifications (1) SCE	Lower Bound	(10) SCE	Lower Bound
Bell R & D Expenditures	1.58	1.47	1.73	1.56
Access to Direct Distance Dialing	1.62	1.48	*	
Access to Modern Switching	1.73	1.54	*	
Bell and Western R & D Expenditures	1.78	1.48	1.91	1.57

* Convergence not achieved.

Table 7

Parameter Estimates for the Variable Cost Functions

Bell R & D Technology

Parameter	Quasi-fixed Capital Specification (1)	Quasi-fixed Capital Specification (10)	Quasi-fixed Capital Allowing for Serial Correlation Specification (1)
α_0	8.834 (.004)	8.830 (.004)	8.860 (.010)
α_Y	.719 (.076)	.773 (.082)	.935 (.097)
α_Z	-.090 (.059)	-.158 (.089)	-.373 (.097)
β_K	.363 (.007)	.406 (.003)	.330 (.022)
β_L	.490 (.007)	.455 (.003)	.487 (.019)
β_M	.147 (.003)	.139 (.001)	.183 (.019)
ω_A	-.172 (.028)	---	-.083 (.042)
γ_{KK}	---	.211 (.014)	---
γ_{LL}	---	.209 (.018)	---
γ_{MM}	---	.033 (.024)	---
γ_{KL}	---	-.193 (.014)	---
γ_{KM}	---	-.018 (.010)	---
γ_{LM}	---	-.016 (.018)	---
η_{YK}	---	.150 (.020)	---
η_{YL}	---	-.100 (.021)	---
η_{YM}	---	.109 (.011)	---
η_{ZK}	---	-.009 (.028)	---
η_{ZL}	---	-.164 (.029)	---
η_{ZM}	---	.014 (.016)	---
δ_{YY}	---	-.112 (.154)	---
δ_{YZ}	---	.167 (.230)	---
δ_{ZZ}	---	-.250 (.345)	---

Table 7 (continued)

Parameter	Quasi-fixed Capital Specification (1)	Quasi-fixed Capital Specification (10)	Quasi-fixed Capital Allowing for Serial Correlation Specification (1)
λ_K	---	-.137 (.123)	---
λ_L	---	-.456 (.014)	---
λ_M	---	.613 (.527)	---
λ_Z	---	-.186 (.234)	---
ρ_C	---	---	.361 (.066)
ρ_S	---	---	.934 (.030)

Parameter	Quasi-Fixed Experienced Labor Specification (1)	Quasi-Fixed Experienced Labor Specification (10)	Quasi-fixed Experienced Labor Allowing for Serial Correlation Specification (1)
α_0	8.733 (.003)	8.723 (.004)	8.747 (.008)
α_Y	.845 (.052)	.773 (.043)	.771 (.071)
α_Z	-.197 (.039)	-.118 (.041)	-.175 (.054)
β_K	.686 (.009)	.723 (.005)	.644 (.026)
β_L	.145 (.009)	.121 (.004)	.098 (.048)
β_M	.169 (.004)	.156 (.002)	.258 (.056)
ω_A	-.096 (.043)	---	-.023 (.057)
γ_{KK}	---	.231 (.031)	---
γ_{LL}	---	.027 (.032)	---
γ_{MM}	---	-.039 (.031)	---
γ_{KL}	---	-.149 (.026)	---
γ_{KM}	---	-.082 (.012)	---
γ_{LM}	---	.122 (.028)	---
η_{YK}	---	-.184 (.036)	---
η_{YL}	---	.180 (.032)	---

Table 7 (Continued)

Parameter	Quasi-Fixed Experienced Labor Specification (1)	Quasi-Fixed Experienced Labor Specification (10)	Quasi-Fixed Experienced Labor Allowing for Serial Correlation Specification (1)
η_{YM}	---	.004 (.015)	---
η_{ZK}	---	.226 (.050)	---
η_{ZL}	---	-.260 (.044)	---
η_{ZM}	---	-.006 (.021)	---
δ_{YY}	---	.478 (.284)	---
δ_{YZ}	---	-.691 (.426)	---
δ_{ZZ}	---	.999 (.607)	---
λ_K	---	.213 (.062)	---
λ_L	---	-.338 (.171)	---
λ_M	---	-.924 (.423)	---
λ_Z	---	.148 (.137)	---
ρ_C	---	---	.233 (.055)
ρ_S	---	---	.954 (.029)

Parameter	Quasi-fixed Management & Experienced Labor Specification (1)	Quasi-fixed Management & Experienced Labor Specification (10)	Quasi-fixed Management & Experienced Labor Allowing for Serial Correlation Specification (1)
α_0	8.707 (.006)	8.717 (.004)	8.742 (.008)
α_Y	.721 (.089)	.729 (.055)	.806 (.078)
α_Z	-.153 (.082)	-.103 (.060)	-.229 (.073)
β_K	.491 (.009)	.724 (.005)	.651 (.025)
β_L	.391 (.009)	.118 (.004)	.084 (.049)
β_M	.118 (.002)	.158 (.002)	.265 (.056)
ω_A	-.084 (.068)	---	-.035 (.057)

Table 7 (continued)

Parameter	Quasi-fixed Management & Experienced Labor Specification (1)	Quasi-fixed Management & Experienced Labor Specification (10)	Quasi-fixed Management & Experienced Labor Allowing for Serial Correlation Specification (1)
γ_{KK}	---	.229 (.030)	---
γ_{LL}	---	.021 (.034)	---
γ_{MM}	---	-.044 (.036)	---
γ_{KL}	---	-.147 (.025)	---
γ_{KM}	---	-.082 (.013)	---
γ_{LM}	---	.126 (.031)	---
η_{YK}	---	-.240 (.049)	---
η_{YL}	---	.232 (.043)	---
η_{YM}	---	.009 (.022)	---
η_{ZK}	---	.364 (.071)	---
η_{ZL}	---	-.351 (.062)	---
η_{ZM}	---	-.013 (.034)	---
δ_{YY}	---	.579 (.433)	---
δ_{YZ}	---	-.876 (.648)	---
δ_{ZZ}	---	1.326 (.973)	---
λ_K	---	.286 (.091)	---
λ_L	---	-.365 (.212)	---
λ_M	---	1.038 (.579)	---
λ_Z	---	.087 (.152)	---
ρ_C	---	---	.228 (.053)
ρ_S	---	---	.954 (.028)

Table 8

Scale Economies for the Translog Variable Cost Function
Technology Variable: Bell R & D Expenditures

Quasi-Fixed Inputs	Without Serial Correlation Adjustment Specification				With Serial Correlation Adjustment Specification	
	(1)		(10)		(1)	
	SCE	Lower Bound	SCE	Lower Bound	SCE	Lower Bound
Buildings COE, LPBX	1.52	1.36	1.50	1.36	1.47	1.33
Experienced Labor (5+ years)	1.42	1.32	1.45	1.36	1.52	1.34
Management and Experienced Labor (5+ years)	1.60	1.39	1.51	1.41	1.52	1.36

Table 9

Parameter Estimates for the Translog Total Cost Function
Using Sub-samples reflecting difference in utilization

Technology Variable Based on Bell R & D Expenditures

Specification (1): First Order Terms

Parameter	High Utilization	Low Utilization
α_0	9.062 (.005)	9.060 (.004)
α_Y	.690 (.031)	.663 (.024)
β_K	.488 (.010)	.482 (.011)
β_L	.391 (.010)	.402 (.011)
β_M	.121 (.003)	.116 (.003)
ω_A	-.124 (.033)	-.095 (.028)

Specification (10: Second Order Prices, Factor Augmenting Technology

Parameter	High Utilization	Low Utilization
α_0	9.065 (.004)	9.063 (.004)
α_Y	.740 (.029)	.688 (.025)
β_K	.514 (.005)	.519 (.007)
β_L	.371 (.004)	.369 (.006)
β_M	.114 (.002)	.112 (.003)
γ_{KK}	.258 (.021)	.283 (.032)
γ_{LL}	.192 (.017)	.183 (.038)
γ_{MM}	-.034 (.022)	-.053 (.039)
γ_{KL}	-.242 (.017)	-.259 (.026)
γ_{KM}	-.016 (.009)	-.024 (.013)
γ_{LM}	.050 (.017)	.076 (.036)
λ_K	.127 (.038)	.172 (.032)
λ_L	-.420 (.066)	-.380 (.057)
λ_M	-.595 (.278)	-.432 (.236)

Table 10

Scale Economies for the Translog Total Cost Function
Using Sub-Samples Reflecting Differences in Utilization

Technology Variable Based on Bell R & D Expenditures

	Specification			
	(1)		(10)	
Sub-Sample	SCE	Lower Bound	SCE	Lower Bound
Years with Relatively High Utilization	1.45	1.33	1.35	1.25
Years with Relatively Low Utilization	1.51	1.41	1.45	1.36

Economic Analysis of Telecommunications:
Theory and Applications
L. Courville, A. de Fontenay and R. Dobell (eds.)

ECONOMIES OF SCALE AND SCOPE IN BELL CANADA

FERENC KISS, SETA KARABADJIAN, BERNARD J. LEFEBVRE
BELL CANADA

1. INTRODUCTION

The main objective of this paper is to establish whether there are internal economies inherent in the process of producing telecommunications services by Bell Canada in large quantities (economies of scale) and great variety (economies of scope).[1]

It is assumed that the services of Bell Canada's productive factors (labour, capital, etc.) can be related to the output (telecommunications service) volumes of the company by a transformation function which exhibits certain useful and economically meaningful mathematical properties (continuous, twice differentiable, strictly monotone and quasi-concave). Further it is assumed that Bell's production technology can be expressed equivalently by a dual cost function, relating exogenous output levels and input prices to the company's total production cost. Technological changes are regarded as shifts in the transformation and cost functions.

The statistical estimation of cost functions is pursued. The translog (TL) and a generalized form of the translog (GTL) cost function are chosen. The estimation effort is guided by deductive reasoning. A very elaborate form of the GTL cost function is estimated first. Then statistical tests on restrictive hypotheses and an analysis of the estimated parameters and economic properties are used to gradually restrict the specification, thereby reducing its degree of generality and making it more reflective of the specific economic characteristics of Bell Canada's technology. This process is pursued until the limits of statistically justifiable and economically meaningful restrictions are approached.

The estimation results obtained from the gradually restricted cost functions are compared in an examination of the robustness of empirical findings on internal economies. To further such an examination, the sensitivity of estimates of internal economies to sample variation and alternative variable measurements is also explored in models which appear to be preferable to other estimated models.

The structure of the paper is as follows. In Section 2, the applied cost model is described. Several forms of internal economies are defined and the statistical testing procedures are established in Section 3. Section 4 contains a description of empirical results. Sub-sections are devoted to multi-output and single output models. The main conclusions of the study are drawn in Section 5, where the empirical results are summarized and compared to the evidence that is available from other studies of the Bell Canada production process.

The following main conclusions are drawn in the last section of the paper. First, the estimated models of Bell Canada's production technology give a robust indication of substantial overall economies of scale. Secondly, the degree of economies of scale was very low in the early 1950's, increased gradually with the introduction of DDD calling and crossbar switching technology during the late 1950's and the entire 1960's, reached high levels by 1970-71 and declined slightly during the 1970's. Thirdly, statistically insignificant economies of scope and cost complementarity are generally indicated by the models. Finally, the models fail to produce any evidence on the output-specific economies of scale of Bell Canada.

2. THE COST MODEL

Increasing exploration of the theory of duality during the 1970's led to the recognition that cost functions were more suitable than production functions for estimating the characteristics of the Bell Canada production process.[2] The neoclassical cost function can be written as

$$(1)\quad C = g(Q_1,\ldots,Q_n;W_1,\ldots,W_m;T)\ ,$$

where $Q_1,\ldots,Q_n$ denote output volumes, $W_1,\ldots,W_m$ denote factor input prices, T is an index of technological change and C refers to the total production cost, defined as the sum of payments to m factor inputs $X_1,\ldots,X_m$, i.e., $C=\sum_i(W_iX_i)$.

The translog (TL) and generalized translog (GTL) forms of the cost function are used to represent Bell Canada's production structure.[3] Each is a class of the flexible second order Taylor series approximation to the general form in equation (1). Both functional forms contain the natural logarithms of $W_1,\ldots,W_m$ and C, but the GTL substitutes the Box-Cox transformation for the natural logarithms of $Q_1,\ldots,Q_n$ and/or T in the translog form.[4] The Box-Cox transformation of output Q_k is $(Q_k^{\lambda_k}-1)/\lambda_k$ $(k=1,\ldots,n;\ \lambda_k\neq 0)$ and the technology variable is transformed in a similar fashion as $(T^{\lambda_T}-1)/\lambda_T$ $(\lambda_T\neq 0)$. The Box-Cox expression reduces to the logarithmic transformation of Q_k and T as the respective λ_k and λ_T values tend to zero. Hence, the GTL specification contains TL as a special limiting case.

To simplify the presentation of the models, equation (2) below contains Q_k^* and T^* variables, which represent both the logarithmic and the Box-Cox transformations; i.e., $Q_k^*=\log Q_k$ and $T^*=\log T$ in TL and $Q_k^*=(Q_k^{\lambda_k}-1)/\lambda_k$ and $T^*=(T^{\lambda_T}-1)/\lambda_T$ in GTL.

The TL/GTL cost function can now be written as

$$(2)\ \log C = \alpha_o + \sum_{i=1}^{m} \alpha_i \log W_i + \sum_{k=1}^{n} \alpha_{Qk} Q_k^* + \beta T^* + (\tfrac{1}{2})\beta_T (T^*)^2$$

$$+ (\tfrac{1}{2}) \sum_{i=1}^{m} \sum_{j=1}^{m} \gamma_{ij} \log W_i \log W_j + (\tfrac{1}{2}) \sum_{k=1}^{n} \sum_{\ell=1}^{n} \delta_{k\ell} Q_k^* Q_\ell^*$$

$$+ \sum_{i=1}^{m} \sum_{k=1}^{n} \rho_{ik} \log W_i Q_k^* + \sum_{i=1}^{m} \beta_i \log W_i T^* + \sum_{k=1}^{n} \beta_{Qk} Q_k^* T^*$$

where the variables are defined as in equation (1) above.[5]

The cost function is constrained to be homogeneous of degree one in the input prices[6] by the following set of restrictions:

$$(3)\ \sum_{i=1}^{m} \alpha_i = 1; \quad \sum_{i=1}^{m} \gamma_{ij} = \sum_{i=1}^{m} \rho_{ik} = \sum_{i=1}^{m} \beta_i = 0 \qquad (j=1,\ldots,m;\ k=1,\ldots,n).$$

The symmetry conditions in a second order approximation together with (3) imply that

$$(4)\ \sum_i \gamma_{ij} = \sum_j \gamma_{ij} = 0.$$

Since the number of parameters to be estimated is usually large, it is advisable to use additional information to construct a more complete model of the cost structure. Assuming that the cost minimizing input levels are chosen to produce the observed output volumes, invoking a lemma ($\partial C/\partial W_i = X_i$) by Shephard (1970) and partially differentiating the cost function with respect to input prices, cost share equations for each input are constructed as

$$(5)\ S_i = \alpha_i + \sum_{j=1}^{m} \gamma_{ij} \log W_j + \sum_{k=1}^{n} \rho_{ik} Q_k^* + \beta_i T^* \qquad (i=1,\ldots,m).$$

Equations (2) and (5) are estimated simultaneously. Since the parameters of (5) are a subset of those of (2), the cost share equations increase the available degrees of freedom and improve statistical precision. Following Christensen and Greene (1976), disturbance terms are added to each cost share equation to reflect random errors in optimization. Since the cost shares sum to unity, their disturbances sum to zero. To preserve the non-singularity of the covariance matrix, one of the m cost share equations is deleted from the estimated model.[7] Maximum likelihood parameter estimates are invariant to the deleted equation. Using the iterative estimation technique for seemingly unrelated equations of Zellner (1962) on a large sample of Bell Canada data ensures that maximum likelihood estimates are obtained if the covariance matrix converges.

Bell Canada is assumed to be a cost minimizer and all the n outputs in equations (2) and (5) are exogenous. This assumption conforms with the single output models of Bell Canada by Denny et al (1979) and Smith and Corbo (1979), and also with single output models of the Bell System by Nadiri and Schankerman (1979) and Christensen et al (1980), but differs from multi-output specifications by Fuss and Waverman (1977, 1978, 1981), Smith and Corbo (1979) and Denny et al (1979), where the local service output is exogenous but the volume of toll service output is assumed to be endogenously set by Bell Canada (through endogenously determined prices) at the point where the marginal cost of toll services equals their marginal revenue.[8]

3. INTERNAL ECONOMIES

Four distinct forms of internal economies are defined below. These are overall economies of scale, cost complementarity between outputs, global economies of scope and output-specific economies of scale.

3.1 Overall Economies of Scale

Overall economies of scale are measured by the inverse of the sum of cost elasticities with respect to outputs. This statistic, denoted by ε and called scale elasticity[9] below, can be derived from the general cost function as

$$(6) \quad \varepsilon = \left[\sum_{k=1}^{n} \frac{\partial \log C}{\partial \log Q_k} \right]^{-1} .$$

The TL form of equation (2) yields the expression

$$(7a) \quad \varepsilon = \left[\sum_{k=1}^{n} (\alpha_{Qk} + \sum_{i=1}^{m} \rho_{ik} \log W_i + \sum_{\ell=1}^{n} \delta_{k\ell} \log Q_\ell + \beta_{Qk} \log T) \right]^{-1}$$

and the scale elasticity is derived from the GTL function as

$$(7b) \quad \varepsilon = \left[\sum_{k=1}^{n} \{ Q_k^{\lambda_k} [\alpha_{Qk} + \sum_{i=1}^{m} \rho_{ik} \log W_i + \sum_{\ell=1}^{n} \delta_{k\ell} \frac{Q_\ell^{\lambda_\ell} - 1}{\lambda_\ell} + \beta_{Qk} \frac{T^{\lambda_T} - 1}{\lambda_T}] \} \right]^{-1} .$$

Local overall economies (diseconomies) of scale are said to exist if

$\varepsilon > 1$ ($\varepsilon < 1$), while the underlying technology is characterized locally by neither economies nor diseconomies of scale if $\varepsilon=1$. The latter is also referred to as constant returns to scale.

At the expansion point, where $W_i=Q_k=T=1$, the cost elasticity with respect to Q_k reduces to α_{Qk} in both functional forms; hence the scale elasticity is

$$(8) \quad \varepsilon = \left(\sum_{k=1}^{n} \alpha_{Qk}\right)^{-1} .$$

The hypothesis of constant returns to scale can be tested by constructing confidence limits for the terms on the right hand side of equations (7a) and (7b) and observing if the value of $\varepsilon=1$ falls within or outside the confidence interval. The procedure can be simplified if the confidence limits are computed for the expansion point only, using equation (8).

Likelihood ratio tests, which are used extensively to test the validity of various hypotheses that result in parametric restrictions, are also utilized in the process of testing for the hypothesis of constant returns to scale. The following parametric restrictions are imposed on the TL/GTL[10] equation by hypothesizing constant returns to scale:

$$(9) \quad \sum_{k=1}^{n} \alpha_{Qk} = 1; \quad \sum_{k=1}^{n} \rho_{ik} = \sum_{\ell=1}^{n} \delta_{k\ell} = \sum_{k=1}^{n} \beta_{Qk} = 0 \qquad (i=1,\ldots,m; k=1,\ldots,n).$$

The number of restrictions in a given specification depends on the number of outputs (n) and inputs (m). Since the parameters are maximum likelihood estimates, the log of the likelihood function for the estimates with restricted and unrestricted parameters (Ω_R and Ω_U, respectively) can be used in likelihood ratios of the form $\lambda=\Omega_R-\Omega_U$. -2λ is distributed asymptotically as Chi-squared, with degrees of freedom equal to the number of independently imposed restrictions on the parameters if the restrictive null hypothesis is true. Normally, the comparison is made between the computed -2λ and the χ^2 value at the .05 level; however, in some cases the χ^2 value at the .01 level is also considered. The null hypothesis cannot be rejected (is rejected) if the critical χ^2 value is less (greater) than the computed -2λ.

3.2 Cost Complementarity

Cost complementarity gives a local estimate of economies of scope at specific output levels. The test statistic for cost complementarity in a twice differentiable cost function is the second order cross-derivative of the cost function with respect to any two outputs:[11]

$$(10) \quad CC_{k\ell} = \frac{\partial^2 C}{\partial Q_k \partial Q_\ell} \qquad (k\neq\ell; k,\ell=1,\ldots,n) .$$

Cost complementarity exists, when CC<0; i.e., when infinitesimal increases (decreases) in the volume of one output make the marginal cost of the other output decline (increase).

In the translog model, the test statistic of equation (10) can be written as

$$(11a) \quad CC_{k\ell} = \frac{C}{Q_k Q_\ell}\left[\frac{\partial \log C}{\partial \log Q_k} \cdot \frac{\partial \log C}{\partial \log Q_\ell} + \delta_{k\ell}\right]$$

and in the GTL model it becomes

$$(11b) \quad CC_{k\ell} = \frac{C}{Q_k Q_\ell}\left[\frac{\partial \log C}{\partial \log Q_k} \cdot \frac{\partial \log C}{\partial \log Q_\ell} + \delta_{k\ell} Q_k^{\lambda_k} Q_\ell^{\lambda_\ell}\right] .$$

At the expansion point, the cost complementarity test statistic in both models reduces to

$$(12) \quad CC_{k\ell} = \alpha_{Qk}\, \alpha_{Q\ell} + \delta_{k\ell} \quad .$$

The null-hypothesis of no cost complementarity can be tested by constructing confidence intervals for terms on the right hand side of equation (11a) or (11b) (or equation (12) when the test is performed at the expansion point) and observing if the value CC=0 falls within or outside the confidence interval.

3.3 Economies of Scope

Economies of scope exist globally (in the entire range of output volumes) when the joint production of an industry's outputs is cheaper than their separate production; i.e., when

$$(13) \quad C(Q_1,\ldots,Q_n) < C(Q_1,0,\ldots) + C(0,Q_2,0,\ldots) + \ldots + C(0,\ldots,0,Q_n),$$
$$(Q_1,\ldots,Q_n) > 0 \quad .$$

This case is referred to as that of overall economies of scope. Output-specific economies of scope exist, when the joint production of an output (Q_k) with the existing combination of other outputs is cheaper than its separate production; i.e. when

$$(14) \quad C(Q_1,\ldots,Q_n) < C(Q_1,\ldots,Q_{k-1},0,Q_{k+1},\ldots,Q_n) + C(0,\ldots,0,Q_k,0,\ldots),$$
$$(Q_1,\ldots,Q_n) > 0 \quad .$$

Global and output-specific economies of scope become equivalent when the number of outputs is two.

Following Panzar and Willig (1979), a test statistic for Q_k-specific economies of scope can be written as

$$(15)\quad SCOPE_k = \frac{C(Q_1,\ldots,Q_{k-1},0,Q_{k+1},\ldots,Q_n)+C(0,\ldots,0,Q_k,0,\ldots)-C(Q_1,\ldots,Q_n)}{C(Q_1,\ldots,Q_n)} .$$

Q_k-specific economies of scope exist when $SCOPE_k>0$. To simplify the procedure, tests of output-specific economies of scope are carried out at the expansion point only. The translog cost function is not well defined for zero output levels; thus, it is not suited to carry out tests of global economies of scope. The economies of scope statistic can be derived from the GTL function as

$$(16)\quad SCOPE_k = \frac{\exp[\alpha_o - \frac{\alpha_{Qk}}{\lambda_k} + \frac{1}{2}\frac{\delta_{kk}}{\lambda_k^2}] + \exp[\alpha_o - \sum_{\substack{i=1\\i\neq k}}^{n} \frac{\alpha_{Qi}}{\lambda_i} + \frac{1}{2}\sum_{\substack{i=1\\i\neq k}}^{n}\sum_{\substack{j=1\\j\neq k}}^{n}\frac{\delta_{ij}}{\lambda_i\lambda_j}] - \exp[\alpha_o]}{\exp[\alpha_o]} .$$

The null-hypothesis of no economies of scope is tested by constructing confidence intervals for the terms on the right hand side of equation (16) and observing whether the value $SCOPE_k=0$ falls within or outside the confidence limits.

3.4 Output-Specific Economies of Scale

Output-specific economies of scale in a multi-product firm result from less than proportional increases (decreases) in the cost that is specific to an output, when the level of <u>that</u> output increases (decreases), while all other output levels remain unchanged. Q_k-specific cost is the addition to the total cost of production that results from Q_k being produced. It is the incremental cost of Q_k, defined as

$$IC_k = C(Q_1,\ldots,Q_n) - C(Q_1,\ldots,Q_{k-1},0,Q_{k+1},\ldots,Q_n) .$$

The average incremental cost of Q_k is defined as $AIC_k=IC_k/Q_k$. AIC_k declines if Q_k-specific economies of scale are present. In this case, AIC_k is greater than the marginal cost of Q_k and their ratio is greater than one.

Following Panzar and Willig (1979), the degree of Q_k-specific economies of scale is defined as the ratio between the average incremental cost and marginal cost of Q_k and can be expressed as

$$(17)\quad S_k = \frac{IC_k/Q_k}{\partial C/\partial Q_k} = \frac{IC_k/C}{\varepsilon_{CQk}} ,$$

where $C=C(Q_1,\ldots,Q_n)$ and $\varepsilon_{CQk}=\partial\log C/\partial\log Q_k$. There are economies of scale specific to Q_k if $S_k>1$.

The translog cost function is not suitable for the determination of IC_k, because it is not well defined for zero output levels. For the GTL function, Fuss and Waverman (1981) show that the degree of output-specific economies of scale can be estimated with relative ease at the expansion point, where $W_i=Q_k=T=1$. The GTL function yields the following expression for Q_k-specific economies of scale:

$$(18)\quad S_k = \frac{\exp[\alpha_o] - \exp[\alpha_o - \alpha_{Qk}/\lambda_k + \delta_{kk}/2\lambda_k^2]}{\alpha_{Qk}\exp[\alpha_o]} \qquad (k=1,\ldots,n)\ .$$

Testing for the null-hypothesis of no Q_k-specific economies of scale is done again by constructing a confidence interval for the terms on the right hand side of equation (18) and by observing if $S_k=1$ falls within or outside the confidence interval.

3.5 The Relationship among Internal Economies

Panzar and Willig (1979) show that overall and output-specific economies of scale and economies of scope are associated by the following relationship

$$(19)\quad \varepsilon = \frac{w_k S_k + (1 - w_k)S_\eta}{1\text{-}SCOPE_k} \qquad \eta=\{1,\ldots,k-1,k+1,\ldots,n\},$$

where

$$w_k = \frac{Q_k \frac{\partial C}{\partial Q_i}}{\sum_{i=1}^{n} Q_i \frac{\partial C}{\partial Q_i}} \geq 0 \quad\text{and}\quad S_\eta = \frac{IC_\eta}{\sum_{\substack{i=1\\ i\neq k}}^{n} Q_i \frac{\partial C}{\partial Q_i}}\ ;$$

ε, $SCOPE_k$ and S_k are defined in equations (6), (15) and (17) respectively; IC_η is the incremental cost specific to the product set η; i.e.,

$$IC_\eta = C(Q_1,\ldots,Q_n) - C(0,\ldots,Q_k,\ldots,0).$$

In the absence of economies of scope, ε is the weighted average output-specific returns to scale. However, when economies of scope are present, the multi-product technology exhibits greater overall returns to scale than the weighted average of the output-specific scale elasticities.

4. EMPIRICAL RESULTS[12]

4.1 Multi-Output Models

Considering that a high degree of multicollinearity may be introduced in the TL/GTL cost model by the large number of second order terms, it was decided that the estimation of the cost function would not be attempted with more than three output and three input variables. The three output volumes are represented by Törnqvist volume indices of Bell Canada's local, directory advertising and miscellaneous service outputs (Q_1), message toll (intra-Bell, Canada, US and overseas) and WATS outputs (Q_2) and private line, TWX and other toll outputs (Q_3). Q_1 is simply referred to as local output, Q_2 as monopoly toll output and Q_3 is called competitive toll output. The three input price variables are those of labour (W_1), capital (W_2) and material (W_3). Labour price is an implicit Törnqvist price index of labour, computed as the ratio of the index of total labour cost to the Törnqvist volume index of labour. Capital price is an implicit Törnqvist price index, obtained as the ratio of the index of total capital compensation (calculated with the user cost or rental price of capital) and the Törnqvist volume index of capital. Material price is the implicit Törnqvist price index of materials, rents, supplies and services. The sample on which the models were estimated contains annual data for the period 1952 to 1978. The variables are normalized around their respective 1967 values. The data, including the indices of technological changes, are described in Kiss (1981).

In addition to the unconstrained GTL function, constrained models were also estimated. It was assumed in one model that the Box-Cox parameters for all outputs were equal ($\lambda_1=\lambda_2=\lambda_3$) and logarithmic transformation was used for the outputs ($\lambda_1=\lambda_2=\lambda_3=0$) in a further constrained model. Another estimated function used logarithmic transformation for the technology variable ($\lambda_T=0$), while one model combined two restrictions ($\lambda_1=\lambda_2=\lambda_3$; $\lambda_T=0$). Finally, the most restricted translog form ($\lambda_T=\lambda_1=\lambda_2=\lambda_3=0$) was estimated. The 3-output TL/GTL model proved to be too general. Many estimated parameters exhibited a high degree of instability and close to 50% of them were insignificant in the constrained models. As a result of instability in the parameters that appear in the test statistics, the estimation of cost complementarity and output-specific economies of scale and scope was not successful.[13] However, the scale elasticities were only slightly influenced by estimation problems. The annual estimates showed a fairly realistic pattern in the two well-behaved cost functions (with the $\lambda_1=\lambda_2=\lambda_3$ constraint) and the expansion point values gave a uniform indication of overall economies of scale in all six estimated models.

With the failure of the 3-output TL cost function to produce realistic estimates of the economic properties of Bell Canada's technology, the possibilities offered by the 3-output model were exhausted and the number of outputs was reduced to two by aggregating the monopoly and competitive toll services. The output-related parameters were generally poorly estimated in the 3-output models (instability, insignificance) and it was hoped that their quality could be improved by reducing their number.

The evidence on overall economies of scale that emerges from the 2-output TL/GTL cost models is remarkably robust. All models produced significant estimates of economies of scale at the expansion point and the estimated scale elasticities fell into a relatively narrow range between 1.44 and 1.66. The annual scale elasticity estimates exhibited a pattern which had realistic features (lower values in the early years and an increase in the degree of scale economies, resulting from the introduction of crossbar and DDD technologies). A marked decline in the technology elasticity of cost after 1970 in all models seems to suggest that the scale elasticities of the 1970's were underestimated to varying degrees. However, this problem led to radical consequences only in the case of the unconstrained model. In the constrained models, the estimated scale elasticities range between 1.18 and 1.49 during the last three years of the observation period.

The 2-output models generated a uniform, but rather weak, indication of cost complementarity between Bell Canada's local and toll service outputs. Uniformity is indicated by the fact that the estimated cost complementarity (CC) statistics were negative in each year of the sample period in four out of six models, while one model yielded negative values for 78% of the observations. Only the unconstrained GTL model produced mostly positive CC statistics. The weakness of the estimates lies partly in their statistical insignificance at the expansion point and partly in the fact that five out of six models produced upward or downward trended estimates of CC. The indication of global economies of scope was unanimous but weak. Each of the three constrained GTL models in which the estimation could be meaningfully accomplished generated positive values for SCOPE in each year of the observation period. Nevertheless, the SCOPE statistics were insignificant at the expansion point and the annual estimates had very strong upward or downward trends with some extremely high values at both ends of the period. Finally, the effort to estimate the degree of output-specific economies of scale failed in all models. A large percentage of the estimated S_k statistics had the a priori incorrect negative sign.

The estimation results with respect to overall economies of scale (ε), cost complementarity (CC) and economies of scope (SCOPE) at the expansion point are summarized in Table 1. The asymptotic standard errors are in parentheses.

Since the number of interaction terms in the 2-output TL/GTL model was still very large, multicollinearity could not be ruled out as a serious concern. In fact, the estimation problems of cost complementarity, economies of scope and especially output-specific economies of scale might be due to the still highly general nature of the specification. A comparison of the parameter estimates of the six 2-output models revealed a certain degree of instability in the first and second order output and the output-technology interaction terms. The parameters associated with these variables showed considerable fluctuation across models, some changed signs and they were generally insignificant. These variables play an important role in the calculation of test statistics for internal economies. The 2-output models seemed to suggest that some improvement with respect to the stability of parameter estimates could be achieved if the number of output-related terms were reduced.

Table 1

Internal Economies in Multi-Output Models

No. MODEL	THREE OUTPUTS	TWO OUTPUTS		
	ε	ε	CC	SCOPE
1. Unconstrained GTL	1.43*(.09)	1.44*(.06)	.02 (.02)	**
2. $\lambda_1=\lambda_2(=\lambda_3)$	1.26 (.15)	1.50*(.15)	-.49(1.88)	1.75(7.26)
3. $\lambda_k=0$	1.67*(.18)	1.64*(.14)	**	N.A.
4. $\lambda_T=0$	1.43*(.10)	1.44*(.11)	-.17 (.47)	.51(1.98)
5. $\lambda_1=\lambda_2(=\lambda_3);\lambda_T=0$	1.22 (.12)	1.44*(.15)	-.45(2.05)	1.27(5.26)
6. $TL(\lambda_k=\lambda_T=0)$	1.67*(.18)	1.66*(.14)	**	N.A.

* Significantly greater than one at the .05 level. (None of the CC or SCOPE estimates are significantly different from zero.)
** Meaningless.

4.2 Truncated Two-Output GTL Models

Several truncated models were estimated. Five of them are referred to in Table 2. The truncated models were systematically arrived at by the exclusion of parameters with insignificant t-statistics, while at the same time ensuring no significant decrease in the log of the likelihood function. Through this approach, those terms in the cost function that were not adding a significant amount of information to the overall model were excluded. The GTL model with the $\lambda_T=0$ constraint was chosen over others, partly because of its favourable economic properties in comparison with the unconstrained GTL model, and partly because likelihood ratio tests rejected three further restricted models at the .05 level.

The first group of three truncated models attempted to eliminate interaction terms associated with the toll output (Q_2), since its respective first order parameter (α_{Q2}) was found to be consistently insignificant. The eliminated parameters were related to the toll-technology interaction term ($\beta_{Q_2}=0$), then to the squared toll variable ($\delta_{22}=0$), and finally, both parameters were excluded ($\beta_{Q_2}=\delta_{22}=0$). The insignificant output interaction term (δ_{12}) was left in the model to make the estimation of cost complementarity possible. The truncated models could not be rejected on the basis of likelihood ratio tests.

Further experiments restricted the output interaction parameter (δ_{12}), along with other insignificant parameters, to zero. Two of these further truncated models are presented in the last two columns of Table 2. The first model restricted the local squared (δ_{11}), the output interaction (δ_{12}) and the toll squared (δ_{22}) parameters and the second model set the toll-technology interaction (β_{Q_2}) parameter, in addition to δ_{11}, δ_{12} and δ_{22}, to zero. These two truncated models are the end products of less restricted attempts and mark the limit to which this type of approach can be taken. Both models were narrowly rejected over the full λ_T=0 model at the .05 level of confidence, but neither model could be rejected at the .01 level.

Even though some of the estimated internal economies varied under different parametric restrictions, the truncated models offered many improvements over the full λ_T=0 2-output cost model. The most consistent indication was that of statistically significant economies of scale in the 1.23 to 1.90 interval at the expansion point, with the point estimates ranging between 1.45 and 1.62. Nearly equally consistent but mainly statistically insignificant was the indication of output-specific economies of scale. Local scale economies appeared to be greater or about the same as toll economies. The truncated models also produced some statistically insignificant evidence on local-toll economies of scope and cost complementarity. The relationship between firm level and output-specific scale elasticities, see equation (19), further suggested the presence of economies of scope in four of the five truncated models.

Table 2

Internal Economies in Truncated Two-Output GTL (λ_T=0) Models

INTERNAL ECONOMIES	β_{Q_2}=0	δ_{22}=0	β_{Q_2}=δ_{22}=0	δ_{11}=δ_{12} =δ_{22}=0	δ_{11}=δ_{12}=δ_{22} =β_{Q_2}=0
Overall scale economies:	1.47* (.11)	1.45* (.11)	1.53* (.14)	1.62* (.14)	1.62* (.13)
Cost complementarity:	-.58* (.34)	.29* (.10)	.13 (.07)	N.A.	N.A.
Scope economies:	6.50 (20.4)	-.03 (.08)	.18 (.10)	.24 (.14)	.09 (.10)
Local scale economies:	-8.29 (32.0)	1.50* (.24)	1.25 (.21)	1.26 (.31)	1.56* (.25)
Toll scale economies:	-5.54 (9.2)	1.31 (.21)	1.28 (.17)	1.10 (.15)	1.09 (.16)

* Significant at the .05 level.

4.3 Models with One Output

Another method of reducing the number of output-related terms in the TL/GTL model, alternative to truncating the function, is the reduction of the number of outputs from two to one. This was achieved by aggregating the local and toll service outputs of Bell Canada. The opportunity to obtain evidence on cost complementarity, economies of scope and output-specific economies of scale was lost in the resulting single output specifications, but it was hoped that greater stability of the parameters and improvement in the precision with which they would be estimated might enhance the estimates of overall economies of scale.

The unconstrained single output GTL function presented significant improvements in the estimation results. The model produced a very good fit, most parameters were significant, the cost function was well behaved at most observation points and the estimated economic properties were generally realistic. A sharp estimate of scale elasticity was obtained at the expansion point. Substantial economies of scale were indicated in each year after 1956. The annual scale elasticities composed a realistic pattern. The parameter estimates of the input-output interaction terms (ρ_{1Q}, ρ_{2Q}, ρ_{3Q}) and the Box-Cox transformation parameter of the output (λ_Q) were insignificantly different from zero in the unconstrained GTL model. This suggested that the specification could be reduced to a homothetic model ($\rho_{1Q}=\rho_{2Q}=\rho_{3Q}=0$) with a logarithmic transformation of the output ($\lambda_Q=0$). Two independently constrained models were estimated. One model eliminated the input-output interaction parameters ($\rho_{1Q}=\rho_{2Q}=\rho_{3Q}=0$), while the transformation of the Q variable was restricted to be logarithmic in the second constrained model. The λ_Q parameter in the first restricted model was insignificantly different from zero. A further model applied the joint hypotheses of $\lambda_Q=0$ and $\rho_{1Q}=\rho_{2Q}=\rho_{3Q}=0$. To ensure that no significant amount of information was lost by the imposed restrictions, likelihood ratio tests were applied at every step. Table 3 contains the -2λ test statistics that were computed.[14] The numbers of restrictions are shown in parentheses.

As we moved from one set of restrictions to the other, the change in the log of the likelihood function proved to be marginal. In order to test whether the technology variable should be logarithmically transformed, all the above mentioned models were estimated under the additional constraint $\lambda_T=0$. The loss in the log of the likelihood function proved to be significant in all cases and these models were rejected at the .05 level.

On the basis of likelihood ratio tests, the homothetic GTL with a logarithmic transformation of the output variable and Box-Cox transformation of the technology variable was chosen as the reduced cost model. It is important to observe that the restrictions thus accepted did not alter the properties of the cost model of Bell Canada. The parameter estimates of the four models were found to be very stable and the technological characteristics, including the scale elasticities shown in Table 4, followed a similar pattern.

The preferred homothetic $\lambda_Q=0$ cost function was well behaved. Monotonicity was satisfied at all observation points and the concavity conditions were met for 78% of the observation points. All parameter estimates, with the exception of the second order output parameter (δ_{QQ}) and the material-technology interaction term (β_3), were statistically significant at the .05 level. The model indicated that technological change was capital using ($\beta_2>0$), labour saving ($\beta_1<0$) and material neutral. Cost elasticities with respect to technology had the expected sign in each year. The absolute values were low for the first two years of the sample, peaked in 1955-56 and fell until 1964, after which they were trended upward. Labour and material price elasticities had the correct negative sign throughout the sample period, but the capital price elasticities were positive for the first 5 years. Demand for all three productive factors was price inelastic, with material demand being the relatively most sensitive (-.5) and capital demand the most insensitive (almost perfectly inelastic at -.05) to price changes. The price elasticity estimates were stable through time. Partial elasticities of factor substitution showed a high degree of labour-material substitutability and, more importantly, a low degree of substitutability between capital and labour. Both were fairly stable throughout the observation period. Capital and material were complementary in the early and mid 1950's, but neither complementarity nor substitutability was indicated during the 1960's and 1970's.

Table 3

Likelihood Ratio Tests for Single Output Models

CONSTRAINED MODELS \ UNCONSTRAINED MODELS	1. UNCONSTRAINED GTL	2. $\rho_{1Q}=\rho_{2Q}=\rho_{3Q}=0$	3. $\lambda_Q=0$	4. $\lambda_Q=\rho_{1Q}=\rho_{2Q}=\rho_{3Q}=0$
2. $\rho_{1Q}=\rho_{2Q}=\rho_{3Q}=0$	.58 (3)	-	-	-
3. $\lambda_Q=0$	.16 (1)	-	-	-
4. $\lambda_Q=\rho_{1Q}=\rho_{2Q}=\rho_{3Q}=0$	.63 (4)	.01 (1)	.43 (3)	-
5. $\lambda_T=0$	10.0 (1)	-	-	-
6. $\lambda_T=\rho_{1Q}=\rho_{2Q}=\rho_{3Q}=0$	14.58 (4)	14.0 (1)	-	-
7. $\lambda_Q=\lambda_T=0$	10.70 (2)	-	10.54 (1)	-
8. $\lambda_Q=\lambda_T=\rho_{1Q}=\rho_{2Q}=\rho_{3Q}=0$	14.64 (5)	14.05 (2)	14.57 (4)	14.05 (1)

Single output GTL cost functions with a logarithmic technology variable were also considered, even though the hypothesis of $\lambda_T=0$ was rejected, in order to obtain further evidence on the sensitivity of scale elasticity estimates to changes in the underlying assumptions. One model used the Box-Cox transformation of the output variable and a second model was a translog. Both models were estimated in non-homothetic and homothetic forms. Only slight changes and no improvements were noticed in the estimation results. As the logarithmic transformation of the technology variable generated some changes (especially towards the end of the period) in the estimates of the technology elasticity of cost, the scale elasticities became lower than in the $\lambda_T\neq0$ models. However, their pattern did not change noticeably.

Table 4

Annual Scale Elasticity Estimates in Single Output GTL Models

YEAR	1. UNCONSTRAINED GTL	2. $\rho_{1Q}=\rho_{2Q}=\rho_{3Q}=0$	3. $\lambda_Q=0$	4. $\lambda_Q=0$; $\rho_{1Q}=\rho_{2Q}=\rho_{3Q}=0$
1952	1.08	1.06	1.07	1.06
1953	1.08	1.07	1.08	1.07
1954	1.09	1.08	1.09	1.08
1955	1.07	1.07	1.08	1.06
1956	1.09	1.09	1.10	1.09
1957	1.13	1.14	1.15	1.14
1958	1.23	1.24	1.25	1.23
1959	1.29	1.29	1.31	1.29
1960	1.39	1.40	1.41	1.40
1961	1.45	1.45	1.47	1.45
1962	1.48	1.49	1.50	1.49
1963	1.60	1.61	1.62	1.61
1964	1.72	1.72	1.74	1.72
1965	1.73	1.73	1.75	1.73
1966	1.74	1.74	1.76	1.74
1967	1.73	1.73	1.75	1.73
1968	1.77	1.77	1.79	1.77
1969	1.78	1.78	1.79	1.78
1970	1.78	1.79	1.80	1.79
1971	1.78	1.80	1.80	1.80
1972	1.77	1.79	1.79	1.79
1973	1.75	1.77	1.77	1.77
1974	1.74	1.77	1.76	1.76
1975	1.72	1.76	1.75	1.76
1976	1.71	1.75	1.74	1.75
1977	1.69	1.75	1.73	1.74
1978	1.71	1.76	1.74	1.75

In summary, the single-output TL/GTL cost functions produced sharp and stable estimates of overall economies of scale in Bell Canada. Based on test results on various hypotheses, the homothetic GTL model in which output was logarithmically transformed appeared to be preferable to other models. The expansion point values of scale elasticity, together with their asymptotic standard errors, are shown in Table 5.

Table 5

Scale Elasticity Estimates in Single Output Models

No.	MODEL	ε
1.	Unconstrained GTL	1.73 (.09)
2.	$\rho_{iQ}=0$ (i=1,2,3)	1.73 (.09)
3.	$\lambda_Q=0$	1.75 (.06)
4.	$\lambda_Q=0$, $\rho_{iQ}=0$ (i=1,2,3)	1.73 (.05)
5.	$\lambda_T=0$	1.61 (.07)
6.	$\lambda_T=0$, $\rho_{iQ}=0$ (i=1,2,3)	1.69 (.09)
7.	TL($\lambda_Q=\lambda_T=0$)	1.62 (.06)
8.	TL($\lambda_Q=\lambda_T=0$), $\rho_{iQ}=0$ (i=1,2,3)	1.69 (.06)

Note: All estimates are significantly greater than one at the .05 level.

4.4 The Sensitivity of the Best One-Output Model

The expansion point estimate of scale elasticity in the preferred homothetic $\lambda_Q=0$ GTL cost model (No. 4 in Table 5) increased only very slightly in the non-homothetic alternative model (No. 3) and remained unchanged when Box-Cox transformation was applied to the output variable, regardless whether the model was non-homothetic (No. 1) or homothetic (No. 2). Further evidence on the robustness of the scale elasticity estimate was obtained by re-estimating the model (1) on slightly different data samples, (2) with alternative assumed forms of technological change, (3) with alternative proxy variables for technological change, (4) with different measures of the cost of capital and (5) with various assumed capital utilization rates.

The sensitivity of scale elasticity estimates to sample variation was tested by omitting years both at the beginning and at the end of the period of observation. In order to prevent a serious loss of information, only a few data points were eliminated. When the function was reestimated for various sub-periods, the scale elasticity estimates changed only very slightly at the expansion point.[15]

Three alternate forms of technological change were considered, viz. factor augmenting, output augmenting and Hicks-neutral. Following Gollop and Roberts (1981), the alternate forms were constructed by restricting the technology-related parameters of the preferred GTL model.[16] The most realistic statistical and economic properties were obtained under the assumption of factor augmentation. However, the scale and technology elasticity estimates were strongly trended in this model. The scale elasticity increased from 1.11 in 1952 to 2.20 in 1978 and the expansion point value (1.56) fell below the lower bound of the 95% confidence interval of the expansion point estimate of the preferred model. The other two models (output augmenting and Hicks-neutral) exhibited signs of serious mis-specification due to the restrictions applied to the technology-related parameters. The magnitude of parameter estimates changed rather noticeably and some parameters switched signs. The downward trended technology elasticities of cost were positive for the first years of the sample and stayed uncharacteristically close to zero in later years when they acquired the correct negative sign. Correspondingly, the annual estimates of scale elasticity were trended upward and were rather high. The expansion point values were well above the upper bound of the 95% confidence interval of the estimate in the preferred model, but some overlap existed between the confidence intervals. Several conclusions can be reached. First of all, factor augmentation lowered, and output augmentation as well as Hicks-neutrality increased, the estimated degrees of economies of scale. Secondly, robustness of scale economy estimates was evident in the sense that the scale elasticities indicated a high degree of statistically significant economies of scale in all three models. Thirdly, other economic and statistical properties appeared to be more sensitive to the variation in the assumed form of technological changes than scale elasticity. Finally, with respect to scale elasticity, the trended nature of the annual estimates appeared to be as important a symptom of mis-specified technological changes as the significant differences in the expansion point estimates.

Sensitivity tests with respect to the technology proxy variable were carried out with the aid of three different models. All estimations presented above used the T2 variable of Kiss (1981), because this measure produced generally superior empirical results. When the T3 and T4 indices were substituted for T2, the scale elasticity estimate at the expansion point remained unchanged (with T3) or increased very slightly (with T4). Another proxy, the FNEW3 variable, also taken from Kiss (1981), represented a dramatic change in the measurement of the technology index, as T2 had grown almost twice as fast as FNEW3 and its pattern was also considerably different (most of its faster growth occurred during the first half of the period). Predictably, the scale elasticity increased significantly at the expansion point, indicating that the estimate would be sensitive to major measurement errors in the technology index. It was also indicated that major errors would destroy the reasonableness of the estimates as the technology elasticity of cost acquired the wrong sign for the majority of the data points and other estimation problems were also encountered.

The next set of sensitivity tests involved the capital price variable. The user cost of capital in Kiss (1981) was altered in three ways: (1) the assumption of zero capital gain ($q_t=0$) was relaxed; (2) the cost of common equity was altered by substituting expected yield for actual yield and changing the growth factor[17] in its formula; (3) the average cost of debt was substituted for the cost of new debt. The alternative user cost of capital measurements resulted in statistically insignificant increases in the expansion point estimate of scale elasticity.

The assumption of full capital stock utilization was relaxed in the last set of sensitivity tests. The unavailability of information and some conceptual difficulties made it impossible to approximate either the level or the annual rates of change of utilization rates. Thus, a low (30)% and a high (70)% level of utilization were arbitrarily chosen for

Table 6

Scale Elasticity Estimates in Homothetic λ_Q=0 Single Output GTL Models

No.	MODEL			ε
1.	Best model:	GTL, $\lambda_Q=0$, $\rho_{iQ}=0$ (i=1,2,3)		1.73 (.05)
2.	Period:	1954-1978		1.72 (.06)
3.		1956-1978		1.79 (.09)
4.		1952-1977		1.73 (.05)
5.		1952-1976		1.68 (.04)
6.		1952-1975		1.70 (.04)
7.	Technology:	Factor augmenting		1.56 (.03)*
8.		Output augmenting		1.93 (.06)*
9.		Hicks-neutral		1.92 (.07)*
10.	Technology proxy:	T3		1.73 (.06)
11.		T4		1.74 (.06)
12.		FNEW3		2.38 (.22)*
13.	Cost of capital:	$\dot{q}_t \neq 0$		1.80 (.06)
14.		alternative equity cost		1.76 (.05)
15.		alternative cost of debt		1.75 (.08)
16.	Cap. utilization:	30% in 1967,	0% growth	1.71 (.08)
17.			1%	1.60 (.09)*
18.			4%	1.68 (.16)
19.			8%	1.52 (.19)*
20.		70% in 1967,	0% growth	1.72 (.06)
21.			1%	1.52 (.06)*
22.			4%	1.23 (.06)*

*The point estimate is outside the 95% confidence interval of the best model's estimate.

the expansion point (1967). The estimated scale elasticities showed only negligible changes when either of the two utilization levels was kept constant for the entire sample period. Since improvement over time in utilization rates might be responsible for some of the estimated economies of scale, a 1% annual improvement was superimposed on the chosen expansion point levels of utilization. In order to carry the experiment to the extreme, annual improvements of 4% and 8% (for the 30% level only) were also assumed. Capital stock was rescaled by the alternatively computed annual utilization rates in the total cost calculation and capital price was kept unchanged. The scale elasticity estimates were significantly reduced but remained high (1.52 to 1.60) at the expansion point, when 1% annual improvement was assumed. The concavity conditions were violated at all observation points and other estimation problems were encountered under higher improvement rates. The expansion point scale elasticity remained significantly greater than 1.0 even in the most extreme cases. It can be concluded that if Bell Canada's capital utilization had indeed improved during the period of observation, then the improvement appears to have been responsible only for a small portion of the estimated scale elasticities.

Table 6 illustrates the sensitivity of expansion point estimates of scale elasticity. The asymptotic standard errors are in parentheses. As they indicate, all estimates are significantly greater than 1.0 at the .05 level.

5. SUMMARY, COMPARISON AND CONCLUSIONS

Twenty-five flexible translog and generalized translog cost models of Bell Canada have been examined in this paper. The most important conclusion that can be drawn from the models is that they offer a robust indication of substantial overall economies of scale in Bell Canada. A certain pattern of economies of scale estimates emerges in Table 7 from the comparison of models with one, two and three outputs.

It is interesting to observe that both the expansion point estimates of scale economies and their 95% confidence intervals substantially overlap in the 2 and 3-output models. The expansion point estimates of scale economies are consistently higher in the single-output models than in the multi-output models and the overlap is narrow (1.61 to 1.67). However, due to their low standard errors, the entire 95% probability range of single-output estimates falls within the confidence interval of the multi-output estimates.

The effect of the Box-Cox generalization of variable transformation can be analysed in Table 8 by following the changes in estimates of economies of scale at the expansion point.

The Box-Cox generalization of the output transformation resulted in a greater reduction of the estimated economies of scale in the 3-output models than in the 2-output models and only a negligible effect is observable in the case of single output models. At the same time, the Box-Cox generalization of the technology variable either left the scale elasticity estimates unchanged or resulted in very small changes (usually increases) in the values of ε.

Table 7

A Summary of Scale Elasticity Estimates

No.	MODEL	ε*	95% Confidence Intervals
1.	3-output	1.22-1.67	.96-2.03
2.	2-output	1.44-1.66	1.14-1.94
3.	2-output, truncated	1.45-1.62	1.23-1.90
4.	1-output	1.61-1.75	1.47-1.91

*At the expansion point (1967).

In order to compare our results to those of other econometric studies, a summary of estimates of economies of scale at the 1967 observation point (or at the mean observation, which closely corresponds to 1967) in seven externally constructed TL or GTL cost functions of Bell Canada is given in Table 9.

With the exception of the GTL function of Fuss and Waverman (1981), all models suggested economies of scale in Bell Canada. Fuss and Waverman found that the Box-Cox transformation of the output variables reduced (from 1.43 to .94) the economies of scale estimate. This finding is consistent with our results, even the magnitude of the reduction is similar. However, our estimates were significantly higher than those of Fuss and Waverman. There are two major differences between the models which might be responsible for the difference in the level of scale elasticity estimates. First, Fuss and Waverman assumed endogenous toll outputs (unhindered monopolistic profit maximization with respect to message and other toll service outputs), while exogenous toll outputs (cost minimizing behaviour) were assumed in our models. Secondly, Fuss and Waverman assumed output augmenting technological changes, while the hypothesis of output augmentation was rejected in our models.

There is some overlap between our scale elasticity estimates and those from externally conducted studies in the case of the 3-output models, but our estimates are generally considerably higher in 3-output models and always higher in models with one or two outputs.

The comparison of scale elasticities in external models with one, two and three outputs is inconclusive, except for Denny et al who found lower estimates in the 3-output model than in the single output case. As mentioned above, our models yielded the same relationship.

Most external studies failed to produce a reasonable pattern for the annual economies of scale estimates. In four of the seven models that are shown above, the estimates were very strongly trended upwards. In contrast, the model by Fuss and Waverman (1981), yielded estimates in the .9 to 1.1 range, which seems to suggest that the underlying technology is

Table 8

The Sensitivity of Scale Elasticity Estimates to Box-Cox Transformation

RESTRICTION	$\lambda_k \neq 0$	$\lambda_k = 0$	RESTRICTION	$\lambda_T \neq 0$	$\lambda_T = 0$
3-output models (k=1,2,3)					
$\lambda_1 \neq \lambda_2 \neq \lambda_3$, $\lambda_T \neq 0$	1.43	1.67	$\lambda_1 \neq \lambda_2 \neq \lambda_3$	1.43	1.43
$\lambda_1 = \lambda_2 = \lambda_3$, $\lambda_T \neq 0$	1.26		$\lambda_1 = \lambda_2 = \lambda_3$	1.26	1.22
			$\lambda_k = 0$	1.67	1.67
$\lambda_1 \neq \lambda_2 \neq \lambda_3$, $\lambda_T = 0$	1.43	1.67			
$\lambda_1 = \lambda_2 = \lambda_3$, $\lambda_T = 0$	1.22				
2-output models (k=1,2)					
$\lambda_1 \neq \lambda_2$, $\lambda_T \neq 0$	1.44	1.64	$\lambda_1 \neq \lambda_2$	1.44	1.44
$\lambda_1 = \lambda_2$, $\lambda_T \neq 0$	1.50		$\lambda_1 = \lambda_2$	1.50	1.44
			$\lambda_k = 0$	1.64	1.66
$\lambda_1 \neq \lambda_2$, $\lambda_T = 0$	1.44	1.66			
$\lambda_1 = \lambda_2$, $\lambda_T = 0$	1.44				
1-output models (k=1)					
$\lambda_T \neq 0$ (homothetic)	1.73	1.73	$\lambda_Q \neq 0$	1.73	1.69
$\lambda_T = 0$	1.69	1.69	$\lambda_Q = 0$	1.73	1.69

linearly homogeneous. (The Fuss-Waverman estimates were also trended after 1958 in their narrow range.) Two models (the single-output and 2-output translog functions of Smith and Corbo, 1979) indicated that scale elasticities gradually, but substantially, increased from 1956 to 1964 and moderately declined after 1964. This pattern appears reasonable to the extent that direct distance dialing and the crossbar switching technology, whose introduction began in 1956, probably increased scale elasticity and, due partly to the increasing sphere of activities of Bell Canada (e.g., intensifying regulatory activities in the 1970's) and partly to the demand slowdown of recent years, some decline in the scale elasticities during the 1970's can be reasonably expected. These characteristics as well as the remaining unrealistic features (viz. high ε values for the first years of the sample period, an inexplicable decline from 1952 to 1956 and the relatively early peak of scale elasticity in 1964) were shared by our multiple output models and were eliminated only in the single output models. The preferred single output homothetic GTL function with logarithmic output shows relatively stable but low values for 1952-55, replaces the 1964 peak with a flat maximum in the 1968 to 1971 period and makes the decline of scale elasticities during the 1970's less pronounced.

The second conclusion that can be drawn from the study is that statistically insignificant economies of scope and cost complementarity were generally indicated by the estimated multi-output cost models. The standard errors of estimates (as well as the annual estimates themselves) improved when the number of outputs was reduced from three to two and this suggests that the insignificance of the estimates is due to a high degree of multicollinearity in the models.

An almost perfectly uniform indication of economies of scope was obtained for all output categories and all years in all models, where the form of the function allowed for the computation of the $SCOPE_k$ statistic. Compared to the two well-behaved 3-output cost functions, the 2-output models reduced the standard error of $SCOPE_k$ rather drastically and resulted in

Table 9

Comparison of Estimates of Economies of Scale in Various Studies of Bell Canada

STUDY	ε
3-output models	
1. Fuss - Waverman (1978)	1.46*
2. Denny et al (1979)	1.46
3. Fuss - Waverman (1981)	.94
2-output models	
4. Smith - Corbo (1979)	1.20
5. Breslaw - Smith (1980)	1.29
1-output models	
6. Smith - Corbo (1979)	1.29
7. Denny et al (1979)	1.58

*Re-estimated in Denny et al (1979). The original estimate was 1.06 and the hypothesis of constant returns to scale could not be rejected in the original model.

generally reasonable values for the annual estimates.

Local economies of scope were indicated by the negative sign of estimates of the cost complementarity statistic between local and toll service outputs. However, none of the computed statistics were significantly different from zero at the expansion point and the annual estimates were generally trended. All constrained 2-output models indicated complementarity (the full model produced a very small positive value) between local and total toll, and the two well-behaved 3-output models gave a further suggestion that complementarity existed between local and both monopoly and competitive toll. It is interesting to observe that the indication of local/competitive toll complementarity was very strong at the expansion point and the annual estimates were not trended.

Only one external study offers estimates of global economies of scope. Fuss and Waverman (1981) found that the annual estimates of economies of scope, specific to other toll services were downward trended and switched signs (from positive to negative) in 1962. The estimates were small and all but one were insignificant. Our comparable other toll-specific economies of scope estimates were positive for the entire period of observation, but the values were trended and fell into the reasonable range only at the beginning of the period.

Table 10 sums up the evidence on cost complementarity in externally conducted econometric studies of Bell Canada. The computed CC values are shown either at the expansion point of the model (Fuss-Waverman, Denny et al) or in 1967 (Smith-Corbo, Breslaw-Smith).

The evidence is inconclusive. The indication from our 3-output models is similar to Denny's results, but there is no further resemblance between our estimates and those shown in the table.

The third conclusion is that <u>our models failed to produce statistically significant and reasonable estimates of output-specific economies of scale with any degree of consistency</u>. In general, the estimates either had the a priori incorrect negative sign, or the positive estimates were unrealis-

Table 10

Comparison of Estimates of Cost Complementarity in Various Studies of Bell Canada

COST COMPLEMENTARITY	Fuss - Waverman (1978)	Denny et al (1979)	Fuss - Waverman (1981)	Smith - Corbo (1979)	Breslaw -Smith (1980)
Local - Message Toll	-.016	-.062*	.099	.000	-.001
Local - Other Toll	.009*	-.037*	.042	-	-
Message - Other Toll	-.002	.017*	-.021*	-	-
Local - Total Toll	-	-	-	-	-

*Significantly different from zero.

tic in magnitude and in pattern in the nontruncated models.[18] On the other hand, encouraging results were obtained from the truncated 2-output models. Two models yielded reasonable and significant estimates of local economies of scale, and fully consistent estimates of overall and output-specific economies of scale and global economies of scope were obtained from three truncated models. These models suggested a higher degree of economies of scale in local than in toll services and also provided an indirect indication of global economies of scope by estimating the degree of overall economies of scale higher than that of either of the two output-specific economies of scale. Based on our estimation results, it appears that the truncation of cost functions (justified by both t-tests and likelihood ratio tests) is a promising method of breaking the multicollinearity of multi-output models and obtaining econometric evidence on economies of scope and output-specific economies of scale.

To summarize the conclusions, a robust indication of substantial overall economies of scale was obtained from the estimated cost models of Bell Canada. The evidence on economies of scope was uniform but not as convincing, either from a statistical or from an economic point of view. Finally, the difficulties associated with the estimation of multiple output cost models and the possible hazards of estimating hypothetical production cost for zero output levels prevented us from producing strong econometric evidence on economies of scale with respect to Bell Canada's local and toll services. However, the statistically justified truncation of cost functions yielded encouraging results.

FOOTNOTES:

[1] The paper represents the views of the authors and not necessarily those of Bell Canada.

[2] One advantage is that output is an exogenous variable in the cost function. This makes cost functions especially suitable for regulated public utilities, whose prices are determined by regulatory agencies, hence their output levels are exogenously set. Another advangage is the relative ease with which the multi-output case can be considered in cost models. Multi-output specifications are required for obtaining evidence on cost complementarity, economies of scope and output-specific economies of scale. Still another advantage of cost functions is that they yield direct estimates of such important properties as marginal costs, cost elasticities and partial elasticities of factor substitution. The most notable disadvantages are the measurement problems of the cost of capital and the lack of direct marginal product estimates.

[3] The translog cost function was first used to estimate economies of scale by Christensen and Greene (1976) and was used in empirical studies of Bell Canada by Fuss and Waverman (1977, 1978), Denny et al (1979), Smith and Corbo (1979), Breslaw and Smith (1980) and also in studies of the Bell System by Nadiri and Schankerman (1979) and Christensen et al (1980). The Box-Cox transformation (Box and Cox, 1962) of variables of cost functions was proposed by Khaled (1978) and the first GTL of Bell Canada was published by Fuss and Waverman (1981).

[4] Total cost and input prices remain in logarithmic form in GTL in order to facilitate easy parameter restrictions to ensure the first degree homogeneity of the cost function in input prices. See the restrictions in (3).

[5] T* is introduced as a variable in the expansion of the production frontier. The specification allows for input-neutral as well as biased changes in technology. The same solution can be seen in a number of sources, e.g., Smith and Corbo (1979). In contrast, Fuss and Waverman (1977, 1978, 1981) and Denny et al (1979) assumed input or output augmenting technological changes in their multiple output models.

[6] The restriction to first degree homogeneity ensures that when all input prices are changed by the same percentage and outputs and technology remain unchanged, the resulting percentage change in total cost will be equal to the equiproportional input price change.

[7] The material cost share equation is deleted.

[8] There are several reasons for regarding exogenous toll output as a more realistic assumption than endogenous toll output for Bell Canada. First, the overwhelming majority of toll rates are set basically exogenously; i.e., influenced to various degrees, but not determined by Bell Canada. This applies to all message toll service categories. Intra-Bell rates are regulated, while TCTS, adjacent member, US and overseas long distance tariffs are set, usually for several years, by bilateral and multilateral agreements and are subject to approval by the CRTC. Secondly, the assumption that Bell Canada reaps maximum monopoly profit on services whose prices are endogeneously set appears to be unrealistic, because of the competitive nature of these services. Thirdly, the internal procedures of Bell Canada reflect the company's ambition to minimize production costs but do not reveal any pursuit of a monopolistic profit maximum. The company's budgeting process and the procedures aimed at determining the level of rate increases do not contain calculations equating the marginal costs and marginal revenues of services. Finally, there are indications (see Bell Canada General Increase in Rates (1980), Exhibit Nos. B-80-200 (Section 5) and B-80-234 that demand at least for the largest categories of toll services may be price inelastic. Monopolistic profit maximum cannot exist in the region of price inelastic demand.

[9] For more on the problems of measuring economies of scale, and some conceptual clarification, see Hanoch (1975).

[10] The restrictions in (9) imply constant returns to scale at each observation point in TL, but the GTL function is restricted by them to exhibit constant returns to scale at the expansion point only, where $Q_k=1$.

[11] See Baumol and Braunstein (1977).

[12] A more elaborate description of empirical results was presented at the conference and is available from the authors on request.

[13] The estimates of output-specific economies of scale and scope were

either negative or had very large positive values. However, the two well-behaved cost functions (in which the Box-Cox parameters of output transformation were constrained to be equal) indicated cost complementarity between local and competitive toll services for each year of the observation period and the annual estimates were not trended. An inconsistent indication of cost complementarity was obtained with respect to local and monopoly toll services, while no complementarity was evident between monopoly and competitive toll services.

[14] The critical χ^2 values are the following:

NO. OF RESTRICTIONS:	$\chi^2_{.05}$	$\chi^2_{.01}$
1	3.84	6.63
2	5.99	9.21
3	7.81	11.34
4	9.49	13.28
5	11.07	15.09

The test results in column 1 suggest that the unconstrained GTL model can be rejected in favour of the homothetic GTL ($\lambda_Q=0;\rho_{1Q}=\rho_{2Q}=\rho_{3Q}=0$) and homothetic TL ($\lambda_Q=\lambda_T=0;\rho_{1Q}=\rho_{2Q}=\rho_{3Q}=0$); the latter at the .01 level only. However, when the homothetic GTL ($\lambda_Q=0;\rho_{1Q}=\rho_{2Q}=\rho_{3Q}=0$) model is tested against the homothetic TL, the latter is rejected at both significance levels.

[15] Small changes were registered in the scale elasticity estimates at the beginning of the observation period. The 1975 estimate of ε increased gradually as the last year, the last two years and the last three years were omitted in successive steps.

[16] Each augmentation function $A_i(T)$ is specified as a second-order expansion of T, indexed to unity when T=1 (i.e. at the expansion point); therefore,

$$A_i(T) = \exp\left[\theta_i T^* + \tfrac{1}{2}\phi_i (T^*)^2\right] \qquad (i=1,2,3,Q),$$

where T^* represents the Box-Cox transformation of T. For the factor-augmented model, $A_i(T)$ augments the price of labour (i=1), capital (i=2) and materials (i=3). In the output-augmented model, $A_Q(T)$ is used to augment output. The Hicks-neutral model is specified by setting $\beta_1=\beta_2=\beta_Q=0$ in the homothetic $\lambda_Q=0$ GTL cost model. F-tests were conducted in order to determine the validity of the restrictive hypotheses in the factor augmenting, output augmenting and Hicks-neutral specifications of technological changes. The F-tests rejected all three of the hypothesized models. The sensitivity analysis was done in order to establish the consequences of mis-specifying the form of technological changes with respect to the statistical and economic properties (especially the scale elasticity estimates) of the cost model of Bell Canada.

[17] Expected yield is the annualized quarterly dividend in percent of the

average market price of common shares, declared in the fourth quarter of the previous year. Actual yield is the declared dividend relative to the actual average market price of common shares in the test year. The 10-year log-linear average growth rate of dividends and earnings per share was substituted for that of dividends alone.

[18] In the 3-output GTL cost model of Fuss and Waverman (1981), the hypothesis of constant returns to scale with respect to other toll services was rejected and the annual estimates of economies of scale of other toll fell in the 2.12 to 2.37 range.

REFERENCES:

[1] Baumol, W.J., and Braunstein, Y.M. (1977) Empirical study of scale economies and production complementarity: The case of journal publication, Journal of Political Economy, Vol 85, no. 5 (October) 1037-48.

[2] Baumol, W.J., Fischer, D., and Nadiri, M.I. (1978) Forms of empirical cost functions to evaluate efficiency of industry structure, Paper no. 30, Centre for the Study of Business Regulation, Graduate School of Business Administration, Duke University.

[3] Bell Canada General Increase in Rates (1980) Part B: Memoranda of Support (February 19).

[4] Box, G.E.P., and Cox, D.R. (1962) An analysis of transformations (with discussions), Journal of the Royal Statistical Society, Series B, 211-243.

[5] Breslaw, J., and Smith, B. (1980) Efficiency, equity and regulation: An econometric model of Bell Canada, Final Report to the Department of Communications, (March).

[6] Christensen, L.R., Cummings, D., and Schoech, P.E. (1980) Econometric estimation of scale economies in telecommunications, Paper no. 8013, Social Systems Research Institute, University of Wisconsin, Madison, Wis.

[7] Christensen, L.R., and Greene, W.H. (1976) Economies of scale in U.S. electric power generation, Journal of Political Economy, Vol 84, no. 4 (August) 655-676.

[8] Corbo, V., Breslaw, J., Dufour, J.M. and Vrljicak, J.M. (1979), A simulation model for Bell Canada: Phase II, Special study No. 77-002, Institute of Applied Economic Research, Concordia University, Montreal (March).

[9] Denny, M., Fuss, M., and Everson, C. (1979) Productivity, employment and technical change in Canadian telecommunications: The case for Bell Canada, Final Report to the Department of Communications (March).

[10] Fuss, M., and Waverman, L. (1977) Multi-product, multi-input cost

functions for a regulated utility: The case of telecommunications in Canada, Presented at the N.B.E.R. Conference on Public Regulation, Washington (December).

[11] Fuss, M., and Waverman, L. (1978) Multi-product multi-input cost functions for a regulated utility: The case of telecommunication in Canada, Draft, Institute for Policy Analysis, (June); Revision of Fuss and Waverman (1977).

[12] Fuss, M., and Waverman, L. (1981) The regulation of telecommunications in Canada, Technical Report No. 7, Economic Council of Canada, (March).

[13] Gollop, F.M., and Roberts, M.J. (1981) The sources of economic growth in the U.S. electric power industry, in Cowing, T.G., and Stevenson, R.E. (eds.), Productivity Measurement in Regulated Industries (Academic Press, New York) 107-143.

[14] Hanoch, G. (1975) The elasticity of scale and the shape of the average costs, American Economic Review, Vol 65, no. 3 (June) 492-496.

[15] Khaled, M.S. (1978) Productivity analysis and functional specification: A parametric approach, Ph.D. dissertation, University of British Columbia, Department of Economics (April).

[16] Kiss, F. (1981) Productivity gains in Bell Canada, Presented at the conference "Telecommunications in Canada: Economic Analysis of the Industry", Montreal (March); included in condensed form in this volume.

[17] Nadiri, M.I., and Schankerman, M.A. (1979) The structure of production, technological change and rate of growth of total factor productivity in the Bell System, New York University, National Bureau of Economic Research.

[18] Panzar, J.C., and Willig, R.D. (1979) Economies of scope, product specific economies of scale, and the multi-product competitive firm, Bell Laboratories (The research paper was written in 1978 and revised in 1979).

[19] Shephard, R.W. (1970) Cost and Production Functions, (Princeton University Press, Princeton, N.J.).

[20] Smith, J.B., and Corbo, V. (1979) Economies of scale and economies of scope of Bell Canada, Institute of Applied Economic Research, Concordia University, Final Report to the Department of Communications (March).

[21] Willig, R.D. (1979) Multi-product technology and market structure, American Economic Review, Vol 69, no. 2 (May) 346-351.

[22] Zellner, A. (1962) An efficient method for estimating seemingly unrelated regressions and tests for aggregation bias, Journal of the American Statistical Association, Vol 57 (June) 348-368.

PART 1
PRODUCTION ANALYSIS

Total Factor Productivity

Economic Analysis of Telecommunications:
Theory and Applications
L. Courville, A. de Fontenay and R. Dobell (eds.)

PRODUCTIVITY GAINS IN BELL CANADA

FERENC KISS
BELL CANADA

1. INTRODUCTION

The Bell Canada productivity study was conceived in the mid-1960's as a consequence of management's realization that, in addition to the multitude of operational efficiency measures that had existed for several decades and were used essentially by lower and middle management as a tool of control and evaluation in their everyday work, all-inclusive aggregate economic measures of performance were needed. The main purpose of total factor productivity measures was the broad evaluation of overall productive performance for executive and regulatory use.

The work has increased in complexity as measurement methods were gradually refined and the analysis of productivity gains and their effect on the company's operations was expanded. For instance, the analysis of productivity gains now utilizes aggregate econometric cost models of Bell Canada in an effort to establish the approximate effect on productivity gains of technological changes, economies of scale and other factors. Further analyses are aimed at determining the impact of productivity gain on the company's output prices, revenues, production costs and profits.

This paper has two objectives. The first objective is to discuss the issues of productivity measurement and analysis that telephone companies must face when developing a study of their productivity improvements. This objective is served by a brief discussion on methodological considerations (Section 2), an elaboration on some of the major measurement problems and an account of how the component variables are measured in Bell Canada (appendix). The second objective is to describe and analyse Bell Canada's productivity performance during the period 1952 to 1980. This is accomplished in Section 3, where it is concluded that Bell's productivity gains have been high, although there were several unfavourable years and periods from the point of view of productivity, and that approximately a quarter of Bell's productivity gains has been generated by improvements in production technology while the remaining three quarters seem to have been generated by growth in demand for the telecommunications services that Bell Canada offers.

2. METHODOLOGY

At the expense of some simplification, the various approaches to productivity measurement that have been proposed in the literature can be classified in two broad categories: the indexing approach and the econometric approach.[1]

The indexing approach measures productivity gains as the difference between the aggregate growth rates of output and input and utilizes index number formulae to obtain the aggregates. The econometric approach uses additional information about the structure of the production process, derived from the parameter estimates of production or cost functions. The indexing approach views the productivity gain as a residual of output growth which cannot be explained by proportional growth in inputs. The econometric approach attempts to measure productivity gains by component.

Both approaches are based on the neoclassical theory of production. For a brief summary of the underlying theoretical issues,[2] let us begin with a general transformation function,

$$H(Q_{1t}, \ldots, Q_{nt}; X_{1t}, \ldots, X_{mt}; t) = 0,$$

in which Q_{it} $(i=1,\ldots,n)$ and X_{jt} $(j=1,\ldots,m)$, respectively, refer to outputs and inputs at time t. Aggregates of output and input at time t can be denoted by Q_t and X_t and their proportionate rates of growth by $\dot{Q}$ and $\dot{X}$.[3] The productivity gain is then defined as

$$(1) \quad P\dot{R} = \dot{Q} - \dot{X}.$$

The condition for measuring productivity gains in this manner is that the transformation function is of the homothetic weakly separable form; i.e., it can be written as

$$\begin{aligned} H(Q_{1t},\ldots,Q_{nt};X_{1t},\ldots,X_{mt};t) &= H'[G'(Q_{1t},\ldots,Q_{nt}),F'(X_{1t},\ldots,X_{mt}),t] \\ &= H''[G''(Q_t),\ F''(X_t,\ t)]. \end{aligned}$$

The traditional production function

$$(2) \quad G(Q_t) - F(X_t,\ t) = 0$$

is obtained, when the homothetic separability is additive.

Productivity gains are regarded as the consequence of technical progress and several authors, e.g., Solow (1957), Jorgenson and Griliches (1967), have established the conditions under which the residual or index number measure of productivity coincides with the effect on output (or cost) of technological changes. The conditions are (1) constant returns to scale and (2) perfect competition in the input and output markets.

Abramovitz (1956), Fabricant (1959), Kendrick (1961), Denison (1962) and others emphasize that residually measured productivity gains capture, in addition to technical progress, a multitude of systematic and random effects and that they are "a measure of our ignorance". There are two sources of other influences, viz. random disturbances in the output or cost of the firm and violations of the above mentioned assumptions. Errors in optimization, economies of scale and measurement errors are all parts of the residual productivity gain. Some monopoly phenomena (cross-subsidized prices and rate of return constraint) were added to the list of components in recent years.[4] There is reason to believe that especially strong violations of the assumptions of perfect competition and constant returns to scale exist in the case of a regulated public utility like Bell

Canada. The product market was largely monopolized over the study period, marginal cost was not a declared "rate setting objective" and social equity considerations appear to have resulted in some instances of cross-subsidization (e.g., between residential and business services). Regarding returns to scale, most econometric cost studies indicate that Bell's technology is characterized by increasing returns to scale.[5]

Attempts to decompose residually measured productivity gains by econometrically estimated components[6] constitute what is called the econometric approach to productivity measurement. This approach estimates production or cost functions in which some of the restrictive assumptions of the indexing approach are relaxed, phenomena such as increasing returns to scale, non-marginal-cost pricing or rate of return regulation are modeled, and the parameter estimates of the functions are used to measure components of productivity gains.

2.1 The Indexing Approach

The Bell Canada productivity study uses the discrete log-change Törnqvist-Theil approximation[7] to the time-continuous Divisia indices of output and input to measure changes in productivity. The index is referred to as the Törnqvist index of productivity.

Output and input prices are utilized in the process of aggregating changes in individual inputs and outputs under the assumption that the revenue share of each output is equal to the cost elasticity with respect to the same output and that the cost share of each input is equal to the output elasticity with respect to the same input.

The index has generally favourable properties. It is invariant, it approximates the factor reversal test very closely and satisfies all the other conventional index tests. Among its limitations, three problems deserve attention. First, the Törnqvist-Theil approximation is not reproductive, unless all the aggregator functions are in linearly homogeneous translog form.[8] Secondly, the discrete index is asymptotic to the continuous index; i.e., the closer the observation points the better the approximation. The use of annual data may generate an approximation error in Bell's productivity index. The third problem is that of possible path dependence of the Divisia index. Hulten (1973) established three conditions for the path independence of the Divisia line integrals. The first condition, the existence of an aggregate, is normally assumed. However, in the case of Bell Canada, Smith and Corbo (1979) tested and rejected the hypothesis of weak separability of inputs from outputs in a multi-output multi-input translog cost function, suggesting that a single output aggregate cannot be formed.[9] The second condition, constant returns to scale, is generally not satisfied in econometric production and cost models of Bell Canada.[5,10] The third condition is no more than the mathematical implication of behaviour optimization.

The Törnqvist formulae of output and input volume indices are:

$$\left[\frac{Q_t}{Q_{t-1}}\right]_T = \prod_{i=1}^{n} \left[\frac{Q_{it}}{Q_{i,t-1}}\right]^{\frac{1}{2}(r_{it} + r_{i,t-1})}$$

$$\left[\frac{X_t}{X_{t-1}}\right]_T = \prod_{j=1}^{m}\left[\frac{X_{jt}}{X_{j,t-1}}\right]^{\frac{1}{2}(s_{jt} + s_{j,t-1})}$$

where r and s refer to revenue and cost shares, respectively; i.e.,

$r_{it} = \frac{P_{it}Q_{it}}{\sum_i(P_{it}Q_{it})}$, $s_{jt} = \frac{W_{jt}X_{jt}}{\sum_j(W_{jt}X_{jt})}$; $r_{i,t-1}$ and $s_{j,t-1}$ are similarly defined.

The Törnqvist productivity index is

$$(3)\quad \left[\frac{PR_t}{PR_{t-1}}\right]_T = \left[\frac{Q_t}{Q_{t-1}}\right]_T \Big/ \left[\frac{X_t}{X_{t-1}}\right]_T .$$

The discrete approximation of the continuous form of productivity gain in equation (1) is the logarithm of the Törnqvist productivity index.[11]

2.2 The Econometric Approach

Following Denny, Fuss, Everson (1979) and Denny, Fuss, Waverman (1979), an econometric estimation of productivity gains through a technology effect and a scale effect is described below. Although the estimation is simplified in that it uses a single output model and does not quantify the effect of non-marginal-cost pricing, it has produced more reasonable empirical results (better statistical properties, more realistic annual cost elasticity estimates) than more elaborate productivity compositions.

The neoclassical production function is

$$Q = f(X_1, \ldots, X_m, t),$$

where Q, X and t denote output, input and technological changes, respectively. It yields a definition of output shifts caused by technological changes. The proportional output shift is $\dot{A} = \partial \log Q/\partial t$.

The "econometric" productivity gain can be defined as the sum of the technology effect and the scale effect; i.e.,

$$(4)\quad (\dot{PR}) = \dot{A} + (\varepsilon - 1)\dot{X},$$

where ε is the scale elasticity, estimated in the production function as $\varepsilon = \sum_j(\partial \log Q/\partial \log X_j)$, and $\dot{X}$ denotes the Divisia input index $\dot{X} = \sum_j \frac{W_j X_j}{\sum_j(W_j X_j)} \dot{X}_j$, where W denotes input prices.

In the dual cost function of the form

$$C = g(W_1, \ldots, W_m, Q, t),$$

the cost shifts that are caused by technological changes are defined as

$\dot{B}=\partial \log C/\partial t$, where $C=\Sigma(W_j X_j)$ is total cost, and the productivity gains are expressed again as the sum of the technology effect and the scale effect; i.e.

$$(5) \quad (\dot{PR}) = -\dot{B} - (\varepsilon_{CQ}-1)\dot{Q},$$

where ε_{CQ} is the cost elasticity with respect to output, estimated in the cost function as $\varepsilon_{CQ}=\partial \log C/\partial \log Q$, and $\dot{Q}$ is the Divisia output index $\dot{Q}=\sum_i \frac{P_i Q_i}{\sum_i (P_i Q_i)} \dot{Q}_i$, where P denotes output prices.

Equations (4) and (5) yield alternative econometric measures of productivity gains. In equation (4), a production function provides the estimates of $\dot{A}$ and the scale elasticity (ε) and the input index is used. In equation (5), the estimates of $\dot{B}$ and the cost elasticity (ε_{CQ}) are derived from a cost function, and the output index is utilized. Since the shifts in the production function and the cost function are related through the cost elasticity, $-\dot{B}=\varepsilon_{CQ}\dot{A}$,[12] and the scale and cost elasticities are also related, $\varepsilon=\varepsilon_{CQ}^{-1}$, it is possible to use an estimated cost function ($\dot{B}$ and ε_{CQ}) both with the input index, as in the equation

$$(6) \quad (\dot{PR}) = -\dot{B}/\varepsilon_{CQ} + (\varepsilon_{CQ}^{-1} - 1)\dot{X},$$

and with the output index, as in equation (5), to arrive at econometric estimates of productivity gains.[13]

The productivity gains of equations (5) and (6) are equal if the actual input growth rate of the company is equal to the rate of growth of the cost minimizing total input. However, favourable or unfavourable production conditions and errors in cost minimization can make the actual input grow either slower or faster than the cost minimizing input.

When the estimated cost function does not include time as a variable, but contains a proxy (T) for technological changes, the intertemporal cost shifts can be obtained as

$$(7) \quad \dot{B} = \frac{\partial \log C}{\partial \log T} \cdot \frac{\partial \log T}{\partial t} .$$

Since the Törnqvist volume indices of output and input that yield the residually measured productivity index in equation (3) represent a discrete approximation to the continuous Divisia indices, the continuous formulae on the right hand side of equations (5) and (6) should be approximated as well. A linear approximation of the technology effect and the scale effect can be obtained with ease from an estimated <u>translog</u> cost function. Taking equation (5) as an illustration,[14] the technology effect $-\dot{B}$ is approximated by component, as they are shown in (7), in the formula

$$-\dot{B}_t = -\tfrac{1}{2} (\varepsilon_{CT,t} + \varepsilon_{CT,t-1}) (\log T_t - \log T_{t-1}),$$

where $\varepsilon_{CT}=\partial \log C/\partial \log T$ is the technology elasticity of cost.

As $\dot{Q}=\partial \log Q/\partial t$ becomes $\dot{Q}_t=\log Q_t-\log Q_{t-1}$, the scale effect is expressed in the discrete case as

$$(8)\quad -\dot{E}_t = -\left[\tfrac{1}{2}\left(\varepsilon_{CQ,t}+\varepsilon_{CQ,t-1}\right)-1\right]\left(\log Q_t - \log Q_{t-1}\right) .$$

When the underlying cost function is a generalized translog (GTL) in which the Box-Cox transformation is applied to the output and technology proxy variables, the approximation becomes more complicated.[15]

The Box-Cox transformation of the technology proxy variable results in $T^* = (T^{\lambda_T}-1)/\lambda_T$ and the cost shift caused by T^*, $\Gamma=\partial \log C/\partial T^*$, is directly estimated. Since Γ is a component of the technology elasticity of cost, (7) can be expressed in a GTL-specific form as

$$(9)\quad \dot{B} = \frac{\partial \log C}{\partial T^*} \cdot \frac{\partial T^*}{\partial \log T} \cdot \frac{\partial \log T}{\partial t} .$$

Linear approximation can be applied to Γ, while the other terms on the right hand side of (9) are readily approximated by

$$\Delta T^* = (T_t^{\lambda_T}-1)/\lambda_T - (T_{t-1}^{\lambda_T}-1)/\lambda_T = (T_t^{\lambda_T} - T_{t-1}^{\lambda_T})/\lambda_T,$$

$$\Delta \log T = \log T_t - \log T_{t-1} \text{ and}$$

$$\Delta t = 1 .$$

Hence the generalized translog cost function yields the following discrete approximation to the technology effect in (5):

$$(10)\quad -B_t = -\tfrac{1}{2}(\Gamma_t + \Gamma_{t-1})\,\frac{T_t^{\lambda_T} - T_{t-1}^{\lambda_T}}{\lambda_T} .$$

The scale effect is treated in a similar manner. As the Box-Cox transformation of the output variable generates $Q^*=(Q^{\lambda_Q}-1)/\lambda_Q$, the cost shift caused by Q^*, $\phi=\partial \log C/\partial Q^*$, is directly estimated and it is a component of the output elasticity of cost, the scale effect in (5) can be written in a GTL-specific form as

$$(11)\quad -\dot{E} = -\left[\frac{\partial \log C}{\partial Q^*} \cdot \frac{\partial Q^*}{\partial \log Q} -1\right]\dot{Q} .$$

The discrete approximation of (11) is

$$-\dot{E}_t = -[\tfrac{1}{2}(\phi_t + \phi_{t-1}) \frac{(Q_t^{\lambda_Q} - Q_{t-1}^{\lambda_Q})/\lambda_Q}{\log Q_t - \log Q_{t-1}} -1] [\log Q_t - \log Q_{t-1}].$$

Productivity gains resulting from the indexing approach contain (1) the effect of scale economies, (2) the effect of technological changes and (3) other effects. Because they exclude other effects, the econometric measures as defined in equations (5) and (6) do not fulfill the role of a single-number measure of changes in overall productive efficiency. Thus, instead of being an alternative to the indexing approach, the econometric approach yields measures which are components of residually measured productivity gains.

With the indices of output, input and productivity given and the cost elasticity derived from an estimated cost function, there are several ways to arrive at the composition of productivity gains. When the technology effect is obtained as the residual of Törnqvist gains after subtracting the scale effect (with either $\dot{Q}$ or $\dot{X}$), the other effects, which are related to the residual term of the estimated cost function, are attributed to technology. They can bias the annual estimates of technology effect either upward or downward but, because they have an approximately zero mean, the long-run (sample period length) average technology effect remains unbiased. Another problem is that the error in the cost elasticity estimate influences not only the scale effect in which it appears, but also the residual technology effect. When the decomposition is done at the bounds of the 95% confidence interval of ε_{CQ}, the results can change rather drastically and the technology effect may become negative in several years at the lower bound of the cost elasticity.[16]

The solution pursued in this paper defines the technology and scale effects as in equation (5) and separates out other effects into a residual ($\dot{R}_t$) of Törnqvist productivity gains;[17] thus, the composition becomes

$$(12)\quad (P\dot{R})_{t,T} = -\dot{B}_t - \dot{E}_t + \dot{R}_t \quad .$$

The underlying cost function is specified as a single-output generalized translog with a logarithmic output variable and Box-Cox transformation of the technology proxy. Hence $-\dot{B}_t$ is defined as in (10) and $-\dot{E}_t$ is approximated according to (8).[18]

3. THE PRODUCTIVITY PERFORMANCE OF BELL CANADA

Table 1 contains the information required for the following description and analysis of the productivity gains of Bell Canada during the period 1952 to 1980. The first column contains the annual productivity index as defined in equation (3). The next three columns show annual productivity gains and their components generated by technological changes and economies of scale, respectively. The last column presents residual productivity gains, due to factors other than technological changes and scale economies. Törnqvist output and input volume indices were used to obtain a measure of actual gains. The output index is that of real gross production, the capital measure is narrowly defined (plant in service) and the user cost of capital is utilized in the process of input aggregation. Table 1 reflects in its structure the components of equation (12). The alternative decomposition formula yields somewhat different numerical results, but the differences do not alter any of the following conclusions.

The cost function from which the estimates of technology shift and output elasticity of cost were obtained is a single-output homothetic generalized translog cost function with logarithmic output and Box-Cox technology variable transformations. It appears in Kiss, Karabadjian and Lefebvre (1981). The sample on which the cost function was estimated contains annual observations from 1952 to 1978. The 1979 and 1980 values of the cost shift (Γ) and the output elasticity of cost (ε_{CQ}) were obtained with

Table 1: Annual Productivity Gains

YEAR	PRODUCTIVITY		TECHNOLOGY EFFECT	SCALE EFFECT	RESIDUAL EFFECT
	INDEX	GAIN			
t	$\frac{Q_t}{Q_{t-1}} \Big/ \frac{X_t}{X_{t-1}}$	$\log\left(\frac{Q_t}{Q_{t-1}} \Big/ \frac{X_t}{X_{t-1}}\right)$	$-\dot{B}_t$	$-\dot{E}_t$	$\dot{R}_t$
1953	1.0302	.0298	.0043	.0050	.0205
1954	1.0068	.0067	.0058	.0055	-.0046
1955	1.0077	.0077	-.0007	.0069	.0014
1956	1.0025	.0025	.0189	.0079	-.0243
1957	1.0401	.0393	.0231	.0101	.0061
1958	1.0212	.0209	.0327	.0114	-.0231
1959	1.0531	.0517	.0155	.0166	.0197
1960	1.0337	.0331	.0180	.0170	-.0019
1961	1.0479	.0468	.0086	.0212	.0169
1962	1.0549	.0534	.0069	.0315	.0150
1963	1.0080	.0080	.0138	.0219	-.0277
1964	1.0330	.0325	.0088	.0285	-.0048
1965	1.0342	.0336	.0024	.0379	-.0066
1966	1.0403	.0395	.0029	.0397	-.0031
1967	1.0742	.0716	.0016	.0373	.0327
1968	1.0487	.0475	.0073	.0339	.0063
1969	1.0341	.0336	.0050	.0435	-.0149
1970	1.0438	.0428	.0047	.0332	.0049
1971	1.0041	.0041	.0033	.0251	-.0243
1972	1.0640	.0620	.0039	.0418	.0163
1973	1.0524	.0511	.0034	.0438	.0039
1974	1.0551	.0536	.0064	.0448	.0024
1975	1.0772	.0744	.0064	.0447	.0233
1976	1.0191	.0189	.0043	.0314	-.0168
1977	1.0080	.0079	.0036	.0282	-.0239
1978	1.0219	.0217	.0055	.0342	-.0180
1979	1.0366	.0359	.0061	.0240	.0058
1980	1.0338	.0332	.0088	.0324	-.0080

the aid of the respective 1979 and 1980 values of the exogenous variables and the estimated parameters, whose constancy was assumed for 1979 and 1980.

The actual average annual productivity gain of Bell Canada was 3.44% during the entire 28-year period of observation. The annual productivity gains appear to be generally very high.[19] Several sub-periods are distinguished by the pattern of annual gains. Table 2 gives a summary of period-average productivity gains.

The productivity gain of 1953 was slightly below the long-term average. Technological changes and scale elasticities resulted only in very small productivity improvement, but other circumstances were favourable, as indicated by the positive residual term in Table 1.

Very low productivity gains were registered in the following three years. Despite large increases in the size of the company's operations, the scale effect was very small, because only negligible scale economies existed in this period. Technological changes did not begin to contribute significantly to productivity gains until 1956. In fact, the technology change indicator declined slightly in 1955. The residual productivity gains indicate that the conditions were generally unfavourable for productivity improvement.

Table 2: Average Annual Productivity Gains

PERIOD	PRODUCTIVITY GAIN	TECHNOLOGY EFFECT	SCALE EFFECT	RESIDUAL EFFECT
1953	2.98%	.43%	.50%	2.05%
1954-56	.56%	.80%	.68%	- .92%
1957-71	3.72%	1.03%	2.72%	- .03%
1972-75	6.03%	.50%	4.38%	1.15%
1976-78	1.62%	.44%	3.13%	-1.95%
1979-80	3.46%	.75%	2.82%	- .11%
1953-80	3.44%	.83%	2.71%	- .10%

The revolution in switching technology, which started in 1956 with the introduction of the first crossbar central offices and customer-dialed long distance telephone calls (DDD), resulted in a suddenly very high direct technology effect on productivity gains. Technological changes also aided productivity improvement in indirect ways by increasing the degree of economies of scale and generating an upsurge in demand for long distance telephone services. As a result, the effect of economies of scale began to increase and reached very high levels by 1962-63. During the four years between 1963 and 1966, the average annual productivity gain generated by scale economies was approximately 3.2%. As the effect of the switching revolution gradually subsided and other circumstances were highly unfavourable (negative residual gain in each year), this 3.2% gain proved to be greater than the actual productivity gain of Bell Canada. The following years witnessed a continuation of small

Figure 1: Annual Productivity Gains

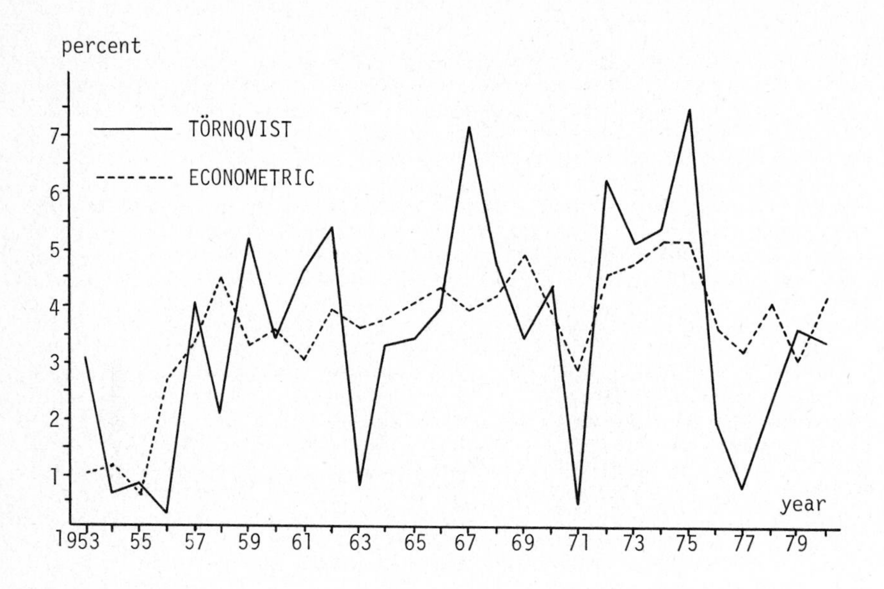

Figure 2: The Composition of Econometric Productivity Gains

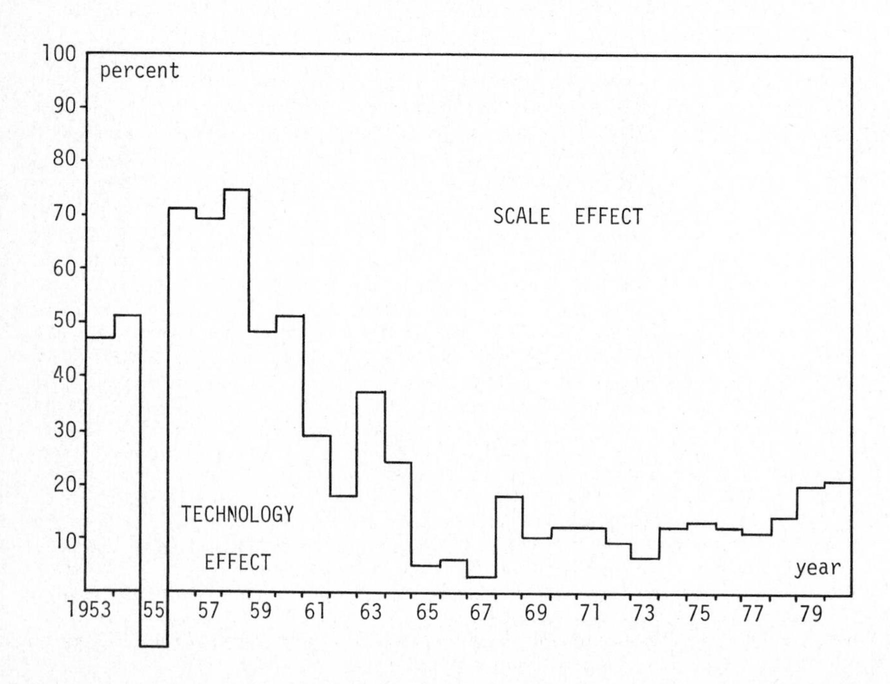

contributions to productivity gains by technological improvements, but the high degree of economies of scale kept productivity gains high. The residual gains show that favourable and unfavourable years alternated between 1966 and 1971. The entire 1957 to 1971 period is characterized by high rates of productivity improvement and fluctuations in the annual gains. The productivity gain was exceptionally high in 1967 (largely because of the high residual effect) and almost nonexistent in 1971.

The highest productivity gains of the period of observation were achieved between 1972 and 1975. Technological improvements contributed to productivity gains only very modestly, but very fast growth in demand for telephone, especially toll, services allowed the existing high degree of economies of scale to generate high productivity gains. The average annual productivity gain due to scale economies was 4.4% between 1972 and 1975. The residual gains were positive in each year and contributed significantly in 1972 and 1975.

The 1976-78 period represents a good example of the demand sensitivity of productivity gains. The actual gains were below the long-term average in all years and the average annual gain slipped to 1.62%. There are two major contributors to the erosion of productivity gains. First, demand for local and other toll services grew more slowly than in any other 3-year period during the observed 28 years and message toll demand also slowed down significantly. As a result, the contribution of scale economies to productivity gains dropped from 4.38% (1972-75) to 3.13% per annum. Secondly, the slowdown in demand for telephone services coincided with, and was in part caused by, worldwide economic problems and some political uncertainties in Canada. Intensifying regulatory activities and some changes in accounting methods may also have had a significant negative impact on Bell Canada's productivity gains. The residual productivity gains have rather large negative values in this period.

The last two years show an average productivity performance in every respect. The productivity gains were very close to their long term average, the technology effect on productivity was strengthened by the introduction of new technologies such as the DMS family of digital electronic switches, the impact of scale economies was higher than in any subperiod, with the exception of the extraordinary years between 1972 and 1975. The residual effect stayed close to zero.

A comparison of the actual and "econometric" productivity gains of Bell Canada in Figure 1 shows that <u>the econometric measure captures the level and the essential features of the pattern of the company's productivity improvement, but leaves a substantial part of the annual variation of productivity gains unexplained</u>. The unexplained variation is helpful in the analysis of productivity gains to the extent that it identifies "favourable" years (e.g. 1967) and periods (e.g. 1972-75) and "unfavourable" years (e.g. 1971) and periods (1963-1966 and 1976-79) for productivity improvements in Bell Canada.

Turning to the composition of the explained ("econometric") portion of productivity gains, Figure 2 shows that the contribution of scale economies was generally much greater than that of technological changes and that a certain pattern of relative contributions prevailed. High contributions were registered from technological changes (48 to 74% in 1956 to 1960) as a result of the introduction of crossbar central offices and DDD.

As the new technologies gradually became dominant, their impact diminished in size and especially relative to the rapidly increasing effect of scale economies. By 1967, the contribution of technological changes dropped to only 4% of the "econometric" productivity gains. During the period 1968 to 1977, the share of the technology effect was fluctuating in the 7% to 13% range. As higher rates of introduction of new technology were registered and the scale effect declined somewhat due to the recent slowdown in demand for telephone services, the share of technological changes in "econometric" productivity gains increased to the 30% level in the last two years.

While the indexing approach to the measurement of Bell Canada's productivity produced an average annual gain of 3.44% for the entire period of observation, the average annual "econometric" gain was 3.54%. Technological progress in Bell Canada was directly responsible for a .83% average annual productivity improvement and scale economies generated a 2.71% average productivity gain per annum. Approximately a quarter of Bell Canada's productivity gains was generated directly by technological changes and three quarters were due to the company's economies of scale. Technological changes were also the ultimate cause of a large part of the scale effect, because technological changes increased the degree of scale economies of Bell Canada and generated demand (hence scale increases) by lowering the cost of production, improving the quality and increasing the variety of telecommunications services.

4. APPENDIX: VARIABLE MEASUREMENT[20]

4.1 Output Categories

Ideally, prices and physical volumes for each individual service should be known[21] and the aggregation of output should be done by applying the chosen index number formula in a consistent manner. However, because of the enormity of the task of obtaining and working with 30 to 40 thousand prices and volumes, the number of outputs has been reduced to ten categories in the Bell Canada productivity study. The output categories can be identified through the revenues they generate. (1) Local service revenues include contract basic and auxiliary charges, message charges, public telephone revenues, private line revenues, installation charges, etc. (2) Intra-Bell message toll revenues contain all revenues derived from long distance calls originating from and terminating in Bell Canada territory. (3) Canada and (4) US and overseas message toll revenues are, in theory, derived from long distance calls in both directions between Bell Canada and other telephone companies in the respective geographical locations. (5) WATS revenues are generated by both INWATS and OUTWATS services. (6) TWX revenues are derived from message charges and equipment rental charges. (7) Private line toll revenues originate from the sale of private line voice and data services (e.g., inter-exchange voice and teletypewriter, radio and TV program transmission, Telpak, Datapac, Dataroute). (8) Miscellaneous other toll revenues are a residual category. Multicom and Voicecom are the most important component services. (9) Directory advertising revenues were derived from Yellow Pages advertising. With the establishment of Tele-Direct as a Bell subsidiary in 1971, this category was discontinued. (10) Miscellaneous revenues include Tele-Direct commission; rents of equipment, poles, buildings, satelite, etc.; general services and licences, e.g., service agreement revenues; Teleboutique/Phone Centre sales and various other revenues.

Although the productivity study uses a single-output aggregate, some two-output and three-output subaggregates are also generated for analytical purposes as well as for various econometric studies. The two-output subaggregates are defined as

- local, directory and miscellaneous (Nos. 1, 9, 10)
- toll (Nos. 2, 3, 4, 5, 6, 7, 8);

and the three-output subaggregates are derived by breaking down the toll category in two ways into

A. - message toll (Nos. 2, 3, 4)
   - other toll (Nos. 5, 6, 7, 8) or

B. - "monopoly toll" (Nos. 2, 3, 4, 5)
   - "competitive toll" (Nos. 6, 7, 8).

The relatively small directory advertising and miscellaneous service categories are aggregated with local services in order to minimize the consequences of not having a price index for miscellaneous services and also to reduce outputs to a manageable number in econometric studies. The "competitive toll" category is a rather crude approximation to the truly competitive toll services.

4.2 Output Volume

Volumes of the ten output categories are represented by deflated or constant dollar revenues in the Bell Canada productivity study. Deflated revenues encounter the usual problems associated with the appropriateness of prices as weights (e.g., cross-subsidization may distort the output aggregates) and, in the case of Bell Canada, flat monthly rates for local services present a special problem in that they reduce the sensitivity of the measure to output changes, because changes in usage are not reflected in deflated revenues.

Deflated revenues also present a number of advantages in measuring output volumes. Perhaps the most important advantage is that the availability of prices for individual services makes a detailed and elaborate aggregation procedure possible, while the theoretically more suitable cost weights are not available and would be difficult to approximate even for large service aggregates. Another desirable aspect of deflated revenues is that they are capable of reflecting changes in the quality of services, when these are accompanied by telephone rate changes.[22]

The output of a firm is generally measured as the aggregate of the volumes of all products being produced. This all-inclusive aggregate is referred to as (real) gross production. If certain input separability conditions are fulfilled in the production function of the firm (i.e., when materials are weakly separable from labour and capital), material input is removed from the output and input variables of the productivity measure. The omitted input is called intermediate input and the resulting output variable is referred to as (real) value added. The available empirical evidence suggests that the material input of Bell Canada is not separable;[23] thus, the measurement and analysis of productivity gains is based entirely on real gross production in this paper.

Törnqvist volume indices of gross production are constructed. In order to maintain the consistency of the chosen index number formula in the process of output aggregation, Törnqvist price and volume indices should be available for each of the ten output categories. However, only Paasche price indices and Laspeyres volume indices or their approximations are available at the present. As a result, the $Q_{it}/Q_{i,t-1}$ coefficients in the Törnqvist output volume index formula of Section 2.1 are represented by ratios of deflated revenues, denoted by $(P_{iB}Q_{it})/(P_{iB}Q_{i,t-1})$, where

$$P_{iB}Q_{it} = \sum_h(p_{ht}q_{ht}) \frac{\sum_h(p_{ht}q_{ht})}{\sum_h(p_{hB}q_{ht})} \qquad (h=1,\ldots,s),$$

p and q represent individual service prices and volumes, h is the number of services aggregated into output category i and B is the base year; $P_{iB}Q_{i,t-1}$ is defined in a similar manner.

4.3 Output Price

Törnqvist price indices are not available. The price indices generally have fixed volume weights in the early years (from 1952 to around 1970) and variable weights for the 1970's. Although the base periods and volume weights have been chosen in various ways, the fundamental features of the procedure of calculating the index series are common. When a change in telephone rates takes place, the appropriately chosen service volumes in the base year (q_{hB}) are priced out at old $(p_{h,t-1})$ and new (p_{ht}) prices. The index formula (the "reprice" effect in Bell jargon)

$$P_{it} = \frac{\sum_h(p_{ht}q_{hB})}{\sum_h(p_{h,t-1}q_{hB})} \qquad (h=1,\ldots,s)$$

is applied, the year-to-year indices are chained and the resulting index series is normalized around its 1967 value (1967=1.0). This procedure yields general price levels, relative to 1967, for the s services included in the given output category i, representing the price of that category in any year t.

4.4 Output Growth in Bell Canada

The average annual rate of growth in gross production is 8.74% for the period 1952 to 1980. The output volume indices illustrate some well known facts; e.g., that local output volumes have grown more slowly than toll, and message toll has grown more slowly than other toll. The growth of local service volumes is characterized by a slowdown, as very high growth rates existed during the 1953 to 1959 period and growth was very slow in recent years. No trend is observable for the period 1960 to 1975. Growth after 1975 was slower than in any other sub-period. There are some rather drastic year-to-year fluctuations in the growth rates of monopoly toll services, with no underlying strong trend. Output grew slowly between 1957 and 1961 and fast growth is observable during the 1972 to 1975 period. Very high growth rates of competitive toll services were achieved in private line on small volumes at the beginning of the observation period. This is a typical new product phenomenon. Later, new services such as TWX and some data services boosted the growth rates several times. It is interesting to observe that competitive toll grew more slowly than monopoly toll after 1970.

The growth pattern of gross production is dominated by year-to-year fluctuations and the sample period cannot be broken down into analytically useful sub-periods. Two periods of below-average growth are 1958 to 1961 and 1976 to 1980, while the only longer period with above-average growth lasted from 1972 to 1975.

4.5 Labour Input Volume

Labour input volume is measured by the total number of hours actually worked by the Bell Canada labour force on the production of telecommunications services. Hours worked by occasional employees are now excluded from the calculations due to the relatively small number of employees (approximately .5% of the total with .3% of total wages and salaries) and high cost of collecting the information. Several measurement issues deserve attention.

First, the Bell Canada labour force is classified in 27 categories. For six full-time occupational groups (operators, plant craftsmen, clerical, other non-management, foremen and supervisors, other management), there is a total of 26 sub-groups distinguished according to the length of service, while part-time employees form a single category. Denison (1962) and Gollop and Jorgenson (1980) include labour classes distinguished according to age, sex and education in their classification of the US labour force. Gollop and Jorgenson suggest that classification according to various demographic characteristics is desirable and Denison argues that different personal characteristics also should be considered. Since the Bell Canada productivity study does not reflect labour characteristics other than occupation and experience in the labour input, these characteristics are captured in the measured productivity changes.

Secondly, in the absence of a suitable measure, hours worked by management and clerical employees on the production of regulatory and other information, not directly related to the production of telecommunications services, are included in the labour input measure. The output of these hours is not accounted for in the output measures of Bell Canada; thus, the inclusion of information-producing hours lowers (increases) productivity gains, when information-producing hours grow faster (slower) than other inputs.

Thirdly, a certain percentage of Bell Canada's labour force is employed in the process of constructing telephone plant rather than directly producing telephone services or managing the company. Since the value of their labour input is included in the value of the resulting telephone plant, the hours they work are excluded from the labour input measure. However, the percentages of expensed to capitalized labour that are derived from company records have been altered on several occasions by changes in accounting procedures and these changes have influenced the measured productivity gains.

Finally, qualitative differences among employees (based on occupation, education, experience, etc.) cause the marginal product of their labour input to vary. Under the conventional assumption that the marginal revenue product of labour is proportional to its rate of compensation, an index of hours worked is obtained by using weights based on the hourly labour cost in each labour category. Since the marginal products are not measured and, at least at the present time, cannot be satisfactorily estimated, it is not known how much distortion the hourly labour cost weights cause in the volume indices of labour input.

Labour input volumes are calculated in four steps. First, the annual average number of employees is obtained by category from the company's records. Secondly, the average number of hours worked per employee per year is calculated for each labour category. Hours worked are obtained as scheduled (basic) hours, minus the hours of vacation days, scheduled days off (SDO's), holidays and sickness leave, plus overtime hours worked. Thirdly, the annual average number of employees is multiplied by the average number of hours worked per employee per year in each of the 27 categories in order to get the total number of hours worked. Both expensed (worked on the production of telecommunications services) and capitalized (worked on the construction of telephone plant) hours are included in this measure. Fourthly, the ratio of expensed to total hours worked is estimated for each category on the basis of ratios of construction employees to total employees. Total hours worked are multiplied with these ratios in order to obtain the total number of expensed hours worked for each of the 27 categories.

4.6 Labour Input Price

Labour price is the hourly rate of total actual labour expense: Employee Expense (wages, salaries, fringe benefits) and five labour-related federal and provincial taxes (Canada Pension Plan, Quebec Pension Plan, Unemployment Insurance, Quebec Health Insurance Plan, Workmen's Compensation).

Wage and salary data are collected together with the number of employees. Year-end levels of weekly wages and salaries per employee are annualized by multiplying them by 52.2. The annualized wage and salary rates are divided by the average number of hours worked per employee per year in order to get the wage and salary portion of the disaggregated labour price for each labour sub-group. Only the total cost of fringe benefits is available from Bell Canada records. Since the rules are different for temporary part-time (and some other) employees, the distribution takes place in two steps. The total cost is established separately for temporary part-time employees and is distributed among them according to their respective shares in the total wage payment. Then the remainder of the total cost of fringe benefits is prorated among the occupational groups according to their shares in total wage payment. Within occupational groups, the prorating is done according to the average number of employees in each sub-group; i.e., it is assumed that the cost of benefits per employee is proportional to the wage rate among occupational groups but it is insensitive to wage rate differentials within the same occupational group. In each sub-group, the total cost of fringe benefits is divided by the average number of employees and the resulting cost per employee is further divided by the average number of hours worked per employee per year in order to obtain the fringe benefit portion of the disaggregated labour price. The sum of the wage-salary portion and the benefit portion of the disaggregated hourly labour cost is adjusted by distributing labour-related non-income taxes among the 27 categories according to their respective shares in Employee Expense. The adjusted hourly labour expense is the disaggregated labour price.

4.7 Labour Input Growth in Bell Canada

During the sample period, total expensed hours worked grew only by 1.5% per year. This slow growth was due mainly to the substantial reduction in the number of telephone operators. The Törnqvist index estimates the growth rate for total labour input at 1.9% p.a., with an implied increase in input per hour of .4% p.a. There are five clearly distinguishable sub-periods.

The first period (1952 to 1957) is characterized by sizable increases in expensed hours. The Törnqvist index shows a 5.5% average annual growth in labour input. In this 5-year period, the fastest growing occupational groups were clerical and other non-management as well as other management. Hours worked by telephone operators on the other hand grew slowly at a rate of 1.6% per year. Hours worked by foremen and supervisors declined slightly.

Between 1957 and 1962, the number of expensed hours worked declined by 3.8% p.a., while the annual labour input decline was 3.2%; therefore, the quality-generated annual increase in input per hour was .6%, representing a fast quality mix improvement. The mix change was the consequence of a very fast decline in operator hours and a moderate increase in other management.[24] Hours worked in all other full-time groups declined, while a small increase can be observed in the part-time employee group. The substantial reduction in operator hours coincided with, and was largely caused by, the introduction of Direct Distance Dialing (DDD).

In the 1962 to 1966 period, expensed hours worked grew again, at a rate of 3.1% per year. The composition of the labour force of Bell Canada shifted toward lower quality and input per hour declined as a result. The decline occured despite slow growth in hours worked by telephone operators and fast growth in other management hours.

During the years 1966 to 1972, a decline in the number of hours worked was accompanied by strong growth in input per hour. Operator hours declined substantially (by 4.3% p.a.) and other management hours continued to increase, though at a lower rate than in the preceding period.

The last eight years (1972 to 1980) produced a 4.1% average annual increase in expensed hours worked. This rate is higher than at any time after 1957 but lower than the rates that prevailed before 1957. Operator hours declined in this period but hours worked in all other full-time occupational groups increased substantially.

4.8 Capital Input Volume

The capital input volume and price measures are conceptually analogous to the volume and price of labour input.[25] Capital input volume is represented by the constant dollar net stock of capital. The measure is often referred to as reproduction cost, signifying that technological changes in equipment manufacturing, resulting in costless quality improvements, are not allowed to lower the price of capital. For more on this subject, see Denison (1957) and Usher (1980).

Capital stock is used instead of utilized capital out of necessity rather than due to theoretical considerations. Suitable capital utilization measures are not available and the adjustment procedures that are recommended in the literature, e.g., Griliches and Jorgenson (1966), Berndt and Wood (1977) or Gollop and Jorgenson (1980), are not applicable to Bell Canada. However, utilization adjustment is a debatable issue. E.g., Kendrick (1973) states that "The degree of capital utilization reflects the degree of efficiency of enterprises ... Hence, in converting capital stocks into inputs, we do not adjust capital for changes in rates of capacity utilization, and thus these are reflected in changes in the productivity ratios".[26]

Capital stock can be defined narrowly as telephone plant in service or more broadly by including plant under construction or even the so-called working capital. Since the volumes of plant under construction and working capital are very small in comparison with telephone plant in service, the difference between the narrow and the broad definitions of capital does not result in a significant alteration of the empirical conclusions. Only the narrow definition is used in this paper.

Plant in service includes land and depreciable plant. It consists of the following six major categories: (1) land, (2) buildings, (3) central office equipment, (4) outside plant, (5) station equipment and (6) general equipment. The constant dollar net stocks of physical capital are obtained from book values by restating their age distribution by appropriately constructed price indices. The calculations require the following information for each year:

1. BG_{ij}, the year-end book value gross plant in service in plant category i and vintage group j;

2. RES_{ij}, the estimated depreciation reserve in plant category i and vintage group j;

3. TPI_{ij}, the Telephone Plant Price Index (1971=100) in class i for year j.[27]

The age distribution of gross plant in service is obtained by approximately 75 categories from the Bell Canada depreciation study. The vintage groups generally go back to 1920. Estimated reserves by category (calculated according to the ELG (Equal Life Group) method, including Bell Canada adjustment procedures) are added up to the account level and are balanced against the accumulated depreciation on each account. The account level actual/estimated reserve ratio is applied to estimated reserves in each component category. Net plant is calculated as $BN_{ij}=BG_{ij}-RES_{ij}$. The Telephone Plant Price Indices yield "translators" which show the rate of change in purchase price level in category i from year j (the year of the purchase) to any arbitrarily chosen base year (c); i.e., $P_{ij}=TPI_{ic}/TPI_{ij}$. Net telephone plant in service in base year dollars by category is given as $KN_i=\sum_j(P_{ij}BN_{ij})$.

Plant in service in each category is restated into current dollars and also into the previous year's dollars in each year of the observation period. The Törnqvist index of capital input volume averages the ratios of each year's constant to the previous year's current dollar value of net plant in service in the six major categories by using average (current and previous year's) category shares of total capital compensation as determined by the user cost of capital.

4.9 Capital Input Price

Capital price is represented by the user cost of capital. Various expressions of the user cost have appeared in the literature; e.g., Jorgenson (1963, 1967), Hall and Jorgenson (1967), Christensen and Jorgenson (1969), Boadway and Bruce (1979), Fuss and Waverman (1981), and Boadway (1980). All measures have been derived from the neoclassical theory of capital accumulation. Differences in the user cost measures arise from variation in the assumptions made in the formulation of the investment problem of the firm.

Cost minimizing behaviour is assumed for Bell Canada. In order to minimize the production cost of a given level of output, the company accumulates physical capital until the price of capital equals its marginal product times the marginal cost of producing the given level of output. Following Boadway (1980), the investment problem can be restated in a dynamic context as one of accumulating physical capital until the unit cost of physical capital equals the present value of the marginal cost of production times the marginal product of capital services.

At any point in time (t), the user cost of capital can be calculated from a perpetual inventory of capital stock model. In the presence of income taxes, the purchase price of capital goods (q_t) is

$$q_t = \int_t^\infty e^{-r(s-t)}\left[(1-u)c_s f_s e^{-\delta(s-t)} + uq_t D_{s-t}\right]ds,$$

where r denotes the discount rate; c_s is the marginal cost (excluding depreciation allowances) of production at time s; f_s refers to the marginal product at time s of the stock of physical capital accumulated at time t; δ is the rate of economic depreciation; $e^{-\delta(s-t)}$ shows the rate at which the marginal product of capital accumulated at time t deteriorates by time s due to economic depreciation; u is the corporate income tax rate; finally, D_{s-t} is the depreciation for tax purposes at time s, relative to the original cost of physical capital of age s-t.

The rate of economic depreciation requires further explanation. The main problem with estimating the economic depreciation of Bell's capital is that the fall in the market value of telecommunications equipment over time, due to simple aging, physical deterioration and obsolescence, is not observable, because there is very limited market for used equipment. However, as a surrogate, accounting depreciation rates are used to represent the degree of economic depreciation.

In Bell Canada, individual depreciation rates for 35 categories of telephone plant are established by the company's depreciation experts. These rates are applied to the corresponding categories of average (12-month simple arithmetic mean) book value depreciable gross plant in order to get annual book value depreciation expenses. The 35 depreciation expenses are summed up according to the five major classes of depreciable gross plant. Book value average depreciable plant is summed up in an identical fashion and the major plant class level depreciation rates are obtained as

$$\delta_i = \frac{\sum_d (\delta_{id}\ BG_{id})}{\sum_d BG_{id}} \qquad (d=1,\ldots,35),$$

where d refers to the number of depreciation categories in plant class i, δ denotes the depreciation rate and BG is average book value depreciable plant.

Denoting the present value of future depreciation deductions for tax purposes allowed on $1 current investment by z, the equation can be written as

$$q_t = \int_t^{\infty} e^{-r(s-t)}\ (1-u)\ w_s e^{-\delta(s-t)}\ ds + uq_t z\ ,$$

where $w_s = c_s f_s$ is the user cost or the cost of capital services at time s.

Differentiating the re-written equation with respect to t and solving for w_t yields

$$w_t = [q_t(r+\delta) - \dot{q}_t]\ \frac{1-uz}{1-u}\ .$$

The $\dot{q}_t$ term represents capital gain from the resale of telephone plant. Its value is assumed to be zero[28] . w_t then becomes

$$w_t = q_t(r+\delta)\ \frac{1-uz}{1-u}\ .$$

This expression is analogous to the rental price of capital formula of Hall and Jorgenson (1967) if tax credit is not assumed.

The user cost of capital is measured for each of the six major classes of physical capital. For classes of depreciable plant (i=2, ...6), the applied formula is

$$w_{it} = q_{it}(r_t+\delta_{it})\ \frac{(1-uz_t)}{1-u_t} + \frac{CROT_t}{KN_t}$$

and for land (i=1) it is written as

$$w_{it} = \frac{q_{it}}{1-u_t}\ r_t + \frac{CROT_t}{KN_t}\ ,$$

where q_{it} is the price of physical capital in category i, measured by an implicit[29] TPI_{it}; δ_{it} and u_t are as defined above; KN_t is the total net stock of physical capital in constant 1967 dollars and $CROT_t$ refers to the sum of capital-related non-income taxes. The discount rate (r_t) and the present value of depreciation for tax purposes (z_t) are described below.

The discount rate is the weighted average of the cost of new long term debt and equity capital; i.e.,

$$r_t = d_t(1-u_t)\ DRATIO_t + e_t(1-DRATIO_t)\ ,$$

where d_t is the cost of new long term debt; $DRATIO_t$ is Bell Canada's debt ratio (debt/debt plus equity); e_t is the expected rate of return on common equity. For simplicity, the relatively small amount of preferred equity is assumed to have the same rate of return as e.

The expected rate of return on common equity is approximated by the expression

$$e_t = \frac{D_t}{P_t} + G_t,$$

where D_t is common dividends declared per common share; P_t is the annual average market price of the common stock; and G_t is approximated by the 10-year log-linear least squares growth rate of dividends per share. z_t depends on the method of depreciation deductions. Bell Canada followed the straight line depreciation method for tax purposes for the years 1952, 1953 and 1958 to 1966, and the declining balance method in other years. For the straight line depreciation, z_t is calculated as

$$z_t = \frac{1}{nr_t}[1 - \frac{1}{(1+r_t)^n}],$$

where n is the average life of the depreciable asset, approximated by taking the arithmetic mean of the reciprocals of composite book depreciation rates for each year between 1958 and 1966. Under the declining balance method, z_t is calculated as

$$z_t = \frac{CCA_t}{r_t + CCA_t},$$

where CCA_t is the composite federal cost of capital allowance rate, obtained from internal sources.

4.10 Capital Input Growth in Bell Canada

The net stock of physical capital grew at an average annual rate of 7.2% during the 1952 to 1980 period. Its growth has slowed down from annual rates in the 9 to 13% range to an average of 3.3% during the last three years of the period. The annual growth rates have a pronouncedly linear downward trend with fairly small annual variation and two bulges. The first bulge appears in the period 1955 to 1962 and the second one during the years 1974 to 1977. The first bulge is associated with the heavy investments necessitated by DDD and crossbar switchers and the second one, while it may be more complex in nature, appears to be associated mainly with the rapid shift to electronic equipment. Both bulges seem to be related largely to significant changes in switching technology. The pattern of central office equipment volume growth supports this conclusion as it approximates closely (albeit with greater variation and a local minimum in 1966) the pattern of total net capital growth.

4.11 Material Input Volume

Material input is measured by the constant dollar value of a great number of miscellaneous inputs such as materials, rents, supplies and services. The following nine categories are distinguised: (1) maintenance material (all non-labour expenses related to the maintenance of station equipment, central office equipment and outside plant done by Bell Canada); (2) contract maintenance (all expenses related to the maintenance of station, central office equipment, outside plant and buildings by contractors such as Northern Telecom); (3) vehicles and tools (including gasoline expenses and vehicle rentals); (4) rentals (real estate, circuits, poles, computers, etc.); (5) house service (electricity, fuel oil, other supplies, contract services, etc.); (6) postage, printing, stationery; (7) travel and transfer; (8) research and development (mostly external, e.g., by Bell-Northern Research); (9) miscellaneous (e.g., advertising, Ontario official telephone service tax).

Current dollar material expenses are deflated in each of the nine categories by price indices to be discussed in the next section. The deflated expenses are expressed in the previous year's dollars. They are divided by the previous year's current dollar material expense in each category. The resulting ratios indicate category level volume changes in material inputs. Törnqvist's formula is used to aggregate the category level volume changes into a volume index of material input.

4.12 Material Input Price

Material prices are represented by price indices. No price index has been developed for the period 1952 to 1968; thus, the GNE Implicit Price Index of Statistics Canada was used as a proxy for the composite price index of all components. Internally developed price indices are available for each of the nine categories from 1969 onward. The individual prices are observed in a number of internal sources (e.g., purchase accounting, contracts) and various price indices from Statistics Canada are used whenever internal prices are not observable. Approximately 80% of the prices are specific to Bell Canada.

4.13 Proxy Variables for Technological Changes

Proxy variables are used to represent the technological progress of Bell Canada in cost models which provide the structural information necessary to calculate econometric productivity gains.

In the early stages of econometric research, the proxy variables were simple ratios, such as the percentage of customer dialed long distance messages. These proxies depicted only a single aspect of developments in switching technology and did not extend for the entire observation period. In an attempt to incorporate several features of technological changes in one proxy variable, Smith and Corbo (1979) and Corbo et al (1979) introduced a variable, written as

$$T = FNEW\ [\tau\ PDPH + (1 - \tau)\ ACCESS],$$

where FNEW is one plus the percentage of main stations switched by XBAR, ESS and SP1; PDPH is the percentage of dial phones; ACCESS is the percentage of telephones with access to DDD[30] and $\tau = Q_L/(Q_L+Q_T)$, where Q_L is local output and Q_T is toll service output.

Bell Canada used several alternative forms of this variable. The following three are referred to in Kiss, Karabadjian and Lefebvre (1981):

- T2: FNEW1 is defined as one plus the percentage of crossbar and electronic central offices;
- T3: FNEW2 is defined as one plus the percentage of telephones attached to crossbar and electronic central offices;
- T4: FNEW3 is defined as one plus the cumulative value of the first differences of the percentage of telephones served by the technologically most advanced switching equipment.

The variable FNEW3 requires some elaboration. It is calculated from the number of telephones attached to different types of central offices. The variable shows the increases in the percentage of telephones attached to the technologically most advanced switching equipment. The following equipments were considered to be technologically most advanced:

1952-55: Step-by-step,
1953-60: Step-by-step and crossbar,
1961-67: Crossbar,
1968-79: Crossbar and electronic,
1980: Electronic.

Although this classification is arbitrary, there are considerations which suggest that it may reflect the different stages of technological development reasonably well. Step-by-step was the leading switching technology before the appearance of the first crossbar equipment in 1956. Between 1956 and 1960, the percentage of telephones attached to step-by-step equipment increased, indicating that step-by-step was still replacing manual equipment in substantial numbers, thereby representing technological progress even in the presence of crossbar.

In 1961, the percentage of telephones attached to step-by-step equipment began to decline. (The number of telephones served by step-by-step continued to increase until 1973.) It would be unrealistic to say that step-by-step was still one of the leading switching technologies after

1960. Crossbar was the only representative of leading switching technology until the first electronic equipment came into existence in 1967. After 1968, crossbar and electronic switching are considered most advanced. The percentage of telephones attached to crossbar has slowed down considerably after 1974 but it was increasing until 1979.

The first differences of percentages are cumulated from 1952 to 1980. One plus the cumulative values are computed for each year. The series is normalized around the 1967 value.

FOOTNOTES:

[1] The two approaches are not separable. The verification of the validity of the notions of productivity utilized in the indexing approach requires econometric hypothesis testing and the econometric approach uses index numbers.

[2] The description generally follows Berndt (1980). The theory of duality is not explored here, but the econometric productivity gains are derived from a cost function as well as from a production function in Section 2.2.

[3] $\dot{Q} = \frac{\partial Q}{\partial t} \frac{1}{Q}$; $\dot{X} = \frac{\partial X}{\partial t} \frac{1}{X}$.

[4] Denny, Fuss and Everson (1979), Denny, Fuss and Waverman (1979).

[5] See Kiss, Karabadjian and Lefebvre (1981) for a summary of evidence. A study by Fuss and Waverman (1981) found approximately constant returns to scale, other studies show various degrees of increasing returns to scale.

[6] Griliches (1963, 1964, 1967) Denny, Fuss and Everson (1979), Denny, Fuss and Waverman (1979), Nadiri and Schankerman (1979).

[7] See Fisher (1922), Törnqvist (1936), Theil (1967), Christensen and Jorgenson (1970), Diewert (1976, 1977) and Jorgenson and Lau (1977). Different index number formulae correspond to different forms of the production function F in equation (2). The index number which is equal to Q_t/Q_{t-1}, derived from a specific F production function, is called an <u>exact</u> index number for F. Diewert (1976) shows that the Törnqvist <u>volume</u> index (used in the Bell Canada productivity study) is exact for a homogeneous translog production function and the Törnqvist price index is exact for a homogeneous translog cost function. Furthermore, Diewert defines an index number as <u>superlative</u> if F for which it is exact can provide a second order approximation to an arbitrary linearly homogeneous production function, and shows that the Törnqvist index is superlative, while the Laspeyres and Paasche indices are not.

[8] Diewert (1976).

[9] More testing is required before any definitive conclusion is drawn.

[10] The estimated single output translog cost model could not reject the

hypothesis of hometheticity of technology. Samuelson and Swamy (1974) and Usher (1974) show that the Divisia index is path independent if the production function is homothetic; however, if constant returns to scale do not exist the Divisia formula does not yield the desirable index.

[11] $(\dot{PR}) = \dot{Q}-\dot{X} = \frac{d \log Q}{dt} - \frac{d \log X}{dt}$ from equation (1) can be approximated as

$$\dot{PR}_{t,t-1} = \Delta \log Q - \Delta \log X = \log\left[\frac{Q_t}{Q_{t-1}} \Big/ \frac{X_t}{X_{t-1}}\right].$$

[12] Ohta (1974).

[13] Nadiri and Schankerman (1979) use a quasi-Divisia index to aggregate input. In their case, equation (6) becomes

$$(\dot{PR}) = -\dot{B}/\varepsilon_{CQ} + (k\varepsilon - 1)\dot{X},$$

where $k = \Sigma(P_iQ_i)/\Sigma(W_jX_j)$; i.e., the revenue/cost or average price/average cost ratio.

[14] The alternative formula in equation (6) can be treated in a similar fashion.

[15] I am indebted to Professor Melvyn Fuss for the following solution.

[16] The technology effect is the residual of Törnqvist productivity gains in Denny, Fuss, Everson (1979) and Nadiri and Schankerman (1979). An alternative would be to estimate the technology effect from a cost function and attribute the other effects to the scale effect. However, this method would exhibit similar sensitivity to the error in the cost elasticity estimate. Further difficulty appears when $\dot{X}$ is used in the scale effect, because differences between the actual and cost minimizing input growth rates distort the residual technology effect.

[17] A separate residual term is shown in Denny, Fuss and Waverman (1979).

[18] I thank my colleague Hoi Xuan Ngo for programming and running the calculations of the econometric productivity gains.

[19] Törnqvist indices of productivity, computed in a very similar fashion for AT&T by Christensen, Cummings and Schoech (1980), make the following comparison of average annual gains possible:

PERIOD	AT&T	BELL CANADA
1957-66	3.1%	3.7%
1967-77	3.2%	4.3%

Denny, Fuss and May (1980) reported average annual productivity gains in the .22 to 2.43% range for twenty two-digit manufacturing industries in Quebec and Ontario during the period 1961 to 1975. Bell Canada's productivity gains averaged 4.45% per annum in the same period.

[20] Due to space limitations, the following description of variable measu-

rement omits many important technical details. A fuller data appendix was presented at the conference and is available from the author on request.

[21] As an alternative, the base year and test year revenue shares and the coefficient of price change $P_{it}/P_{i,t-1}$ (if the price index of output is calculated and the implicit volume index is taken) or the coefficient of volume change $Q_{it}/Q_{i,t-1}$ (if the volume index of output is calculated) can be obtained for each individual service i (i=1,...,n). The measurement of revenue shares and price coefficients has great practical advantages for regulated public utilities, because price regulation often results in uniform percentage changes in the prices of large portions of total output. In such cases, the task of measurement is reduced to observing the price increase and obtaining the combined revenue share data for all individual outputs which share the same price coefficient.

[22] Telephone exchange upgrouping is a reflection of qualitative improvement in local service output, resulting from increases in the size of the local calling area. Since the local service price index does not reflect rate increases which accompany exchange upgroupings, the resulting increases in revenues are shown as output volume increases.

[23] Smith and Corbo (1979), Denny, Fuss and Everson (1979).

[24] The term "quality" refers to the marginal revenue product of the factor in question. The marginal revenue product of labour is approximated by the hourly rate of total labour cost. The marginal revenue product of capital is approximated by the user cost of capital.

[25] Gollop and Jorgenson (1980), p. 67.

[26] Kendrick (1973), p. 26.

[27] Price indices and age distribution are not available for land. It is assumed that land has the same age distribution and is subject to the same price changes as buildings.

[28] The annual average productivity gain for the 1954 to 1980 period increases from 3.46% to 3.6% if this assumption is relaxed.

[29] I.e., weighted by volumes of plant in service, as opposed to gross additions.

[30] The ACCESS series was discontinued in Bell Canada's reports in 1978. The 1978 to 1980 values of the variable were obtained by imposing the annual growth rates of the percentage of customer dialed long distance calls on the 1977 value of ACCESS.

REFERENCES:

[1] Abramovitz, M. (1956) Resource and output trends in the United States since 1870, American Economic Review 46, no. 2 (May) 5-23.

[2] Berndt, E.R. (1980) Comment on U.S. productivity growth by industry,

1947-73 by F.M. Gollop and D.W. Jorgenson, in: Kendrick, J.W. and Vaccara B.N. (eds.), New Developments in Productivity Measurement and Analysis (National Bureau of Economic Research, Chicago) 124-36.

[3] Berndt, E.R., and Wood, D.O. (1977) Engineering and econometric approaches to energy conservation and capital formation: A reconciliation, MIT Energy Laboratory Working Paper.

[4] Boadway, R.W. (1980) Corporate taxation and investment: A synthesis of the neo-classical theory, Canadian Journal of Economics XIII, no. 2 (May) 250-67.

[5] Boadway, R.W., and Bruce, N. (1979) Depreciation and interest deductions and the effect of the corporation income tax on investment, Journal of Public Economics 11, 93-105.

[6] Christensen, L.R., Cummings, D., and Schoech, P.E. (1980) Productivity in the Bell System, 1947-1977, Presented at the Eight Annual Telecommunications Policy Research Conference, Annapolis, Maryland.

[7] Christensen, L.R., and Jorgenson, D.W. (1969) The measurement of U.S. real capital input, 1929-1967, Review of Income and Wealth 15 (December) 293-320.

[8] Christensen, L.R., and Jorgenson, D.W. (1970) U.S. real product and real factor input, 1929-1967, Review of Income and Wealth, ser. 16, no. 1, 19-50.

[9] Corbo, V., Breslaw, J., Dufour, J.M., and Vrljicak, J.M. (1979) A simulation model of Bell Canada: Phase II, Special Study No. 79-002, Institute of Applied Economic Research, Concordia University, Montreal.

[10] Denison, E.F. (1957) Theoretical aspects of quality change, capital consumption, and net capital formation, in: Capital formation, Studies in Income and Wealth, vol. 19. (National Bureau of Economic Research, New York).

[11] Denison, E.F. (1962) Sources of economic growth in the United States and the alternatives before us, Supplementary Paper 13, Committee for Economic Development.

[12] Denny, M., Fuss, M., and Everson, C. (1979) Productivity, employment and technical change in Canadian telecommunications: The case for Bell Canada, Final Report to the Department of Communications.

[13] Denny, M., Fuss, M., and Waverman, L. (1979) The measurement and interpretation of total factor productivity in regulated industries, with an application to Canadian telecommunications, Presented at the Conference on Productivity Measurement in Regulated Industries, University of Wisconsin, Madison, Wis.

[14] Denny, M., Fuss, M., and May, J.D. (1980) Inter-temporal changes in regional productivity in Canadian manufacturing, Working paper no. 8017 (Institute for Policy Analysis, University of Toronto).

[15] Diewert, W.E. (1976) Exact and superlative index numbers, Journal of Econometrics 4, 115-45.

[16] Diewert, W.E. (1977) Aggregation problems in the measurement of capital, Discussion paper 77-09, Department of Economics, University of British Columbia.

[17] Fabricant, S. (1959) Basic facts on productivity change, Occasional Paper 63, (National Bureau of Economic Research, New York).

[18] Fisher, I. (1922) The making of index numbers (Houghton Mifflin, Boston).

[19] Fuss, M., and Waverman, L. (1981) The regulation of telecommunications in Canada, Technical Report No. 7, Economic Council of Canada.

[20] Gollop, F.M., and Jorgenson, D.W. (1980) U.S. productivity growth by industry, 1947-73, in Kendrick, J.S. and Vaccara, B.N. (eds.), New Developments in Productivity Measurement and Analysis (National Bureau of Economic Research, Chicago) 17-124.

[21] Griliches, Z. (1963) The sources of measured productivity growth: U.S. agriculture 1940-1960, Journal of Political Economy, August 331-46.

[22] Griliches, Z. (1964) Research expenditures, education and the aggregate agricultural production function, American Economic Review (December) 961-74.

[23] Griliches, Z. (1967) Production functions in manufacturing: Some preliminary results, in: Brown, M. (ed.), The Theory and Empirical Analysis of Production (National Bureau of Economic Research, New York) 275-322.

[24] Griliches, Z., and Jorgenson, D.W. (1966) Sources of measured productivity change: Capital input, American Economic Review 56, no. 2 (May) 50-61.

[25] Hall, R.E., and Jorgenson, D.W. (1967) Tax policy and investment behaviour, American Economic Review 57, no. 3, 391-414.

[26] Hulten, C.R. (1973) Divisia index numbers, Econometrica 41, no. 6 (November) 1017-26.

[27] Jorgenson, D.W. (1963) Capital theory and investment behaviour, American Economic Review, Papers and Proceedings 53, no. 2 247-59.

[28] Jorgenson, D.W. (1967) The theory of investment behaviour, in: Ferber, R. (ed.), Determinants of Investment Behaviour (National Bureau of Economic Research, New York).

[29] Jorgenson, D.W. and Griliches, Z. (1967) The explanation of productivity change, Review of Economic Studies 34, no. 99, 249-83.

[30] Jorgenson, D.W. and Lau, L.J. (1977) Duality and technology (North Holland, Amsterdam).

[31] Kendrick, J.W. (1961) Productivity trends in the United States (Princeton University Press, Princeton).

[32] Kendrick, J.W. (1973) Postwar productivity trends in the United States, 1948-1969 (National Bureau of Economic Research, New York).

[33] Kiss, F., Karabadjian, S., and Lefebvre, B. (1981) Economies of scale and scope in Bell Canada: Some econometric evidence, Presented at the conference "Telecommunications in Canada: Economic Analysis of the Industry", Montreal (March). Included in a condensed form in this volume.

[34] Nadiri, M.I. and Schankerman, M.A. (1979) The structure of production, technological change and the rate of growth of total factor productivity in the Bell System, Presented at the Conference on Productivity Measurement in Regulated Industries, University of Wisconsin, Madison, Wis.

[35] Ohta, M. (1974) A note on the duality between production and cost functions: Rate of returns to scale and rate of technical progress, Economic Studies Quarterly 25, 63-65.

[36] Samuelson, P.A. and Swamy, S. (1974) Invariant economic index numbers and canonical duality: Survey and synthesis, American Economic Review 64 566-93.

[37] Smith, J.B., and Corbo, V. (1979) Economies of scale and economies of scope of Bell Canada, Final Report to the Department of Communications.

[38] Solow, R.M. (1957) Technical change and the aggregate production function, Review of Economics and Statistics 39, no. 3 (August) 312-20.

[39] Theil, H. (1967) Economics and information theory (North Holland, Amsterdam).

[40] Törnqvist, L. (1936) The Bank of Finland's consumption price index, Bank of Findland Monthly Bulletin, no. 10, 1-8.

[41] Usher, D. (1974) The suitability of the Divisia index for the measurement of economic aggregates, Review of Income and Wealth 20, 273-88.

[42] Usher, D. (1980) Estimation of capital stock in the United States, in: Usher, D. (ed.), The measurement of capital (National Bureau of Economic Research, Chicago).

Economic Analysis of Telecommunications:
Theory and Applications
L. Courville, A. de Fontenay and R. Dobell (eds.)

COMPARING THE EFFICIENCY OF FIRMS: CANADIAN TELECOMMUNICATIONS COMPANIES

M. Denny, A. de Fontenay and M. Werner

University of Toronto, Government of Canada
and Telecommunications Consultant

1. Introduction*

A study of the efficiency of individual firms is seldom possible due to data restrictions. This paper reports on a unique empirical investigation of the efficiency of three telephone companies in Canada. Most of the data, originally developed for their own separate productivity studies, has been made publicly available by the telephone companies. Without their considerable effort this paper would not be possible.

These data bases are not entirely comparable. (An appendix to this paper clarifies the major differences.) Part of our task is to evaluate the sensitivity of our comparisons to alternative measures of the variables, in order to investigate possible errors arising from the limited comparability of data. Moreover, possible advantages and disadvantages of alternative definitions of economic variables is a problem broader than simply determining the veracity of the measured variables. Telecommunications firms offer a wide variety of services through their networks. There are alternative sensible definitions of economic variables whose use will alter the magnitude and perhaps ranking of the firms' efficiency. While not wishing to obscure the results, we believe that the complexity introduced by exploring the alternatives provides a much better understanding of the detailed changes of efficiency within and across firms.

Given a set of data on the prices and quantities of inputs and outputs, the methods we use to compare efficiency have been discussed elsewhere by us (Denny, de Fontenay and Werner (1980a,b), Denny and Fuss (1980) and by Caves, Christensen and Diewert (1980)). In this paper, we will apply these methods without extensive discussion.

The three companies included in this study are Bell Canada (BELL), Alberta Government Telephones (AGT) and British Columbia Telephone (B.C. Tel). In 1978 these three companies provided 70% of the dollar value of domestic telecommunications services.

2. Labour Productivity

To begin our comparison, we have estimated labour productivity for each company and compared the results. Output is the aggregate of the output disaggregation provided by the firms and discussed in the appendix. For reasons of comparability, labour is measured as unweighted man-hours of labour worked in each company.

In Table I, indexes of labour productivity for AGT, BC Tel. and Bell are shown. Labour productivity in AGT and BC Tel. has grown at approximately 8% a year since 1972 compared to about 4.5% in Bell. Prior to 1972, labour productivity was growing at an annual rate above 10% at AGT and 7.7% at Bell Canada.

Output growth was higher at BC Tel. than at Bell after 1972. Labour input must have grown faster at Bell than at BC Tel. during this period in order to convert BC Tel.'s 2% advantage in output growth into a 3 1/2% difference in labour productivity growth. AGT had the fastest rate of growth of output after 1972 but this was not translated into a higher labour productivity growth relative to BC Tel. Given the rates of growth of output, BC Tel. has managed a superior performance relative to Bell and AGT in achieving labour productivity growth.

The levels of labour productivity are presented in Table II, Bell Canada's labour productivity level is normalized to 100 in 1972 and the results for the other companies are relative to this normalization. While Bell has had the highest level of labour productivity, the other two companies have reduced the gap since 1972. After 1975, the change in the relative levels has slowed down as each company has had increasing difficulty in raising its labour productivity level.

3. Total Factor Productivity

We will measure total factor productivity for AGT, Bell and BC Tel. using a common methodology and data which is partially standardized.

Define the rate of growth of productivity,

$$\dot{TFP} = \dot{Q} - \dot{F}$$

where the aggregate output growth rate $\dot{Q}$ is defined by,

$$\dot{Q} = \sum_j r_j \dot{q}_j$$

and the aggregate input growth rate, $\dot{F}$ is defined by,

$$\dot{F} = \sum_j s_i \dot{x}_i \quad .$$

The disaggregate output $(\dot{q}_j)$ and input $(\dot{x}_i)$ growth rates are weighted by the revenue (r_j) and cost (s_i) shares respectively. This procedure standardizes the methodology for the three companies.

The data are partially standardized by the choice of input variables. For each company, labour input is measured as man-hours worked without any adjustment for skill levels. Capital is measured as the gross capital stock which is an aggregate of detailed physical assets. Material inputs are not completely comparable but this is not believed to be a problem. Finally, the assumption is made that the value of capital services can be measured as a residual component in total realized costs.

	Table I			Table II		
	Labour Productivity (1972 = 100.0)			Levels of Labour Productivity (Index, Bell 1972 = 100.0)		
	AGT	BCT	Bell	AGT	BC Tel.	Bell
1967	61.7	-----	66.3	43.6	-----	66.2
1968	70.7	-----	74.4	50.0	-----	74.6
1969	76.7	-----	80.8	54.3	-----	80.6
1970	81.4	-----	86.2	57.5	-----	86.2
1971	88.2	-----	92.5	62.5	-----	92.6
1972	100.0	100.0	100.0	70.9	82.0	100.0
1973	107.2	104.2	105.4	75.6	84.7	105.2
1974	121.8	111.9	109.7	86.2	91.7	109.9
1975	143.8	131.4	122.3	102.0	107.2	121.9
1976	149.3	150.8	125.5	105.3	123.4	125.0
1977	164.1	159.9	129.6	116.3	129.8	129.8
1978	159.3	157.1	131.7	112.3	128.2	131.6
1979	-----	149.2	133.9	-----	121.9	133.3

Source: See data appendix.

For the three companies, the rates of growth of total factor productivity are shown in Table III and a productivity index (1972 = 100) appears in Table IV. AGT has had a faster rate of growth of TFP than Bell and BC Tel during any time period when comparable data is available. From 1972-78, AGT's productivity grew at an average annual rate of 6.6% compared to a rate of 3.9% for Bell and for BC Tel.

Recall that AGT and BC Tel. had almost identical rates of growth of labour productivity. The TFP results indicate that BC Tel. achieved the labour productivity results through faster rates of growth of the capital-labour and the materials-labour ratio relative to AGT. The later company was more successful at achieving high rates of labour productivity growth via high rates of TFP growth.

Bell had a substantially lower rate of growth of labour productivity than BC Tel. but TFP grew at least as quickly. Relative to Bell as well as AGT, BC Tel. must have had a faster rate of growth of capital and materials to labour intensities in order to achieve the results portrayed above.

4. Relative Efficiency

Relative efficiency will be measured using the methodology originally proposed by Jorgenson and Nishimizu (1978). This methodology has been developed more extensively by Denny and Fuss (1980, 1981), Caves, Christensen and Diewert (1980) and Denny, de Fontenay and Werner (1980a, b,).

One can provide an intuitive interpretation of the method. It would be straightforward to compare the efficiency of the firms if we observed them using the same vector of inputs. Then, the relative output level would measure the relative efficiency levels. As Caves, Christensen and Diewert (1980) have shown, the relative efficiency measure we use can be interpreted as the average of the relative efficiency levels of the firms measured as the relative output levels at each firm's input level. That is, it is equal to the average of the relative output levels when both firms use the observed input levels of one firms. A similar interpretation may be given to the cost efficiency measure. These interpretations imply that the differences in the prices and the quantities of inputs and outputs across firms are accounted for in the relative efficiency measure.

	Table III			Table IV		
	Annual Rates of Growth of TFP			TFP Indexes (1972 = 100)		
	BC Tel.	AGT	BELL	BC Tel.	AGT	BELL
1967	----	----	5.9	----	74.9	86.8
1968	----	5.3	4.3	----	78.9	90.6
1969	----	5.5	2.9	----	83.4	93.3
1970	----	4.6	3.7	----	87.3	96.8
1971	----	4.2	-0.5	----	91.1	96.3
1972	----	9.3	3.7	100.0	100.0	100.0
1973	2.9	7.7	4.7	102.9	108.0	104.8
1974	5.9	11.9	4.4	109.1	121.7	109.5
1975	6.0	8.3	6.9	115.9	132.3	117.3
1976	4.4	3.3	1.0	121.0	132.8	118.5
1977	-2.2	6.6	0.7	118.4	141.8	119.4
1978	3.0	2.0	2.3	122.0	144.8	122.2
1979	2.5	----	2.2	125.1	-----	124.9

Source: See data appendix.

Using the date underlying our calculations of total factor productivity, an initial comparison of the firm's relative levels of efficiency was made. Define the relative total factor productivity level, of firm k relative to firm h, E_{kh}

$$\log E_{kh} = \log (Q_k/Q_h) - \tfrac{1}{2} \sum_i (s_{ik} + s_{ih}) \log (X_{ik}/X_{ih}) \quad ,$$

where s_{ik} is the cost share of factor i in firm k and X_{ik} is the equivalent quantity.

From the cost function, one may define a relative cost efficiency level, CE_{kh}

$$\log CE_{kh} = \log(C_k/C_h) - \tfrac{1}{2} \sum_i (s_{ik} + s_{ih}) \log (w_{ik}/w_{ih}) - \log (Q_k/Q_h) \quad ,$$

where C_k is the total cost and w_{ik} the price of input i in firm k.

Tables V and VI present the results, E_{kh} and CE_{kh}, of measuring both of these relative efficiency measures for the three companies. Consider the results of comparing Bell and AGT in Table V. In 1967 Bell's relative TFP level was 124.8 compared to AGT's 100. Alternatively, one may state that the quantity of output produced by Bell was approximately 25% greater than that produced by AGT after accounting for differences in input quantities. For the companies to be equally efficient, the E value for Bell would have to be 100.

The results are roughly equivalent when measured from the cost side. Bell's cost efficiency in 1967 was 80.2 relative to AGT's 100. Bell's costs were only 80.2% of AGT's after accounting for differences in input prices and output levels.

Through time AGT has eliminated the relative efficiency gap. In 1979, AGT had a 7% relative efficiency advantage. In our explorations below we will try and indicate what led to this sharp improvement in AGT's relative efficiency.

In Table VI, AGT and Bell are compared to BC Tel. for the years 1972-78. In 1972, BC Tel. and Bell had approximately equal efficiency and BC Tel. was 10% more efficient than AGT. Since BC Tel. and Bell had equal average productivity growth during this period there is no substantial change in their relative efficiency levels during the '70's. Since AGT had a very rapid growth in TFP relative to the other companies, the initial efficiency disadvantage of AGT relative to BC Tel. had been sharply reversed. AGT began in 1972 with a 10% cost disadvantage and finished with a 7% cost advantage.

5. Interpreting the Results

Our investigation is limited by the data that we have available publicly. The results suggest that in 1978 Bell and BC Tel. used more real resources to produce a given output level than AGT. To clarify this possibility, we will study the use of each factor and the production of outputs for the three companies. To begin, consider the indexes of the input-output ratios for each factor and company presented in Table VII. The indexes are normalized to 100 for Bell Canada in 1972.

For Bell Canada, the labour to output ratio has declined throughout the period. However the decline was more rapid prior to 1972 than after. BC Tel. had a much larger labour-output coefficient in 1972 but the ratio declined more quickly for BC Tel. than Bell after 1972. There was still a slightly lower labour coefficient in Bell in 1979. AGT had a very high labour coefficient relative to Bell in 1967 but this coefficient has declined more rapidly for AGT than Bell. Most of the large difference had disappeared by 1979. For the input labour, both BC Tel. and particularly AGT have done better than Bell. Notice that this ranking corresponds to the ranking of the output growth rates among the companies. To what extent the output measures are biasing the results will be discussed below.

Table V

Relative Efficiency of Bell Compared to AGT

	Productivity		Cost Efficiency
	Bell	AGT	Bell
1967	124.8	100	80.2
1968	123.9	100	80.7
1969	120.9	100	82.7
1970	120.4	100	83.1
1971	115.6	100	86.5
1972	109.7	100	91.2
1973	106.4	100	93.9
1974	98.8	100	101.2
1975	98.3	100	101.7
1976	98.9	100	101.1
1977	93.3	100	107.1
1978	93.4	100	107.1

Table VI

Relative Efficiency of AGT and Bell Compared to BCT

	Productivity			Cost Efficiency	
	AGT	Bell	BC Tel.	AGT	Bell
1972	89.6	98.8	100	111.7	101.2
1973	94.1	100.7	100	106.3	99.4
1974	100.0	99.5	100	100.0	100.5
1975	102.4	101.0	100	97.6	99.0
1976	98.6	98.1	100	101.4	102.0
1977	108.2	101.2	100	92.4	98.8
1978	107.5	100.5	100	93.0	99.4

The capital-output ratio has fallen for Bell but the temporal pattern is the opposite of the labour coefficient. Prior to 1970 the capital coefficient fell very slowly and after 1972 its rate of decline increased. The rate of decline was always much slower than the decline in the labour coefficient. The capital-labour ratio has increased in Bell throughout this period.

In 1972, the capital coefficient of BC Tel. was lower than at AGT or Bell. The very slow reduction in the BC Tel. capital coefficient has eliminated the gap relative to Bell and AGT at the end of the period.

Table VII

Input-Output Ratios

Indexes: BELL 1972 = 1.00

	Labour			Capital			Materials		
	AGT	BCT	BELL	AGT	BCT	BELL	AGT	BCT	BELL
1967	2.29	-	1.51	1.25	-	1.06	0.92	-	0.97
1968	2.00	-	1.34	1.25	-	1.05	0.91	-	0.94
1969	1.84	-	1.24	1.19	-	1.02	0.87	-	1.01
1970	1.74	-	1.16	1.15	-	1.00	0.84	-	0.94
1971	1.60	-	1.08	1.13	-	1.01	0.81	-	1.05
1972	1.41	1.22	1.00	1.06	.92	1.00	0.72	0.81	1.00
1973	1.32	1.18	0.95	0.98	.90	0.96	0.64	0.79	0.96
1974	1.16	1.09	0.91	0.87	.87	0.91	0.58	0.70	0.91
1975	0.96	0.93	0.82	0.83	.88	0.88	0.60	0.65	0.81
1976	0.95	0.81	0.80	0.82	.88	0.88	0.66	0.66	0.82
1977	0.86	0.77	0.77	0.80	.90	0.87	0.57	0.84	0.86
1978	0.89	0.78	0.76	0.74	.88	0.84	0.61	0.72	0.86
1979	-	0.82	0.75	-	.83	0.82	-	0.66	0.82

At AGT, the capital coefficient has fallen throughout the period at a rate faster than either of the other companies. The large (50%) gap relative to Bell that existed in 1967 has been substantially reduced by 1978. While the capital to labour ratio increased sharply prior to 1972, its growth has been much slower absolutely and relative to the other companies after 1972.

For materials the pattern is different since at the beginning of the period Bell did not have a substantially lower materials coefficient. Instead it was modestly higher. At Bell, the materials coefficient has fallen by less than the other coefficients. The other two companies have maintained their lower materials coefficient throughout the period and after 1972 there has been little change in the relative coefficients. Prior to 1972 AGT's materials coefficients did fall more than Bell's coefficient. The advantage held by BC Tel. and AGT over Bell does not result in a very large impact on the comparison for two reasons. Materials are the least important input due to their smaller cost share, and the differences across companies is smaller than the differences in the other two inputs.

These results suggest that in relative terms AGT has improved its efficiency level through improved utilization of labour particularly. The same pattern is observed for BC Tel. AGT has also improved its capital utilization but this has not been as spectacular.

6. Alternatives

There is an extensive literature on international comparisons of productivity. We have adapted the methodology reported in Kravis (1976) and in Kravis et al. (1975, 1978) to our situation and find that major results are identical and even the numerical magnitudes remain very close. This has convinced us that our results are not sensitive to quite large changes in the methodology used to measure relative efficiency.

We have also attempted to assess the impact of alternative measures of the price and flow of capital services. This required the development of relatively simple measures of user costs of capital. These replaced the implicit user costs inherent in our comparison discussed above. Once again, the results did not change. However, we are developing a more detailed specification of the cost of capital for each company which will provide more accurate estimates.

Finally, we should note that the relative efficiency differentials reported in this paper include all the effects of regulation, non-competitive behavior and scale economies. Any separation of the relative efficiency levels into these types of causes requires econometric analysis. The data series were not long enough to permit this type of study.

7. Conclusions and a Warning

This paper contains an attempt to compare the efficiency of the telephone companies using only the aggregate publicly available data. Our major conclusion is that AGT has made major strides in improving its relative efficiency level compared to BC Tel. and Bell Canada. The latter two companies have had roughly equivalent efficiency levels with no major changes in their relative efficiency.

We do not believe that our results will change until we have better data. However, we do expect that some changes will occur as we are able to incorporate more disaggregate and accurate data. Consequently, we would recommend that these results be viewed as the best that currently exist but ones that may change with the further work that we are currently doing. At the end of the data appendix two sources of data problems are identified. These should be improved in the future.

It must be remembered that neither profit nor efficiency levels explain themselves. One may know that efficiency or profits are high or low but it is a more extended task to ascertain why these results occurred. One should not use the results given here to imply any particular line of causation since we have not developed any causes for the differences in relative efficiency.

8. **Data Appendix**

The comparisons that have been made are based on the public data bases of the three companies. In a small but crucial number of incidents the companies have provided extra data which was very helpful. The purpose of this section is to identify the exact public data series which were used.

For Bell Canada, the data were taken from the most recent productivity submission to the CRTC:

> Bell Canada, Information Requested by National Anti-Poverty Organization, March 30, 1981, Bell (NAPO) 30 Mar. 81-612, CRTC.

For BC Telephone the data were taken from the submission to the CRTC:

> BC Telephone, Total Factor Productivity Study: Data Description and Methodology, by J.T.M. Lee, BC Tel. (NAPO) 80-08-01-406 CRTC.

For AGT, data in current dollars was supplied by the company and the corresponding constant dollar data appear in the CRTC submission by AGT, Saskatchewan Telecommunications and Manitoba Telephone Systems in the CNCP-Bell Canada inter-connect case:

> Some Economic Aspects of Interconnection, Evidence in Chief, H. Harries, economic witness.

9. **Bell Canada**

Labour

The quantity equals the unweighted man-hours (MH) (unadjusted man-hours from Table 6 of NAPO 30 Mar. 81). The price index, PL, is generated by dividing total labour compensation (TLE) (Table 6 NAPO 30 Mar. 81) by unadjusted man-hours.

Capital

The output quantity is a divisia index with the output price = 1.0 divided by constant $ series (Table7 NAPO 30 Mar. 81) yields the asset price series. This asset price series was re-normalized in 1972 and the re-normalized price was divided into current $ total average gross stock of capital to yield a constant dollar gross capital series in 1972 $.

The value of capital services was generated residually by subtracting total labour compensation (Table 6 NAPO) and current $ cost of materials (Table 3 NAPO) from Total Revenue (Table 1 NAPO 30 Mar. 81)

$$VK = TR - PM \cdot M - PL \cdot L$$

The service price of capital was arrived at by dividing the 1972 constant $ gross capital series into the value of capital services.

Materials

The current $ cost of materials, services, rents and supplies is divided by constant $ cost of materials, etc. (from Table 3 NAPO 30 Mar. 81) to arrive at a price index. This price series is re-normalized in 1972. The re-normalized series is divided into the current $ cost of materials to provide a constant $ material series.

Output

The output quantity is a divisia index with the output price = 1.0 in 1972. The components in the divisia index are the prices and quantities of local service, message toll, other toll, directory advertising and miscellaneous. Current and constant $ amounts for these categories appear in Tables 1 and 2 of NAPO 30 Mar. 81. The price series for each classification were found by dividing current $ series by the corresponding constant $ series.

10. B.C. Telephone

Labour

Table A-13 of (BC tel. NAPO 80-08-01-406) provides expensed labour hours and expensed wages, benefits and taxes for the following classifications; management, clerical operators, occupational, engineers, salesmen, service rep., technicians and draftsmen. The quantity of labour is the simple, unweighted sum of the expensed labour hours of all these categories. The price of labour was found by dividing this quantity of labour into the unweighted sum of the expensed wages of all the categories.

$$PL = \frac{\sum_i \text{wages}_i}{\sum_i \text{labour hours}_i}$$

Capital

The value of capital services is defined as the sum of the financial charges (Total line in Table A-4), depreciation (Total line in Table A-5), property tax (Total line in Table A-6) for Okanagan Tel. and the financial expense (Total line in Table A-7), depreciation expense (Total line in Table A-8) and property taxes (Total line in Table A-9) for B.C. Tel.

The capital stock series is defined as the reproduction cost of capital in Table A-11, adjusted to 1972 $.

The price of capital services was generated by dividing the value of capital services series by the capital series.

Materials

The value of materials is generated residually. It is found by substracting total expensed wages (see above) and the value of capital services (see above) from total revenue (see above).

This value of materials series is deflated by a re-normalized (1972) materials price index equal to the Stats Can GNE deflator to yield a constant 1972 $ series for materials.

Output

The output price and quantity series is a divisia index (price = 1.0 in 1972) of the disaggregated output categories given in Tables A-1 and A-2. The quantity series are given in Table A-2 while the corresponding revenues are given in Table A-1. A price series is generated for each category by dividing the quantity series into the revenue series.

11. **Alberta Government Telephones**

Labour

Current $ value of labour (from Harries testimony) is divided by the man-hour series (Interconnection Evidence, App. 4, Table 1) to arrive at a price series for labour. No normalization is performed on these series.

Capital

The value of capital services in current $ (from Harries Testimony) is divided by constant 1972 $ average gross capital series to yield a price of capital services. This series is constructed by dividing the current $ gross capital series (Harries testimony) by the constant 1971 $ gross capital series (Interconnection Evidence) which yields an asset price series. The asset price series is re-normalized in 1972 and then divided into current $ gross capital to arrive at the constant 1972 $ gross capital series.

Materials

The current dollar value of materials (in Harries letter of Dec. 4, 1980) is divided by the constant 1971 $ value of materials (provided in Interconnection Evidence Appendix 4, Table 1) to arrive at a price series. This price series is re-normalized in 1972 and a constant $ material series is found by dividing current $ value materials by the re-normalized price series.

Output

The output quantity series is produced by dividing gross revenue in current $ (Harries Testimony) by gross revenue in constant 1971 $ (Inter. Ev.) to yield an output price series. The output price is re-normalized in 1972 then divided into current $ gross revenue to yield a constant 1972 $ output series.

12. Non-Comparable Data

It was not possible to change the public data bases to eliminate some difficulties. Two areas require further improvement. First, BC Tel. measures aggregate output as the Divisia index of disaggregate quantities. The other two companies use a finer disaggregation of output prices to calculate an aggregate output price index and an implicit quantity index. Due to the more aggregate BC Tel. procedure, the growth rate of BC Tel. output is undoubtedly underestimated. We do not know the magnitude but we can be certain of the direction. Second, the differences in the relative output price levels are underestimated for AGT. This affects the level of AGT output and tends to depress it. Consequently, we have undoubtedly underestimated the level but not the growth rate of AGT's output. Correcting this will reduce AGT's disadvantage in relative efficiency during the early years.

REFERENCES:

[1] Caves, R., Christensen, L. and Diewert, E., Multilateral Comparisons of Output, Input and Productivity Using Superlative Index Numbers, SSRI, W.P. 8008 (revised), University of Wisconsin, Madison (1980).

[2] Denny, M., Fuss, M. and Everson, C., Productivity, Employment and Technical Change in Canadian Telecommunications: The Case for Bell Canada, Final Report to the Department of Communications, Ottawa (1979).

[3] Denny, M., de Fontenay, A. and Werner, M., Comparative Efficiency in Canadian Telecommunications: An Analysis of Methods and Uses, Report the the Department of Communications, Ottawa (1980).

[4] Denny, M., de Fontenay, A. and Werner, M., Total Factor Productivity (TFP) in a Telecommunications Utility as a Measure of Efficiency and as a Regulatory Tool, Analysis, Forecasting and Planning for Public Utilities, Vol. 1, N. Curien, ed., Paris, PTT (1980).

[5] Denny, M. and Fuss, M., Intertemporal and Interspatial Comparisons of Cost Efficiency and Productivity, W.P. 8018, Institute for Policy Analysis, University of Toronto (1980).

[6] Denny, M., Fuss, M. and May, J.D., Intertemporal Changes in Regional Productivity in Canadian Manufacturing, Canadian Journal of Economics (August 1981) 390-408.

[7] Department of Communications, Statistiques Financieres sur les Societes Exploitantes de Telecommunications du Canada, 1978, Ottawa (1980).

[8] Diewert, E., Exact and Superlative Index Numbers, Journal of Econometrics 4 (1976) 115-145.

[9] Gilbert, M. and Kravis, I., An International Comparison of National Products and the Purchasing Power of Currencies, Paris, OEEC (1954).

[10] Gilbert, M. and Assoc., Comparative National Products and Price Levels, Paris, OEEC (1958).

[11] Jorgenson, D. and Nishimizu, K., U.S. and Japanese Economic Growth, 1952-74: An International Comparison, Economic Journal 88 (1978) 707-726.

[12] Kravis, I., Kenessey, Z., Heston, A. and Summers, R., A System of International Comparisons of Gross Product and Purchasing Power, Baltimore (Johns Hopkins Press, 1975).

[13] Kravis, I., A Survey of International Comparisons of Productivity, Economic Journal 86 (March 1976) 1-44.

[14] Kravis, I., Heston, A. and Summers, R., United Nations International Comparison Project: Phase II International Comparison of Real Product and Purchasing Power, Baltimore (Johns Hopkins Press, 1978).

[15] Walters, D., Canadian Income Levels and Growth: An International Perspective, Staff Study No. 23, Economic Council of Canada, Ottawa (Queen's Printer, 1968).

[16] Walters, D., Canadian Growth Revisited, 1950-67, Staff Study No. 28, Economic Council of Canada, Ottawa (Queen's Printer, 1970).

[17] West, E.C., Canada-United States Price and Productivity Differences in Manufacturing Industries, 1963, Staff Study No. 28, Economic Council of Canada, Ottawa (Queen's Printer, 1971).

PART 1
PRODUCTION ANALYSIS

Management Applications

Economic Analysis of Telecommunications:
Theory and Applications
L. Courville, A. de Fontenay and R. Dobell (eds.)

GLOBAL FACTOR PRODUCTIVITY (GFP) AND EDF'S MANAGEMENT

J.N. Reimeringer

Electricite de France

1. Brief Account of the Method

1.1 Labor productivity and overall productivity

In everyday language, the word productivity has become synonimous with labor productivity, itself being defined as the ratio of output or added value[1] to labor quantity. However, improved labor productivity does not necessarily mean that the company has become more "efficient", within the meaning with which we are here concerned. Let us take for example a workshop with 10 workers which has a daily output of 100 units of a given item of goods, let us say, chairs. A change of machinery could allow 200 chairs to be produced using the same quantity of raw materials per chair. Labor productivity has indeed doubled, but the collective gain has not doubled. The machinery itself has involved a certain number of work hours - which we shall refer to as an indirect labor and a certain quantity of raw materials. If we wish to have an overall or "global" idea of the workshop productivity, all the factors involved in the production process must be included in the denominator.

1.2 Principle of global productivity evaluation

The output/sum of factors ratio, at first sight, would appear simple to calculate. Would it not suffice to formulate the product/charge ratio on the basis of the amounts shown on the credit side and on the debit side of the general operating account? The answer obviously is "no". The accounting figures are indeed values, that is quantities of goods multiplied by prices. If, by producing the same quantities of good with the same factors, the company doubles its prices while the price of the factors remains unchanged, the product/charge ratio will be multiplied by two, whereas the company's efficiency will remain the same.

Productivity must be assessed on the basis of a quantity to quantity (or still, volume to volume) ratio, but not on a value ratio.

However, if values can be easily added, (since they are all expressed in francs), it is far more difficult to work out the summation of quantities of goods as dissimilar as work hours, tons of petroleum, machinery and kWh. A set of equivalence factors which must be consistent at lowest cost, is needed.

At the outset, the first thought which very naturally comes to mind is market prices. Indeed, market prices are imperfect in certain ways. No price in the literal sense, can be set to a large number of goods, the so-called non-marketable goods, such as for instance: human life, pollution, disrupted kWh... For the rest, there are many causes for biais: power relationship between countries, structural rigidity patterns, governmental action, the policy of privately held companies in a monopolistic or oligopolistic position... However, such prices have a clear advantage. When the company's management takes a decision, they represent the environment within which it fits. By taking market prices as factors, the quality of the company's management may therefore, hopefully, be comprehended throught global productivity. Since the calculation elements for the criterion are the same as those used for decision making, the surplus productivity determined during a given year, (let us say one year), will be calculated by taking as an equivalent system the prices prevailing at the beginning of the year (the effect of a decision being felt only with a certain lag), except in regard to capital charges, which are dependent on long-term decisions (see chapter 2.2).

2. Calculation of Global Productivity Rates

2.1 Calculation principle[2]

Let us note that:

q^1_i : quantity of goods i produced year 1.

q^0_i : quantity of goods i produced year 0

p^0_i : price of goods i year 0

f^1_j : quantity of factor j consumed year 1

f^0_j : quantity of factor j consumed year 0

p^0_j : price of factor j year 0

Consequently, calculation may then be made of the

Production growth index

$$P_1/P_0 = \frac{\Sigma_i \, p_i^0 \, q_i^1}{\Sigma_i \, p_i^0 \, q_i^0}$$

Consuption growth index of factors F_1/F_0 =

$$F_1/F_0 = \frac{\Sigma_j \, p_j^0 \, f_j^1}{\Sigma_j \, p_j^0 \, f_j^0}$$

The global productivity growth indes is inferred therefrom:

$$1 + \pi = \frac{P_1/P_0}{F_1/F_0}$$

π is called the global productivity rate.

- Comparison with partial productivity figures

The consumption growth index of factor j is merely the ratio of the quantities consumed during year 1 to the quantities consumed during year 0. The partial productivity growth index of factor j is therefore

$$1 + \pi_j = \frac{P_1/P_0}{f_j^1/f_j^0}$$

Therefore, if the weight of factor j in the global volume of teh factors for year 1 is referred to as

$$\phi_j = \frac{p_j^0 \, f_j^1}{F_1}$$

We can note that:

$$1 + \pi = \Sigma_j \, \phi_j \, (1 + \pi_j)$$

The global productivity growth index is the sum of the partial productivity growth indexes of the factors weighted by their respective weights in the volume of factors for year 1.

It can easily be ascertained that if the quantities of the factor always evolve in the same manner (it is then stated that there is no substitution of the factors) $\pi_j = \pi$, whatever be the factor j.

2.2 Specific probem of capital charges

Capital charges are broken down as follows:

- financial costs : these costs represent the interest on loans to be taken out by the company in order to purchase the various assets (such as real property, machinery, inventories, ...) which constitute the capital ;

- depreciation charges : such charges represent the loss in value of the capital owing to wear and tear and to the evolution of technology (such capital becoming obsolete).

Now, in relation to the fixed target, which is to measure the company's efficiency by using the conventional measures adopted for economic calculation, the accounting charges provide the indirect means to do so for the following three main reasons :

- first of all, such charges are expressed in current francs. A capital of F. 1 000 does not represent the same quantity of goods according as to whether it was invested in 1960 or 1978, due to monetary erosion. The first adjustment will therefore consist in reasoning in constant francs, that is, in reevaluating all of the assets ;

- secondly, the financial charges depend on the interest rates of the various loans contracted by the company which can be highly variable. Now, what we are attempting to do is to measure the efficiency of a public corporation ("entreprise publique"), one of the characteristics of which is its capacity to decide as arbitrator in accordance with the public interest between a capital expenditure (and therefore an increased effort) in the immediate present and economies (and therefore increased satisfaction) in the future.

Consequently, the collective preference between an expenditure today and a saving in the future need be ascertained. A "social preference for the future" rate, that is, the "discounting rate" must be determined. It should be noted that such rate (fort instance a) is such that it makes no difference to the community whether goods of an x value be used today, or whether goods of an x (1 + a) value (after readjustment for inflation) be used next year.

Such rate (which is presently 9%) is fixed by the "Commissariat General au Plan". It constitutes one of the basis of the investment decisions taken by E.D.F., and one of the essential conventional measures of economic calculation. For the sake of consistency, that same rate is taken into account for economic performance evaluation. Finally, the accounting depreciation of the capital does not actually represent the loss in value of the equipment, within the meaning of economic calculation. Consequently, certain adjustments to ensure consistency with the objectives aimed at need be effected.

The capital charges thus recalculated are termed "normative", as opposed to the charges entered in the accounts, and which are termed to be actual charges.

2.3 Example : calculation of G.P.F. (Global Productivity Factor) of E.D.F. between 1977 and 1978

Expressed in 1978 francs, the operating accounts for 1977 and 1978 are shown as follows, after adjustment for capital charges.

Table 1

Operating Accounts in Value (1978 Million Francs)

	1977	1978
HV sales	6 371	6 455
MV sales	10 926	11 475
LV sales	19 961	21 516
Total products	37 258	39 446
Fuel	8 683	8 426
Purchases of energy	3 137	3 430
Personnel and social services	7 824	8 289
Miscellaneous expenses	4 963	5 362
Capital charges (normative)	21 682	22 186
Total charges	46 289	47 693

Table 1 (continued)

	1977	1978
Operating result	+ 46	+ 131
Variance between actual capital charges and normative capital charges	9 077	8 378

As from such data, the operating accounts in volume are reconstituted : the values for the 1978 accounts are expressed, heading by heading, with the prices of the 1977 accounts (expressed in 1978 francs) :

Table 2

Operating Accounts in Volume (MF 1978)

	1977	1978
HV sales	6 371	6 549
MV sales	10 926	11 614
LV sales	19 961	22 400
Total products	37 258	40 563
Fuel	8 683	9 060
Purchases of energy	3 137	3 616
Personnel and social services	7 824	8 061
Miscellaneous expenditures	4 963	5 345
Capital charges (normative)	21 682	22 186
Total charges	46 289	48 268

The G.F.P. rate is inferred therefrom :

$$1 + \pi = \frac{\frac{40\ 563}{37\ 258}}{\frac{48\ 268}{46\ 289}} = \frac{1\ 089}{1\ 043} \Longrightarrow \pi = 4{,}40\%$$

2.4 Global factor productivity progress at E.D.F. since 1960

The productivity growth between two consecutive years is difficult to interpret partly because it results from numerous effects (see Chapter 4) some of which may be due to the economic conditions then prevailing, but mostly because investments decisions are taken on a long-term basis and therefore their consequences on the G.F.P. are felt over several years.

For all these reasons, it is worthwhile going over the evolution of the G.F.P. rates over a rather long period, and analysing the tendencies which are determined therefrom.

The two diagrams attached hereto show, in addition to the evolution of the G.F.P. rates, the evolution of product volume and factor volume[3].

Diagram 1

Evolution of G.F.P. Rates

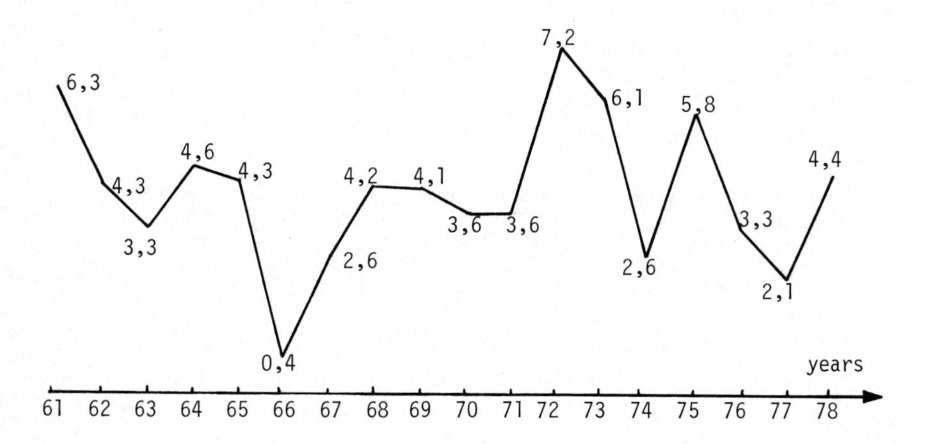

Diagram 2

Evolution of Respective Volume of Products and Factors

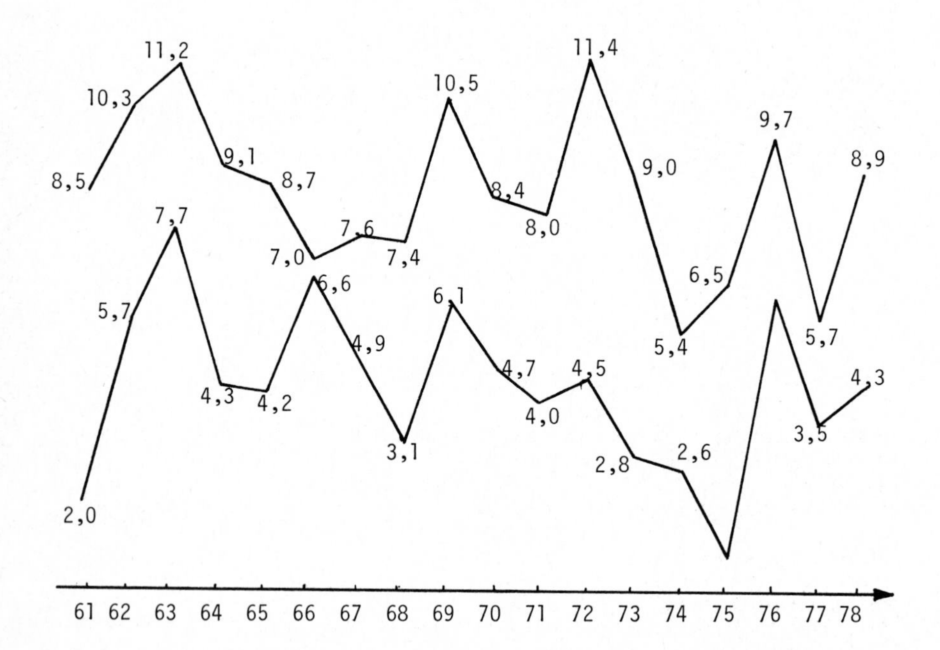

We can note that, from one couple of exercices to the other, the G.F.P. rate varies considerably, thereby making for a serrated curve. However, the simultaneous reading of the two graphs brings to light certain phenomena :

- early in the 1960's the annual consumption growth was substantial, there-by allowing for the obtention of G.P.F. rates of about 4,5%, despite a substantial increase in charges ;
- from 1964 to 1968 approximately, the increase in charges was substantially the same ; however, the product increase was less marked. There followed a deterioration of the G.F.P. which is mainly accounted for by the slow-down in consumption ;
- as from 1969 and until 1975, even though the increase in consumption ranged around 8% per annum, the growth rate of charges had been continuouly declining : consequently, there progressively followed a recovery of the G.F.P. ;
- during these past few years, a further growth of charges has brought back the G.F.P. rate to a lower level.

On the average the G.F.P. has been progressing by 4% since 1960. Notwithstanding the variations due to the prevailing economic condition, it may therefore be stated that E.D.F. is a compagny whose productivity is substantially growing.

3. Productivity Surplus and its Allocation

3.1 Principle and method of calculation

By increasing its productivity, the company has increased the variance between the volume of goods it restitutes to the community and the volume of the production factors it absorbs. Such increase may therefore be expressed as follows :

$$S = \underbrace{(P_1 - F_1)}_{\text{Product-charge variance for year 1}} - \underbrace{(P_0 - F_0)}_{\text{Product-charge variance for year 1}}$$

S is referred to as Productivity Surplus.

We can now proceed to examine the allocation of the surplus wealth thus generated. To do so, we are going to examine the manner in which price variations occur.

We know that prices undergo a general rise due to inflation as measured by the price index of the Gross Domestic Product (G.D.P.). For each specific price, we can evaluate its comparative upward or downward drift, in relation to the average of the other prices : to do so, it will suffice that the reasoning be formulated in constant francs (by dividing the values of year 1 by the price index of the G.D.P. between year 0 and year 1).

When the price of a product diminishes or when the price of a factor increases (in constant francs), this means that a portion of the surplus has been used to that effect : it is then stated that a use has been made. In the opposite case, we have a resource.

The calculation there of is quite simple. Let us take for instance p as being the G.D.P. price index.

For a product : $q_i^1 \left(\frac{p_i^1}{p} - p_i^0\right)$ is a use if it is negative
is a resource if it is positive

For a charge: $f_j^1 \left(\frac{p_j^1}{p} - p_j^0\right)$ is a use if it is positive
is a resource if it is negative

p^1_i and p^1_j being the prices of the goods and factors for year 1.

Let us look for the existing ratio, surplus to uses and resources.

We know that :

$s = (P_1 - F_1) - (P_0 - F^0)$

Which we can also write as :

$$s = \sum_i p_i^0 q_i^1 - \sum_j p_j^0 f_j^1 - \sum_i p_i^0 q_i^0 + \sum_j p_j^0 f_j^0$$

By introducing the prices for year 1, we obtain :

(1)
$$s = - \sum_i p_i^1 \left(\frac{p_i^1}{p} - p_i^0\right) + \sum_j f_j^1 \left(\frac{p_j^1}{p} - p_j^0\right) \}$$

(2)
$$+ \sum_i \left(q_i^1 \frac{p_i^1}{p} - p_i^0 p_i^0\right) + \sum_j \left(f_j^1 \frac{p_j^1}{p} - f_j^0 p_j^0\right) \}$$

We can recognize in the first term (1) of the equation the total sum of the uses less the sum of the resources ; let us now take a look at the second term ; it may also be written out as :

(2)
$$= \underbrace{\frac{1}{p} \left(\sum_i p_i^1 p_i^1 - \sum_j (f_j^1 p_j^1\right)}_{\text{profit for year 1}} \quad \underbrace{\left(\sum_i q_i^0 p_i^0 - \sum_j f_j^0 p_j^0\right)}_{\text{profit for year 0}}$$

The constant franc variation of the company's profit is thus formulated.

We now arrive at the following remarkable equality :

surplus = uses - resources + profit variation

We can thus ascertain that a company which increases its productivity does not necessarily increase its profit: through uses, it may cause its customers, its personnel or its suppliers to benefit therefrom. Productivity and profit are in no way synonymous.

3.2 Example : the productivity surplus determined by E.D.F. between 1977 and 1978 and its allocation.

The productivity surplus is obtained by comparing the 1978 and 1977 accounts in volume. Uses and resources are calculated by a substraction from the 1978 accounts in value and in volume.

Capital lenders are included among the recipients of uses (or the allottees of resources). Consequently, the reasoning must be formulated in terms of actual, and not normative capital charges.

Table 2

1978 and 1977 operating accounts
(in volume and in value, 1978 MF)

	(1)	(2)	(3)	(4)	(5)
	1977	1978	1978	resources	uses
	(volume-value)	(volume)	(value)		
HV sales	6 371	6 549	6 455		94
MV sales	10 926	11 614	11 475		139
LV sales	19 961	22 400	21 516		884
Toatal products	37 258	40 563	39 446		
Fuel	8 683	9 060	8 426	634	
Purchases of energy	3 137	3 616	3 430	186	
Personnel and social services	7 824	8 061	8 289		228
Miscellaneous expenses	4 963	5 345	5 362		17
Financial charges	4 568	4 685	4 741		56
Depreciation	8 037	8 216	9 067		851
Total charges	37 212	38 983	39 315		
			Total	820	2 269

We thus obtain :

- the S productivity surplus through substraction of the 1978 and 1977 product-charge differences, in volume:

$$S = (40\,563 - 38\,983) - (37\,258 - 37\,212) = 1\,534 \text{ MF}$$

Products-charges for year 1 — Products-charges for year 0

the constant franc profit variation through substraction of the 1978 and 1977 product-charge differences, in value :

$$B = (39\,446 - 39\,315) - (37\,258 - 37\,212) = 85 \text{ MF}$$

Profit for year 1 — Profit for year 0

Let us now ascertain that we have duly reached equality of terms :
s = uses - resources + Δ B

In effect :

1 534 = 2 269 - 820 + 85

Such equality may be represented so as to cause the relationship between uses and resources to come out (see diagram), thereby facilitating its interpretation.

Surplus allocation - interpretation

Sum of surplus and resources : 1 534 + 634 + 186 = 2 534 MF

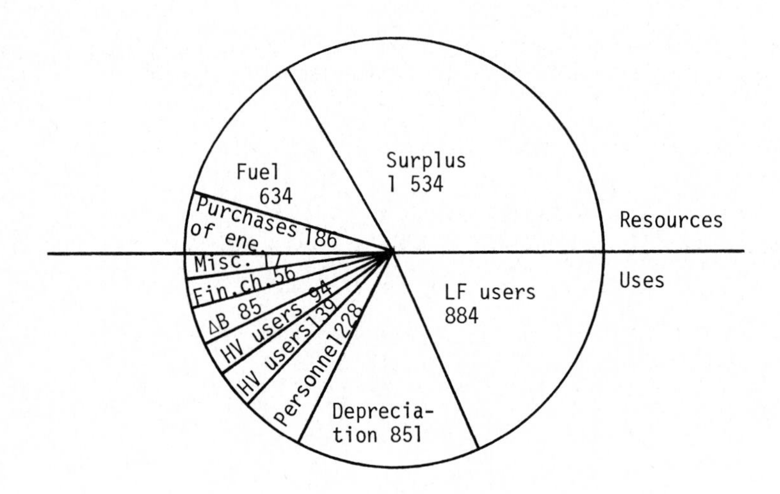

Sum of uses and of profit variation:
94 + 139 + 884 + 228 + 17 + 56 + 851 + 85 = 2354

In addition to the productivity gains made by E.D.F., we also have an inherited fuel resource (due to the diminution in the fuel price in constant francs) and an inherited energy purchase resource (likewise caused by the decline in fuel prices).

Resources are mainly used for increased depreciation (a little more than 10%). E.D.F. may thus generate liquid assets to be used for supplemental self-financing (the very small increase in the profit contributing to the same end).

Users have benefited from such resources to a lesser extent (in regard to their weight). Low voltage users are the most favored (+ 4%). As for employees, the costs borne in their respect by the company have increased by approximately 3% in constant francs, given an equal number of units of personnel with the same qualifications, a figure which can compare with the average drop in the constant franc sale price of electricity.

4. Precautions to be Taken in Interpreting G.F.P. Calculations

We have seen that the G.P.F. rate constitutes a weighting of partial rates representing various factors. Likewise, the product volume index constitutes a weighting of the growth rates of the different goods and services produced by the company. When the factor and product relative weights vary little, the G.F.P. rate directly reflects the increase of partial productivity. However, this does not apply in periods where the prices of factors vary substantially in relation to one another, or where technological or structural pattern evolutions of consumption involve a change in product and factor relative volumes. By definition, a G.F.P. rate "covers" all the phenomena which may bear on such rate and it is sometimes difficult to segregate their specific action (their so-called "effect").

4.1 Factor substitution effect

Economic rationality would require an enterprise to combine its production factors so as to minimize its total production cost. To a given price ratio of the factors, there will be a corresponding given combination. If such ratio should change, it would be in the company's interest to change its combination so as to use a lesser quantity of the factor whose price has increased.

Such adaptation to price evolution is effected more or less rapidly according as to whether it requires more or less substantial investments. Thus it happened that further to the quadrupling of the 1973/1974 petroleum price, E.D.F.'s reaction was expressed in three stages.

- first of all, by a more intensive use of coal-fired plants involving immediate effects ;

- then by the conversion to coal of a certain number of fuel-operated plants ;

- finally, by a progressive substitution of nuclear energy for conventional fuels which has now begun to bear fruit.

Each substitution diminishes the quantity of the factors (weighted by means of the new price system), even though the techniques used may be the same (or even - if related to each factor- more expensive : the total cost of a fuel-fired plant reconverted to coal is higher than that of a plant built for the purpose of being coal-fired).

The G.F.P. increases. This is termed a factor substitution effect. Such effect cancels out when readaptation to the new price structure occurs.

4.2 Expansion effect

One of the most salient characteristics of electricity supply is that the volume of capital and personnel required to supply users increases less rapidly than consumption : given identical technical requirements, the greater the density of power to be supplied, the lower the supply cost. It can then be stated that returns are increasing.

At the production level, returns are increasing but at a much lower rate in thermal production due to size effects and to expansion (the greater the number of production units, the lesser would be the consequences of a unit failure). Such returns decrease in hydraulic production by reason of the saturation of usable sites (it is then stated that an adverse "external factor" occurs).

Nevertheless, having regard to the high proportion of low and medium voltage supply, a consumption growth will reflect an automatic productivity increase : this is the expansion factor.

A product structural effect (similar to the factor substitution effect) will be superimposed on the expansion effect. The fact that the low voltage should increase more rapidly than the high voltage, is not without consequences for the G.F.P. rate : indeed the expansion effect is greater on the former than on the latter.

4.3 Effect of factor weights

We have seen that the G.F.P. rate could be expressed as a sum weighted by the partial productivity factor weights. If such weights be very different according to the factors, the weights are going to have a very substantial influence. Now, differences do exist : considering the present state of technology, it is virtually no longer possible to improve the efficiency of thermal plants. Partial productivity of the "fuel" factor is therefore constant. The weight of such factor is directly related to its price which is in constant variation. The G.E.P. rate is therefore going to be directly affected thereby. This is the effect of factor weight.

4.4 Service quality effect

This is a specific case relating to the consequences of the non-valorisation of non-marketable goods. Black-out time and voltage drops will be more or less substantial in relation to the generaling, transmission and supply equipment for a given consumption requirement. The fact that such quality aspect should not be valorized leads to the absurd conclusion that within the meaning of the G.F.P. calculations, there would be greater advantage in consumption. Factors increase less rapidly while the incidence on the products-due to the simple fact that we are confronted with a potential loss in profit when power cut-outs must be made - is small. Consequently, a normative price representing the hindrance thereby involved for the community should be applied to thek kWh which should fail to be supplied (and to voltage drops). Up to now, the G.F.P. calculation at the national level has not taken this into account, thereby implicity assuming that the reasoning rests on a nearly constant service quality factor. There is the risk that such assumption may not lend itself to ascertainment in the next forthcoming years.

A similar remark could be made in regard to expenses undertaken by E.D.F. for the environment. Insofar as they increase the costs of structures and the advantage accruing therefrom for the community cannot be expressly valued in the products, such costs have a downward effect on the productivity rate calculated, which it would be advisable but difficult to correct.

Consequently, the G.F.P. rate reflets these various effects which are difficult to segregate and also the results of technological progress and a better work organization. Efficient use of such composite management tool would therefore require a careful interpretation of the results.

FOOTNOTES:

[1] Added value = value of products - cost of raw materials (including fuels).

[2] The closeness of the definitions - based on a mathematical formulation - must not be confused with the required conventional measures and approximations which always intervene in the search for global figures.

[3] The homogeneity of the figures may call for some reservations due to changes in the calculations over that period. However, the orders of magnitude and the evolutions are correct.

Economic Analysis of Telecommunications:
Theory and Applications
L. Courville, A. de Fontenay and R. Dobell (eds.)

NET INCOME AND PRODUCTIVITY ANALYSIS (NIPA) AS A PLANNING MODEL

A. Chaudry

American Telephone and Telegraph Company

1. Introduction

The purpose of this paper is to describe a productivity-based planning model which is in current use at a major U.S. corporation and which explicitly recognizes the role total factor productivity plays, along with other factors, in determining the bottom line net income. Using hypothetical data for XYZ Corporation, we illustrate how an existing planning and budgeting process can be enhanced by introducing the contribution of productivity and other factors in terms of dollars and cents to which decision-makers can relate easily.

For this purpose, a Net Income and Productivity Analysis (NIPA) Model (Chaudry, Burnside and Eldor (1980)) has been developed to explain the growth in net income in terms of total factor productivity (TFP), growth of capital, price changes, inflation in input costs, depreciation, taxes and other financial factors[1]. We refer to these as NIPA factors. The NIPA model assumes total product exhaustion by inputs and other specific factors in each period and attributes dollar values to all inputs and factors. By regrouping them according to the way they affect net income, the net sum of these factors accounts for the total change in net income. By separating price and quantity components, changes in the efficiency of the enterprise can be identified.

These factors can be grouped into the following categories:

Income-Augmenting Factors - those directly contributing to growth in net income: productivity, or the improvement in efficiency of the firm; growth in the physical capital stock; changes in product prices; and 'other income' (not directly associated with the physical operations of the firm).

Income-Absorbing Factors - those inversely related to growth in net income: changes in prices of materials and services purchased from other firms; changes in labor input costs due to changes in wages and benefits; changes in non-income (indirect) taxes due to change in the tax rates; changes in depreciation expenses; and changes in income taxes and financial factors. (The financial factors included in this category are interest charges, uncollectibles, miscellaneous deductions from income and extraordinary and delayed items - net. In the detailed model output, these factors are analyzed individually. See Table 1.)

This model can be used by a corporation for corporate planning, budgeting and for strategic targetting to develop and analyze its budget or its corporate plans in such a way that (1) all relationships between productivity and costs, and those between prices and volume of business are treated consistently and (2) all financial factors are fully accounted for. In other words, nothing is allowed to fall through the cracks, as is common with many financial models that rely on ad hoc ratios of selected variables. For example, these ratios are usually calculated in terms of current prices and thus reflect the effects of both price changes and volume changes. NIPA, on the other hand, decomposes each key variable into its price and quantity components and ensures that the total change in the variable is accounted for by the sum of the changes in the respective components.

We describe the theoretical framework of the model in Section II followed by a discussion in Section III of how the basic results might be analyzed in terms of the dollar impact of changes in prices and quantities separately, on the change in net income.

This feature of NIPA is the key to its usefulness as a planning model. The user can vary any of the assumptions explicitly in terms of projected productivity gain, price changes, inflation in various costs and other factors such as tax rates or depreciation rates. For illustrative purposes, Section IV describes three alternative scenarios based on different productivity projections. Some concluding remarks are presented in Section V.

2. The Model

By the usual broad definition, net income is simply the difference between total sales revenue and total expenses or costs. Defining revenue as all financial inflows and expenses as all financial outflows, including taxes, etc., and treating net income as a return to equity holders (thus a claim against revenue and a component of total cost), one has as an accounting identity the requirement of total revenue exhaustion, or

$$R = NI + C^* = C$$

where NI = net income

R = total sales revenue

C^* = total operating cost (excluding return to equity capital)

C = total costs.

Since the changes in revenue and costs, and therefore in net income, reflect the combined effect of price and quantity changes, we need to further decompose the total change in each variable into its price and quantity components. Only then can we measure productive efficiency in terms of the real output and real inputs and account for the price effects separately.

It should be noted that the productivity calculation in NIPA is different from the fixed base-year methodology which is used in standard TFP measurement. Since we are attempting to account for the growth in net income from one year to the next, we are dealing with the quantities (or volumes of output and inputs, respectively), and their corresponding prices for two consecutive years. Thus, the measures of output and inputs for the current year (t) must be constructed in terms of prices of the previous year (t-1). Similarly, any effects attributable to changes in output prices or input prices must be measured with reference to the previous year. This means that all computations of this type to be made within the NIPA framework employ a changing base year as contrasted with the fixed-base-year indexes of traditional productivity studies.

Differencing equation (1), with P(Q) representing the base year price of base year year output Q; and P(X) a vector of prices corresponding to the input vector X, one has

(2) $P(Q).dQ + dP(Q).Q = P(X).dX + dP(X).X$

where P(Q).dQ and P(X)dX represent the real physical effects and dP(Q).Q and dP(X).X represent the effect of changes in output and input prices respectively. By definition, total productivity (TP[2]) is

TP = change in Real Output - change in Real Inputs or

(3) $TP = P(Q).dQ - P(X).dX$

Separating the real and price components in the accounting identity (1), and using (3), we can write

(4) $TP + dP(Q).Q - dP(X).X = 0$

Substituting individual input prices explicitly, the identity (4) can now be written as

(5) $TP + dP(Q).Q = dP(K).K + dP(L).L + dP(M).M$

where K is total capital, L is labor and M is total materials, rents and services.

Equation (5) states that the total productivity gain and the value of the output price changes are absorbed by the three inputs in the form of remuneration to the respective factors of production.

The foregoing exposition has been simplified by explicitly including only the quantities and prices of the three major input factors. We also need to take into account indirect taxes and a number of financial factors for completeness. In defining output for calculating TP, deflated indirect non-income taxes (NIT) are generally substracted from deflated revenues. These include (a) Property taxes, (b) Capital Stock taxes, (c) Gross Receipts taxes, and (d) other Non-income taxes. The first two categories are related to the real investment in plant and equipment, while the last two are related to sales revenue and these relationships are used to make the calculations.

The total change in these taxes (dNIT) consists of real change (dNITR) and the "price change" effect which, in this case, means the change resulting from a change in the tax rate (dNITP), i.e.,

(6) $dNIT = dNITR + dNITP.$

The real effect, dNITR has been implicitly accounted for in the definition of output and of TP[3], as given above, but we also need to account for the "price effect." This is done by expanding the (dP(X).X) vector to add dNITP to the right hand side (RHS) of equation (5).

Since the capital input change is a deduction in the TP calculation, but is included (in part) in the net income, we need to reflect this in our model.

Adding P(K).dK to both sides of the equation, we get

$$(7)\quad TP + dP(Q).Q + P(K).dK = (dP(K).K + P(K).dK) + dP(L).L + dP(M).M + dNITP$$

where P(K).dK is the growth of physical capital input and dP(K).K + P(K).dK = d(P(K).K) is the current undeflated value of the change in capital input. For the present expository purpose the latter may be interpreted as comprising depreciation (DEP), interest charges on debt (INT), income taxes (FIT + SLIT), and the return to equity investors (i.e., the net income (NI) - including other income (OI)) and other miscellaneous financial factors such as uncollectibles (UNC), miscellaneous deductions from income (MDI) and extraordinary and delayed charges and credits-net (E&D).

Substituting these factors for d(P(K).K), we obtain the fundamental equation underlying NIPA as

$$(8)\quad TP + dP(Q).Q + P(K)\,dK + dOI = dP(L).L + dP(M).M + dNITP + dDEP + dFIT + dSLIT + dINT + dUNC + dMDI - dE\&D + dNI$$

This equation is an alternative definition of the change in net income which we present on the following page, along with the traditional definition contained in the income statement. This equation is also the basis for all simulation results generated by NIPA which we discuss in subsequent sections of this paper. The complete set of relationships described in this section is shown schematically in Figure 2, with all variables as defined in the text.

The whole model can be thought of as consisting of four interrelated modules, namely, Productivity Module, Capital Growth Module, Price Effects Module and finally Tax and Financial Module. We will return to the relationships among these modules when we discuss the alternative planning scenarios in Section IV.

Relationship Between

NIPA and the Income Statement

	NIPA	INCOME STATEMENT
	Change In net Income =	Change in Net income =
	Change in Deflated Revenue	Change in Revenues[1]
	- Change in Def. Non-Income Taxes	- Change in Non-Income Taxes[1]
TP	- Change in Labor Input	- Change in Labor Costs[1]
	- Change in Capital Input	- 0.0[2]
	- Change in Materials Input	- Change in Materials Costs[1]
	+ Output Price Changes	(See Footnote 1)
	+ Capital Expansion	+ 0.0[2]
	- Inflation in Materials	(See Footnote 1)
	- Inflation in Labor	(See Footnote 1)
	- Inflation in Non-Income Taxes	(See Footnote 1)
	+ Change in Other Income	+ Change in Other Income
	- Change in Depreciation	- Change in Depreciation
	- Change in Federal Income Tax	- Change in Federal Income Tax
	- Change in State & Local Income Tax	- Change in State&Local Income Tax
	- Change in Interest	- Change in Interest
	- Change in Uncollectibles	- Change in Uncollectibles
	- Change in Misc. Deductions	- Change in Misc. Deductions
	+ Change in Extra. & Del. Items-Net	+ Change in Extra.& Del. Items-Net

Notes:

1. These items are in nominal terms and thus include price changes.

2. In the income statement, there is no deduction for capitalized investment expenditures. Thus the return to capital is a part of net income.

3. TP = Total Productivity. See text for definition.

3. Analyzing The Relative Contribution of NIPA Factors

The basic NIPA model yields an estimate of the dollar contribution of each factor to the growth in net income. An example of the model output is shown in Table 2. However, these dollar estimates can be affected by a number of factors and thus could vary substantially from year to year. Furthermore, a change in the magnitude of a given factor from year to year is hard to interpret because of the continuously shifting base year. For example, the deflated quantities such as revenues in any two years are expressed in terms of the prices of the previous year. Thus the difference between the deflated revenues in two years within the basic NIPA framework cannot be treated merely as a change in the "real" or physical volume of business.

This difficulty can be avoided by comparing the percentage contribution of each of the factors relative to the subtotals for the "income augmenting" and the "income absorbing" factors respectively (% f(AUG,i)) and % f(ABS,i)), computed as follows:

$$\% f(AUG,i) = Fi/TAUG \qquad i = 1, 2, 3, 4$$

$$\% f(ABS,i) = Fi/TABS \qquad i = 5, 6, ..., 14.$$

where Fi are the various NIPA factors and

TAUG = Subtotal for the Income-Augmenting Factors; and

TABS = Subtotal for the Income-Absorbing Factors.

The resulting percentage distribution is shown in Table 2.

While these percentage factors are a little more stable over time, compared with the dollar contributions, they are also subject to several influences whose importance can vary from year to year and which are not explicitly quantified in NIPA. Moreover, the two subtotals themselves are arbitrary and do not bear any direct relationship to any of the financial variables that financial planners have to work with. Thus we propose the following normalized NIPA factors, using the level of net income in the previous year as the normalizing variable.

$$Gi(t) = Fi(t)/NI(t-1)$$

where

Fi(t), i = 1, 2, ..., 14 are the dollar contributions of the NIPA factors in the current year; and

NI(t-1) = Level of Net Income in the previous year.

By definition,

$$\sum_{i=1}^{14} Fi(t) = dNI(t).$$

Thus

$$\sum_{i=1}^{14} Gi(t) = dNI(T)/NI(t-1)$$

Gi(t) can be interpreted as the percentage contribution of the ith factor in year t to the growth rate of net income in that year. That is, each Gi(t) is a proportionate growth factor and that all of them combined account for the total growth in net income during the year.

This normalization procedure is appealing because,

(a) it uses the level of net income in the previous year as the normalizing variable, which is independent of the current year's distribution of NIPA factors themselves; and

(b) it directly shows the importance of a given factor in determining the growth rate of net income.

Table 3 shows a three-year history of these proportionate growth factors (along with projected results for 1980-1985 which will be discussed in Section IV) for the XYZ Corporation. For example, in 1979, productivity accounted for 32% of previous year's net income, earnings on capital 16%, rate changes 12% and other income 1%. The combined contribution of all income augmenting factors in that year was 61%. This means that if there had been no inflation in MR&S, Labor, etc., and no increases in taxes, depreciation or interest charges, etc., net income would have grown by 61%. But unfortunately, all of these factors were present. For example, inflation in MR&S amounted to 11% of previous year's net income and inflation in Labor costs (including Social Security taxes) another 30%. However, there was some relief from the Property and Other Non-income Taxes (-3%) and from Federal Income Taxes (-4%). The combined negative effect of the income absorbing factors was to reduce the net income growth by 55%. Hence the actual percent growth in net income of 6% (=61-55) in 1979.

As shown in Table 3, the relative importance of productivity over the three year historical period is fairly stable around 30%. Capital expansion ranges between 16% and 18% while the contribution of price changes varies widely. The latter is a reflection of the irregular price adjustment process in this particular case where price changes come in lumps.

On the negative side, inflation in MR&S ranging between 8% and 11% is a direct result of the changes in the general price level for materials and services the firm buys from other firms. Similarly, inflation in labor costs broadly reflects the increases in wage rates and related benefits, as well as changes in the Social Security tax rates. This factor varies widely between 18% and 30% and it partly reflects the effects of a three-year bargaining cycle and changes in Social Security tax legislation. Depreciation is very stable around 12%, whereas income taxes and other financial factors show considerable variation.

It should be noted that we are not necessarily implying that the normalized NIPA factors should remain stable over time. But when a particular factor shows a significant change, it should be regarded as a signal of a fundamental shift that should be investigated further.

4. Alternative Planning Scenarios

We are now ready to investigate the behavior of net income under varying assumptions for the future, and to use NIPA to analyze how various factors contribute to changes in net income. In this section we present three illustrative cases. The first case assumes that labor productivity increases in the planning period at the same average rate as it did in the last five years (Simulation or Sim A); in the second we hold the absolute level of labor productivity constant, i.e., zero growth in labor productivity (Sim B); and in the third, it grows at a rate 20% faster than the average growth in the past five years (Sim C). All other variables for the three simulations are projected according to the assumptions sketched out in the table on the next page.

While it is possible to generate many other scenarios with a model like this, we have focused on the effect of varying productivity on the bottom line. Given certain assumptions about the behavior of productivity, we first derive a complete income statement. Then by solving the NIPA equation system, the resulting changes in net income are analyzed in terms of the NIPA factors described in the earlier sections.

In terms of the schematic in Figure 2, our key assumptions about labor productivity primarily affect the Productivity Module through impact on hours while other assumptions impact the Price Effects module. The Capital Growth module is affected only in Sim C where we make the additional assumption that the capital-labor ratio be held constant.

ASSUMPTIONS FOR SOME ALTERNATIVE FUTURE SCENARIOS FOR 1980-1985

Key Variables	Sim A	Sim B	Sim C
Output Volume	Average Rate of Growth	Average Rate of Growth	20% Above Average Growth in Volume
Output Prices	Average Rate of Increase	Average Rate of Increase	Average Rate of Increase
Employee Hours	Average Labor Productivity Gain	Zero Labor Productivity Gain	20% Above Average Labor Productivity Gain
Hourly Compensation	Average Rate of Increase	General Economy's Increase for Hourly Comp.	General Economy's Increase for Hourly Comp.
Capital Stock	Average Rate of Increase	Average Rate of Increase	Constant K/L Ratio at Average Level
Prices of Plant & Equipment	Average Rate of Increase	General Inflation in PDE Prices	General Inflation in PDE Prices
Materials	Average Rate of Increase	Avg. Vol. Growth, General Inflation in Prices	Avg. Vol. Growth, General Inflation in Prices

All Other Variables Are Projected At Average Rates.

This means that once the required hours have been determined, we must further determine the level of capital which is consistent with those hours. Note that while our assumptions were centered around labor productivity which directly affected the required hours (or labor input), NIPA still utilizes the total productivity concept in making all calculations shown in Tables 1 thru 3. By focusing on labor productivity, we are able to isolate the impact of this key factor on the financial performance of the firm. Of course, similar analyses could be conducted on any of the other variables of interest to the planners.

As an example of the complete set of analytical results currently generated by NIPA, we show the following data in Tables 1 thru 3 for Sim A for the planning period 1980 to 1985, along with actual results for 1976 to 1979.

Table 1: NIPA Summary

Table 2: Percentage Distribution of NIPA Factors

Table 3: Relative Importance of NIPA Factors

For comparative purposes, we have plotted the key variables from the three simulations in Figures 3 thru 7. For example, Figure 3 shows the three assumptions about labor productivity; Sim A with average growth in output per hour; Sim B with zero growth and Sim C showing 20% higher than the average growth rate. We see the dramatic impact of zero productivity on net income which takes a nose dive starting in 1980 and ending up negative in 1984 and 1985 (see Table 4). Figure 4 shows this phenomenon in terms of changes in net income, contrasting Sim B with the alternative scenarios. For instance under Sim B, the net income loss increases by nearly $4 billion in 1985 whereas A and C show positive gains.

If we had looked at the nominal net income data in Table 4 and the total revenue and total expenses alone, we would not be able to easily identify the source of the decline in net income. It could have resulted from any number of causes including higher wages, materials costs and other inflationary pressures. NIPA on the other hand, provides a vivid analysis of the true picture. With no growth in labor productivity, the number of employee hours required to produce the growing output increases rapidly resulting in substantial increases in the dollar value of labor input, which reduces the contricution of total factor productivity turning it into losses in 1983-1985. This is evident in Figure 5 showing the dollar contribution of TFP under the three scenarios. We can further examine the implications of zero labor productivity by looking at the dollar and relative impacts of inflation in labor (Figures 6 and 7 respectively). In both dollar and relative terms, we see that Sim B is substantially higher than either A or C, even though increases in the hourly compensation in B and C are the same. This difference occurs because the increase in the hours in B is so much larger than in C (where productivity rises 20% faster than the average of the last 5 years). The relative importance of the underlying productivity growth is especially highlighted in Figure 7 which shows the resulting increase in labor costs (due to the the increases in hourly compensation coupled with labor productivity behavior) as a percent of previous year's net income.

5. Conclusions

We have described a planning model which provides a great deal more information than most financial models offer. NIPA permits the user to account for productivity and many other underlying factors explicitly and in terms of dollars and cents as they affect the bottom line. Because of this feature, managers find NIPA easier to understand than most models offered to planners. Thus they are more inclined to use it as a planning tool and be able to put the dollar productivity estimate in proper perspective of the income statement. It is worth noting that most managers have shyed away from the use of traditional productivity measures expressed in terms of percentage growth rates, because they could not use such data in any direct way. The most some planners were able to do was to compare the projected productivity growth with the past record and make a qualitative judgement as to whether the budget based on that projection was a reasonable one. The only control these managers could exercise was to demand an explanation of the poor productivity built into the budget and ask the operating entity anticipating lower than averge productivity to redo its budget with some target productivity growth. With NIPA, the planner is able to see the dollar contribution of productivity to growth of the bottom line and is thus able to set a specific quantitative target the entity must achieve if it is to meet its budget goal. Moreover, looking at the standard NIPA Summary results, the planner is able to see whether it is poorer output growth, faster input growth or worse inflation beyond the control of the managers which is the culprit. In other words, while NIPA cannot come up with solutions, it can at least point out the problem areas which should be investigated in search of management options to intervene. Moreover, it provides enough disaggregation of the underlying factors to allow a distinction to be made between what management can and cannot control and thus act accordingly.

NIPA is currently operational in an interactive mode and allows the user to provice a variety of inputs and exercise many options in terms of generating alternative scenarios and select desired results. However, this means that the necessary budget data must be prepared in advance through whatever budgeting process may be in use. We plan to extend the model to include target setting by the user and having the model solve for all the endogenous variables before making the standard NIPA calculations. For instance, we could set a target for net income or rate of return and then given specific operating rules, demand conditions and appropriate constraints, solve for the necessary inputs to NIPA.[4]

This would permit the user to construct a budget, analyze it and alter it with the help of NIPA to achieve prespecified management goals. For example, such a model could also be used for determining the change in hours or other resources that would be required if a certain change in the budgeted net income is to be made.

FOOTNOTES:

[1] Werner (1979) has also developed a similar productivity-based model which is designed to calculate a theoretical budget, subject to appropriate constraints facing the corporation. NIPA, by contrast, works with the proposed budget and recasts it in terms of productivity and other factors discussed in the text.

[2] TP as used here refers to the total productivity of all inputs, namely capital, labor and materials. The more commonly known concept of total factor productivity (TFP) is sometimes used interchangeably with TP. However, some authors prefer to limit the use of TFP to the combined productivity of capital and labor only.

[3] In terms of tax-adjusted output and real inputs, total productivity is defined as:

TP = P(Q).dQ -	dNITR -	P(K).dK -	P(L).dL -	P(M).dM.
Total Deflated Revenue	Deflated Non-Income Taxes	Real Capital Input	Labor Input	Materials Input

[4] A similar model has been proposed by M. Werner (1979), using a somewhat different alternative rationale.

REFERENCES:

[1] Chaudry, M.A., Burnside, M. and Eldor D., NIPA: A Model for Net Income and Productivity Analysis, forthcoming.

[2] Craig, C.E. and Harris, R.C., Total Productivity Measurement at the Firm Level, Sloan Management Review (Spring 1973) 13-28.

[3] Davis, B.E., Caccapollo, G. and Chaudry, M.A., An Econometric Planning Model for American Telephone and Telegraph Company, Bell Journal of Economics (Spring 1979).

[4] Denny, M., Fuss, M. and Waverman, L., The Measurement and Interpretation of Total Factor Productivity in Regulated Industries, with an Application to Canadian Telecommunications, a paper presented at the NSF Conference on Productivity Measurement in Regulated Industries, Madison, Wisconsin (May 1979).

[5] Faraday, J.E., The Management of Productivity, Management Publications Limited, Tonbridge, Kent, Great Britain (1971).

[6] Kraus, J., Productivity and Profit Models of the Firm, Business Economics (September 1978) 10-14.

[7] Ross, T.L. and Bullock, R.J., Integrating Measurement of Productivity into a Standard Cost System, Financial Executive (October 1980) 34-40.

[8] Steffy, W., Zearley, T. and Strunk, J., Financial Ratio Analysis, An Effective Management Tool, Industrial Development Division, Institute of Science and Technology, the University of Michigan, Ann Arbor, Michigan (1974).

[9] Tilanus, C.B. (ed.), Quantitative Methods in Budgeting, Martinus Nijhoff, Leiden (1976).

[10] Werner, M., Productivity Based Planning Model for Teleglobe Canada, Proceedings of the International Telecommunications Conference (1979).

Table 1

NET INCOME AND PRODUCTIVITY ANALYSIS (NIPA)
XYZ CORP - AVG GROWTH IN O/H
(MILLIONS OF DOLLARS)

YEARS IN STUDY	1977	1978	1979	1980	1981	1982	1983	1984	1985
CHG. IN DEFL REVENUES	$1,565.8	$2,060.5	$2,223.1	$2,157.2	$2,427.0	$2,732.1	$3,076.8	$3,466.5	$3,907.4
- CHG IN DEFL PONIT	79.3	98.2	93.8	78.0	64.1	69.8	76.0	82.8	90.3
= CHG IN OUTPUT	1,486.5	1,962.3	2,129.3	2,079.2	2,362.9	2,662.3	3,000.8	3,383.7	3,817.1
- CHG IN CAPITAL INPUT	382.4	496.5	504.8	407.9	239.9	268.1	299.7	335.0	374.6
- CHG IN LABOR INPUT	328.0	378.6	351.0	54.2	250.3	276.4	314.3	357.8	407.6
- CHG IN MR&S INPUT	215.5	264.0	247.8	187.3	240.0	275.4	316.1	362.9	416.7
= PRODUCTIVITY GAIN	560.6	823.2	1,025.7	1,429.8	1,632.7	1,842.5	2,070.8	2,328.1	2,618.2
EARNINGS ON CAPITAL EXPANSION	382.4	496.5	504.8	407.9	239.9	268.1	299.7	335.0	374.6
RATE CHANGES	568.2	619.8	379.8	994.0	1,115.9	1,250.6	1,402.7	1,574.4	1,768.6
OTHER INCOME	159.0	84.2	18.2	64.4	71.4	79.2	87.9	97.5	108.2
TOTAL POSITIVE FACTORS	1,670.2	2,023.7	1,928.4	2,896.0	3,059.9	3,440.4	3,861.0	4,335.0	4,869.5
INFL IN MR&S	184.3	245.8	355.5	447.4	488.5	560.7	643.5	738.4	847.3
INFL IN LABOR INCL SS TAXES	407.2	607.8	943.5	989.5	1,100.5	1,238.8	1,391.5	1,559.3	1,743.5
INFL IN PONIT	41.1	36.7	- 96.2	30.4	52.6	56.0	59.7	63.5	67.7
CHG IN DEP DUE INFLATION	98.0	82.5	170.2	221.3	320.4	354.6	392.4	434.3	480.6
CHG IN DEP DUE TO OTHER EFFECTS	224.1	201.9	171.2	155.4	96.4	106.7	118.0	130.6	144.5
CHG IN FEDERAL INCOME TAXES	180.3	307.5	-142.4	-100.5	-379.6	-539.5	-732.6	-140.2	-
CHG IN STATE AND LOCAL INCOME TAXES	35.0	20.4	11.6	-33.1	-37.1	-52.8	-71.6	-13.7	-
CHG IN INT EXP DUE TO INT RATES	-10.6	41.4	136.6	29.6	64.2	69.7	75.7	82.1	89.2
CHG IN INT EXP DUE TO DEBT VOLUME	34.7	68.8	120.4	122.5	100.9	109.5	118.8	129.0	140.0
CHG IN UNCOLLECTIBLES	20.5	46.4	67.9	52.3	62.6	74.9	89.6	107.2	128.3
CHG IN MISC DEDUC FROM INCOME	-	-	-	-	-	-	-	-	-
LESS: CHG IN EXTRA & DEL ITEMS-NET	-31.3	41.0	6.7	-1.3	-.4	-.1	-.0	-.0	-.0
TOTAL NEGATIVE FACTORS	1,246.0	1,544.8	1,731.5	1,916.0	1,869.8	1,978.7	2,085.0	3,090.7	3,641.1
ESTIMATED CHG IN NET INCOME	424.2	478.9	196.9	980.1	1,190.2	1,461.7	1,776.0	1,244.3	1,228.5
ACTUAL CHG IN NET INCOME	424.2	478.9	196.9	980.1	1,190.2	1,461.7	1,776.0	1,244.3	1,228.5

Table 2

NET INCOME AND PRODUCTIVITY ANALYSIS (NIPA)
XYZ CORP - AVG GROWTH IN O/H
(PERCENTAGE DISTRIBUTION)

YEARS IN STUDY	1977	1978	1979	1980	1981	1982	1983	1984	1985
PRODUCTIVITY GAIN	34	41	53	49	53	54	54	54	54
EARNINGS ON CAPITAL EXPANSION	23	25	26	14	8	8	8	8	8
RATE CHANGES	34	31	20	34	36	36	36	36	36
OTHER INCOME	10	4	1	2	2	2	2	2	2
TOTAL POSITIVE FACTORS	100	100	100	100	100	100	100	100	100
INFLATION IN MR&S	15	16	21	23	26	28	31	24	23
INFLATION IN LABOR INCL. S.S. TAXES	33	39	54	52	59	63	67	50	48
INFLATION IN PONIT	3	- 2	- 6	2	3	3	3	2	2
CHG IN DEPRECIATION EXPENSE	26	18	20	20	22	23	24	18	17
CHG IN FEDERAL INCOME TAXES	14	20	- 8	- 5	-20	-27	-35	- 5	-
CHG IN STATE & LOCAL INCOME TAXES	3	1	1	- 2	- 2	- 3	- 3	-	-
CHG IN INTEREST CHARGES	2	7	15	8	9	9	9	7	6
CHG IN UNCOLLECTIBLES	2	3	4	3	3	4	4	3	4
CHG IN MISC DEDUC FROM INCOME	-	-	-	-	-	-	-	-	-
LESS: CHG IN EXTRA & DEL ITEMS-NET	- 3	3	-	-	-	-	-	-	-
TOTAL NEGATIVE FACTORS	100	100	100	100	100	100	100	100	100

Table 3

RELATIVE IMPORTANCE OF NIPA FACTORS XYZ CORP - AVG GROWTH IN O/H (PERCENT OF PREVIOUS YEAR'S NET INCOME)

YEARS IN STUDY	1977	1978	1979	1980	1981	1982	1983	1984	1985
PRODUCTIVITY GAIN	25	31	32	42	38	33	30	27	26
EARNINGS ON CAPITAL EXPANSION	17	18	16	12	6	5	4	4	4
RATE CHANGES	25	23	12	30	26	23	20	18	18
OTHER INCOME	7	3	1	2	2	1	1	1	1
% CHG IN TOTAL POSITIVE FACTORS	74	75	61	86	70	62	55	49	49
INFLATION IN MR&S	8	9	11	13	11	10	9	8	8
INFLATION IN LABOR INCL. S.S. TAXES	18	23	30	29	25	22	20	18	17
INFLATION IN PONIT	2	- 1	- 3	1	1	1	1	1	1
CHG IN DEPRECIATION EXPENSE	14	11	11	11	10	8	7	6	6
CHG IN FEDERAL INCOME TAXES	8	11	- 4	- 3	- 9	-10	-10	- 2	-
CHG IN STATE & LOCAL INCOME TAXES	2	1	-	- 1	- 1	- 1	- 1	-	-
CHG IN INTEREST CHARGES	1	4	8	5	4	3	3	2	2
CHG IN UNCOLLECTIBLES	-	-	-	-	-	-	-	-	-
CHG IN MISC DEDUC FROM INCOME	1	2	2	2	1	1	1	1	1
LESS: CHG IN EXTRA & DEL ITEMS-NET	- 1	2	-	-	-	-	-	-	-
% CHG IN TOTAL NEGATIVE FACTORS	55	57	55	57	43	36	30	35	36
EST % GROWTH IN NET INCOME	19	18	6	29	27	26	25	14	12
ACTUAL % CHG IN NET INCOME	19	18	6	29	27	26	25	14	12

Table 4

ALTERNATIVE NET INCOME SCENARIOS
(Millions Of Dollars)

	Sim A	Sim B	Sim C
1976	2,268	-	-
1977	2,692	-	-
1978	3,171	-	-
1979	3,368	-	-
1980	4,348	3,244	3,940
1981	5,538	2,934	4,596
1982	7,000	2,405	5,370
1983	8,776	1,562	6,269
1984	10,020	-613	7,322
1985	11,249	-3,818	8,533

Notes on key assumptions:

Sim A: For 1980-1985, all variables are assumed to grow at the average rate for the period 1974-1979.

Sim B: Assumes zero growth in output per hour (labor productivity) for 1980-1985.

Sim C: Output and output per hour are assumed to grow at a 20% faster rate but capital is derived by holding the capital labor ratio constant as of 1979.

Figure 1

FACTORS AFFECTING CHANGE IN NET INCOME

XYZ Corporation—1979

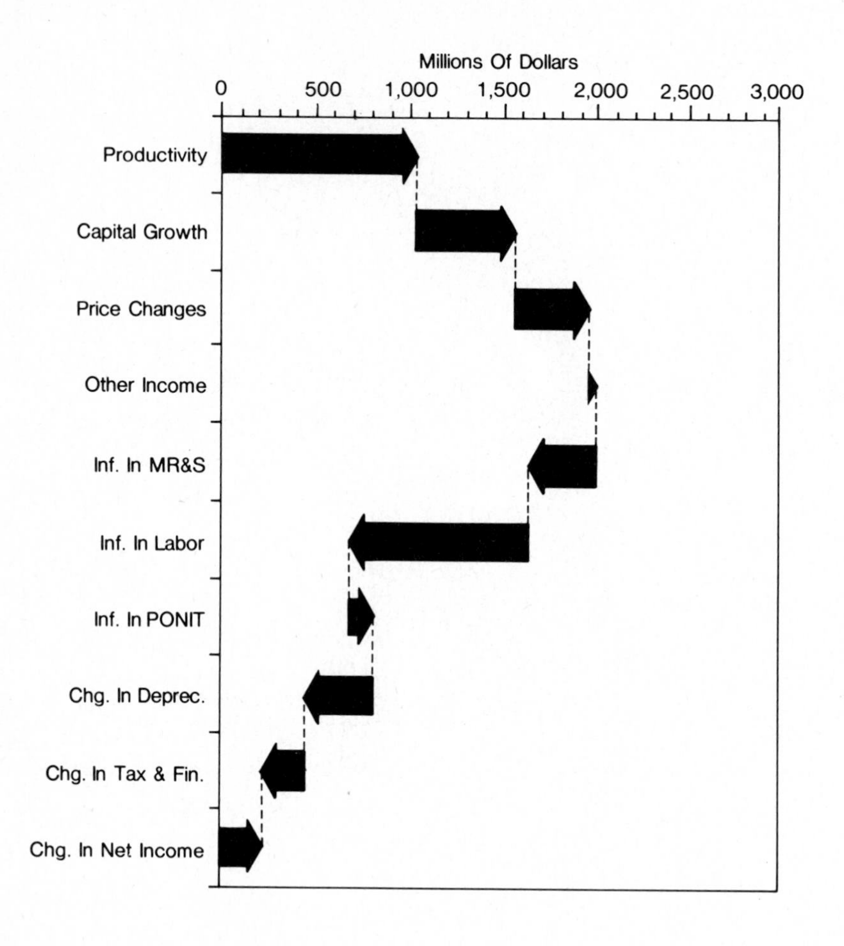

Figure 2

NET INCOME AND PRODUCTIVITY ANALYSIS (NIPA) MODEL

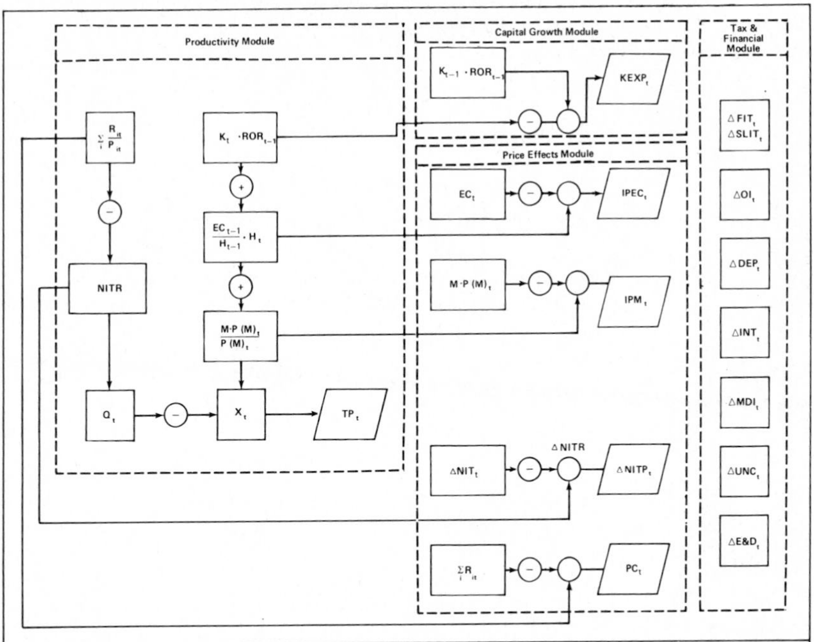

$\Delta NINC = TP + KEXP + PC + \Delta OI - IPM - IPEC - \Delta NITP - \Delta FIT - \Delta SLIT - \Delta DEP - \Delta INT - \Delta MDI - \Delta UNC + \Delta E\&D$

Note: H = L = Hours.

Figure 3

LABOR PRODUCTIVITY

UNDER THREE ALTERNATIVE SCENARIOS

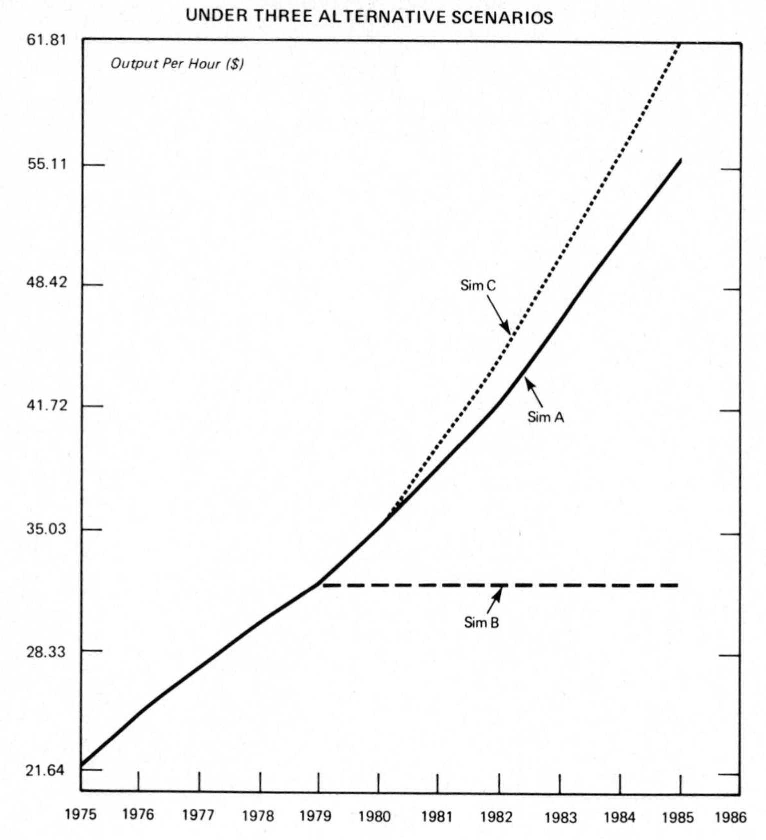

Figure 4

CHANGE IN NET INCOME

UNDER THREE ALTERNATIVE SCENARIOS

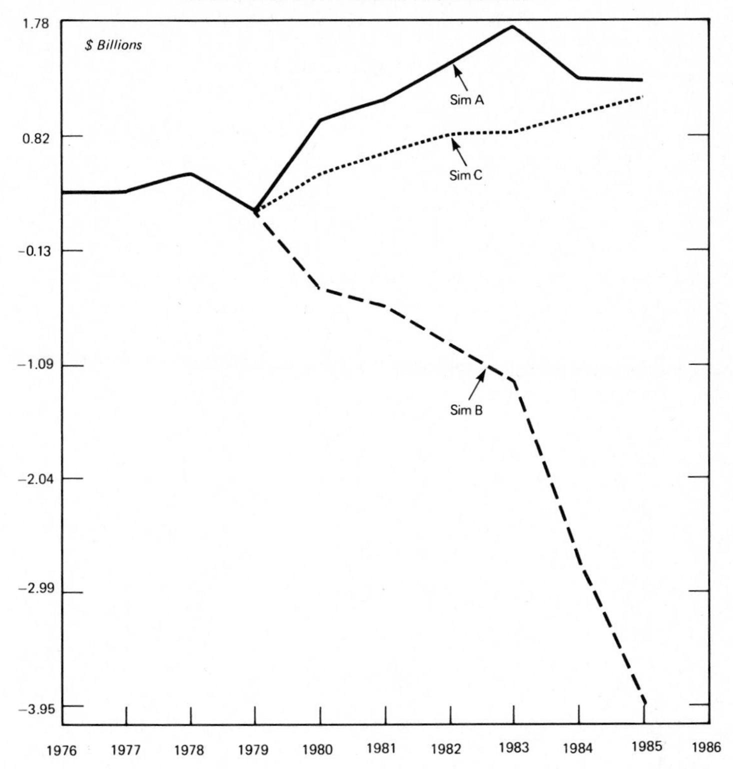

Figure 5

TOTAL FACTOR PRODUCTIVITY CONTRIBUTION

UNDER THREE ALTERNATIVE SCENARIOS

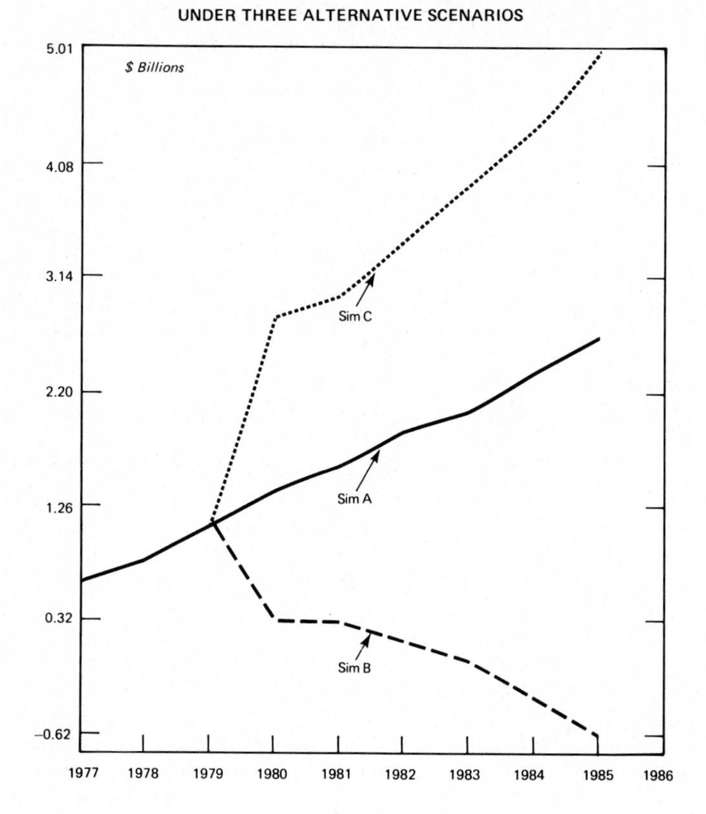

Figure 6

DOLLAR IMPACT OF INFLATION IN LABOR

UNDER THREE ALTERNATIVE SCENARIOS

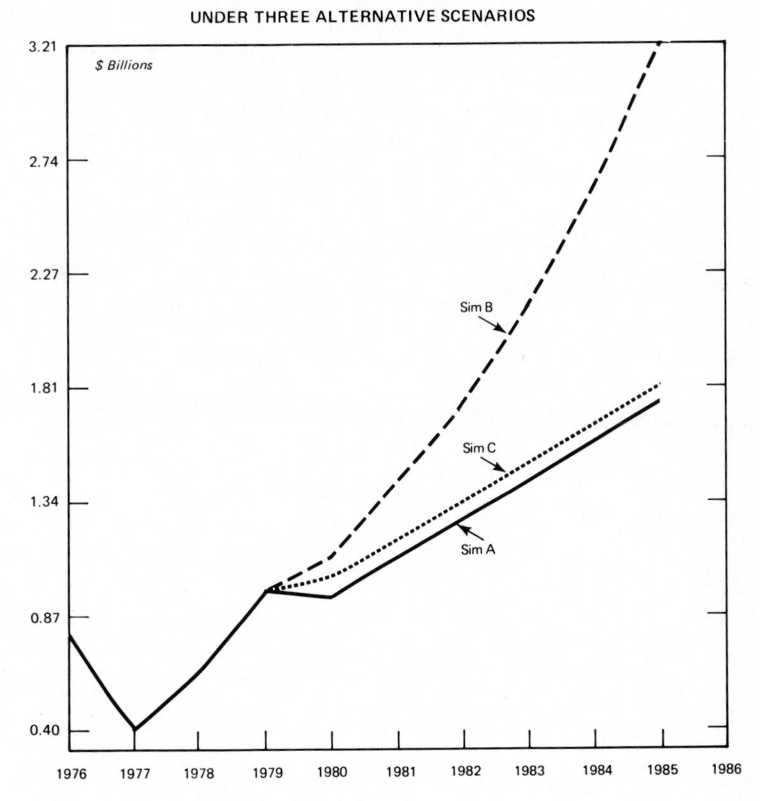

Figure 7

RELATIVE IMPACT OF INFLATION IN LABOR

(% OF NET INCOME IN THE PREVIOUS YEAR)

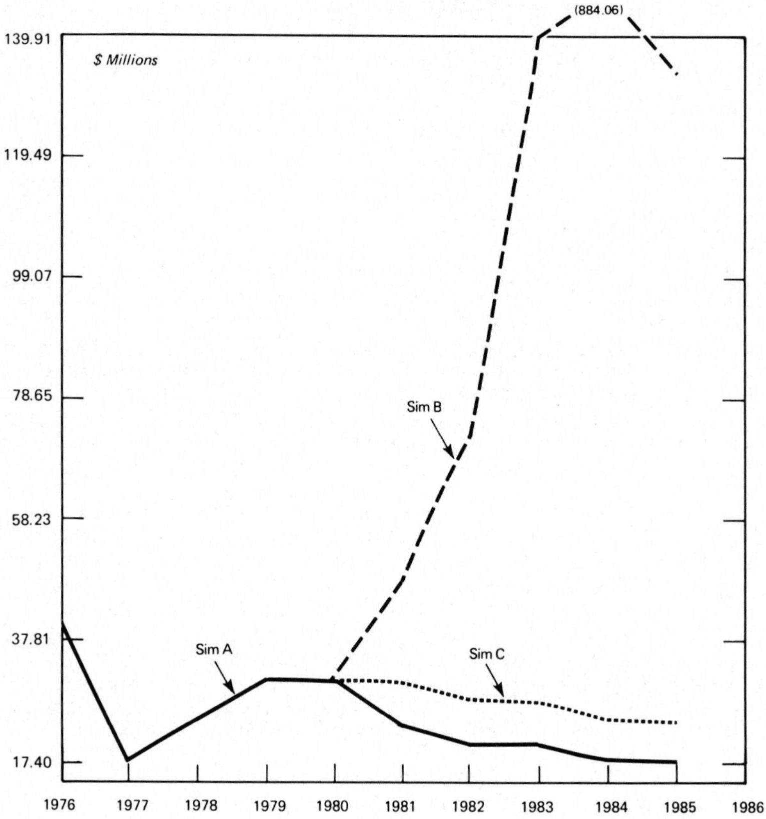

Economic Analysis of Telecommunications:
Theory and Applications
L. Courville, A. de Fontenay and R. Dobell (eds.)

TOTAL FACTOR PRODUCTIVITY FOR MANAGEMENT: THE POST-MORTEM AND PLANNING FRAMEWORKS

M. Denny, A. de Fontenay and M. Werner

University of Toronto, Government of Canada
and Telecommunications Consultant

1. Introduction

Concern with productivity is nothing new. Considerable effort has been devoted to measuring, monitoring and improving efficiency and productivity and most firms collect data on a variety of physical indicators specifically for this purpose. Such partial measures offer excellent tools for control of costs at various levels in the organization. But their significance for overall corporate performance, or for corporate planning, is difficult to determine. They cannot be sting together to yield a global productivity measure useful for corporate budgetary planning.

As an alternative, this paper centers an analytical and planning models that can be integrated into the planning process through use of a Total Factor Productivity (TFP) measure, a measure which synthesizes all the market related activities of the firms. NIPA (Net Income Productivity Analysis Model) provides the starting point. We then develop an unconstrained version (UNIPA) which relates the firm's realized rate of return to the capital market, thus permitting one to study the firm without assuming that its objectives are automatically achieved - that is, that the realized return to equity is adequate. Finally, we introduce PAP (Productivity Analysis for Planning model) which a firm can use as a "top-down" guideline to control its corporate budgeting and planning.

The first model, NIPA, is oriented, in particular, towards explaining the growth in Net Income, which, for the management of the firm, is the most important single statistic which they monitor, and the variable which most clearly mirrors performance. Profits and consequently revenues and costs are decomposed into the basic price and quantity components of the income statement and rearranged to emphasize the distinction between price and quantity-based changes in profits. Many different presentations of the same data are possible. We will begin by a summary and brief commentary on the version developed independently at AT&T and Teleglobe Canada.[1] Following that overview of the NIPA, we will introduce, as a tool to internalize the firm's expected earnings performance, the UNIPA model, which removes the identity between revenue and cost (used in the NIPA analysis to define residually the cost of capital) and thus permits economic profits or losses.

In particular, our version resembles a combination of the analytical models at Electricite de France (Reimeringer (1981)) which do not constrain the return to capital to always equal its cost, and NIPA, which does not admit the possibility that planned and actual costs and revenues may not always be equal. Finally, we shall show how the UNIPA models can be used as post-mortem or quasi-planning models, supporting an analysis of historical performance and review of future plans with a view to identifying implicit productivity gains (or losses) and their impact on Net Income growth. In their present form NIPA and UNIPA models do not, in contrast to the PAP model also presented below, actually generate the plan.

The PAP model is a pure planning model designed to develop a complete budgetary/corporate plan, at a fairly aggregate level, where the components of the various financial/accounting summaries all embody certain key management and corporate targets. More succinctly we may view this as something of a guideline budget generated for top management as a framework for the development of a full-blown, bottom-up corporate budgetary plan. With the results of the planning model the process becomes far less arbitrary. The planners are in a position to quantify upper limits consistent with corporate objectives for all items of the key financial statement, including labour and other expenses and the size of the capital budget. They are armed with the knowledge that any overshooting of these benchmark expense and expenditure figures will ensure that some or all of the present targets will not be attained. While there are a whole array of possible targets, our model is built around what we believe to be the most important of these: the required return to invested capital (r), the forecast demand for the firm's production and the desired growth in productivity.

2. NIPA

As explained by Chaudry and Burnside (1982), the focus of the NIPA models is on changes in net income based on underlying changes in the prices of inputs and outputs and changes in productivity. We can summarize their approach in our own notation as follows.

Assume there exists aggregate quantity and price for output, Q and P and for input W and X. Then revenue $R = P \cdot Q$ and total costs $C = W \cdot X$. The proportional rate of change of any variable will be indicated by placing a dot over that variable. The growth in revenue ($\dot{R}$) may be written

(1) $\dot{R} = \dot{Q} + \dot{P}$

and the growth in costs ($\dot{C}$),

(2) $\dot{C} = \dot{W} + \dot{X}$.

The growth in revenue or cost simply equals the sum of the growth in their relevant price and quantity components.

Letting $\dot{Q}$, $\dot{P}$, $\dot{W}$ and $\dot{X}$ be the weighted sums of the individual rates of growth of the respective component quantities/prices, and using as weights the shares of revenue/cost, then these will be Divisia indexes.

Productivity has been defined as $TFP = Q/X$ and the rate of growth of productivity ($\dot{TFP}$) is therefore written

(3) $\dot{TFP} = \dot{Q} - \dot{X}$

In the NIPA model revenues equal total cost and consequently,

we may write

(4) $\dot{TFP} = \dot{Q} - \dot{X} = -(\dot{P}-\dot{W})$

The rate of growth of productivity ($\dot{TFP}$) must equal the negative of the difference between the rate of growth of output prices and the rate of growth of input prices. That is, to the extent that input prices are rising more quickly than output prices then, given the R = C constraint, it is the result of the productivity gains which are being achieved. There is no other way to achieve this outcome.

To relate the rate of growth of productivity to the rate of growth of profits we will have to make our model more detailed. Let us assume that the firm uses labour, (L), capital, (K) and materials (M) to produce output. Then

(5) $\dot{W} = s_\ell \dot{w} + s_m \dot{m} + s_k \dot{r}$

where s_i, $i=\ell,m,k$ is the share of input i in total costs and $\dot{w}$, $\dot{p}_m$ and $\dot{r}$ are the growth rates of the prices of labour, materials and capital. This method simply says that aggregate input prices grow at a rate determined by the rates of growth of individual input prices. The rate of growth of any input prices has a larger impact on the overall rate of growth of the aggregate the larger is its importance in total costs. This explains the use of the cost shares as weights.

Define gross profits, $\Pi = rK$. The rate of growth of gross profits $\dot{\Pi} = \dot{r}+\dot{K}$. Substituting for $\dot{r}$ in (5) and substituting (5) into (4) we obtain

(6) $\dot{P} + \dot{TFP} + s_k\dot{K} = s_\ell\dot{w} + s_m\dot{m} + s_k\dot{\Pi}$

On the LHS of equation (6) are three terms whose positive growth contribute to the growth in gross profits. The items are output price increases, productivity growth and capital accumulation (or investment). Offsetting any tendency for profits to grow is the growth in the prices of labour and materials.

Using equation (6) one can calculate the dollar values of the components that lead to changes in gross profits. That is, one may easily see how important are the various price components, capital accumulation and productivity growth in determining the changes in gross profits. This analysis is done by both AT&T and Teleglobe and is more fully discussed in Denny et al. (1980).

To provide an analysis of net income it suffices to break down gross profits into its major components: depreciation (D), debt service (B), taxes (T) and net income (NI). As in (6) above, the rate of growth of gross profits (Π) may be written,

(7) $\dot{\Pi} = s_D\dot{D} + s_B\dot{B} + s_T\dot{T} + s_{NI}\dot{NI}$.

The share weights in (7) are the shares of each component in gross profits. Even though one may provide evidence on the underlying components of net income growth using equation (7) alone, it is only in combination with equation (6) which links net income growth directly to productivity growth that one obtains an overview of the whole firm. In this model, the rate of depreciation, the interest rate on the debt and the tax rate are given. The equality between revenues and expenses is maintained by assuming that the net income is a residual.

While the above model provides an extremely useful disaggregation of the financial/accounting income statement, it must be noted that nowhere in the model is anything said about the adequacy of the NI, upon which the relative impact of all the other items is being measured. Given that it is a residual in the cost of capital after payments to depreciation, debt service and taxes, we are led to believe that, within the context of the model, the return to invested capital, i.e., NI, is in fact also identically equal to its cost. Until now, the cost of capital has been defined residually, but this may not be useful in the long run, since it does not reflect the option the firm has to invest its internally generated funds in the capital market. Nevertheless, despite that drawback, this type of income statement presentation can only be a major improvement over the standard format since above all, it isolates the impact of inflation. In addition, while it presents the crucial information to be garnered from a knowledge of the relative impacts of TFP and individual price movements, it preserves all the key information normally found on an income statement including, of course, the critical net income results, now decomposed into inflationary price movements and productivity increases.

3. UNIPA(Unconstrained NIPA)

The cornerstone of the NIPA model is the equality of revenues and costs. However, once the cost of capital is defined exogenously, say, as the opportunity cost of capital, then it does not necessarily follow that actual revenues equal actual costs. The cost of capital in the NIPA model is whatever balances costs and revenues, and nothing in the NIPA analysis prevents this measure from being very high or very low or even negative, reflecting a very good or a very poor performance on the part of the firm (where the concept of good or poor performance has a commonsense meaning formalized in economic analysis as the opportunity cost). To relax this restriction, we define the cost faced by the firm to include $\rho_t K_t$, the opportunity cost of capital, and thus have

$$(8) \quad C' = C(\ ) = wL + mM + \rho K$$

and

$$(9) \quad U = R - C$$

where U is the profit or loss due to the unanticipated returns (positive or negative).

The UNIPA model is ideally applied to post-mortem analysis in which we recognize that deviations from plan are an unavoidable phenomena which will generate positive or negative unanticipated earnings. Whereas ex ante the firm will plan to earn as a "desired" return the opportunity cost of capital, ex post realities will usually differ from anticipations. It should be noted that when we refer to "desired" returns we mean those amounts required to exactly offset all costs, including labour, capital and materials. The firm plans for revenues which, after paying labour, suppliers of intermediate goods and services, depreciation expenses, financial obligations and taxes, will leave a residual to "adequately" compensate the providers of equity capital. However, as is the nature with any residual, in situations of uncertainty, it will equal its planned level, in the short run, only by coincidence. Given the definition of productivity,

$$(10) \qquad \dot{TFP} = \dot{Q} - \dot{X}$$

then

$$(11) \qquad \dot{TFP} = \dot{W} - \dot{P}$$

if, and only if U = 0. That is, U is the repository of all deviations from plan. (Note that the plan was based on U = 0, i.e., $R^* = C^*$.)

The exogenous return on capital is defined in terms of its opportunity cost to the firm assumed equal to the rate of return the firm expected to reach when it developed its plan. This rate will be denoted by ρ such that

$$(12) \qquad C^* = C(\rho) = wL + mM + \rho K.$$

For simplicity, let $C^* = w^*x^*$ where w^* and x^* are the appropriate planned price and quantity input indexes. Then from (12), we can derive:

$$(13) \qquad \left[\frac{\dot{U}}{C}\right] = (\dot{TFP}-\dot{TFP}^*) + (\dot{P}-\dot{P}^*) + (\dot{w}-\dot{w}^*) + (\varepsilon_{CQ}-1)\dot{R}$$

where we have used $\dot{U} = 1$ and where ε_{CQ} is defined as the cost elasticity of output which is equal to the inverse of the scale elasticity. The proportionate change in adjusted unanticipated earnings is the difference between the planned and the realized values, plus a scale term. The first term in brackets is that proportion of the unanticipated earnings due to the difference between planned and actual productivity growth while the second and third terms reflect the degrees to which planned and actual price recovery differs. If the scale elasticity is unity then the last term on the right in zero. The entire expression of course reflects the degree to which the productivity divergence and price recovery divergence offset each other. These can be broken down into all the same elements as the actual NIPA statement since the UNIPA analysis can be seen as the difference between two NIPA analyses.

The post-mortem analysis adds a new dimension in that it enables one to study the impact of the various forecasting errors, be they of exogenous variables such as w, p, ... or of endogenous terms such as L, P, ... through costs and revenues on the income statement. For instance the impact of a strike which might significantly lower L but which may be associated with an unforeseen wage settlement which, in turn, might increases w significantly can now be traced.

By decomposing as in the NIPA analysis, we obtain

$$U = \tfrac{1}{2}\{[(PdQ-WdX) - (P^*dQ^*-W^*dX^*)] + [rdK-\rho dK^*] + [QdP-Q^*dP^*]$$

$$(14)\ -[(L\Delta w-L^{*}\Delta w^{*}) + (M\Delta m-M^{*}\Delta m^{*})] - [\sum_{i=1}^{3} (r_i K-r^{*}_i K^{*})]\} .$$

Here we recognize the familiar NIPA terms in deviations form: the positive factors of productivity, output prices and capital growth minus the negative terms of input inflation and financial charges. The deviations are the results of events which include errors in forecasting wages or the price of materials. Each individual item from the NIPA statement can be matched with its own unique variance. In essence we would have

Plan	Actual	Variance
Positive Factors		
TFP^{*}	TFP	Due to TFP
+Output Price Changes	+Output Price Changes	Due to Output Price Changes
+Capital Growth	+Capital Growth	Due to Capital Growth
Negative Factors		
-Input Price Changes	-Input Price Changes	Due to Input Price Changes
-Capital Cost Changes (excluding NI)	-Capital Cost Changes	Due to Capital Cost Changes
$=NI^{*}$	=NI	U= NI plan - NI actual $\neq 0$

4. Integrated Planning Model

The two versions of NIPA, presented above, while providing a good analytical framework for the intelligent evaluation of budgetary plans, are essentially ex post models. NIPA intervenes in the budgetary process in a sequential manner, taking an active role only after the laborious planning exercise produces its game plan. At that juncture NIPA analyses the budget's implicit productivity performance, which may or may not justify another round of the planning process. Given the scope of the budgetary process in any large firm, it is unlikely that a bad productivity picture, along with good built-in financial results, will move the planners to modify an already overly complex structure. The most natural solution to this dilemma would be to ensure that NIPA results are always favourable by including productivity as an explicit consideration during the planning process. Such a model is the subject of this section. It can be used to develop a complete corporate plan (budgetary and otherwise), explicitly incorporating all essential physical and financial targets such as return to investment and productivity. In this way, top management, who ultimately have to approve any budget, will have available a set of guidelines, incorporating all essential corporate objectives, through which to guide the development of the actual budgetary process.

It is a mixed model which uses econometrics to estimate elements of the planning process, such as the desired factor shares, which cannot be known to the planner.

The major advantage of the following model lies in formally incorporating the productivity objective of the firm in the initial step, thus eliminating to a great extent the need to recycle the budget. In most purely financial models the distinct identifiable input sector is, to a large extent, independently sized and then fitted into the framework of certain corporate constraints, which include the financial rate of return. It is of course only by coincidence that such a process will be satisfactory after a first attempt. Some of the items will be recycled and returned for a new round of integration. In this process the firm's fundamental goal in productivity is approached only indirectly through financial indicators and partial efficiency measures.

From Figure 1, which assumes a capital intensive firm, thus placing a large importance on the capital budgetary process, we can trace the evolution (in very general terms) of a corporate budgetary plan. The most important driving forces are prior and present period demand forecasts. The former creates a requirement for ongoing capital projects, pretty well divorced from present demand conditions, while the latter determines present and longer term capital projects as well as, to a certain extent, replacement requirements. "Other" reasons for increasing the capital budget vary from industry to industry. In telecommunications, for example, international standards and interface exigencies would play significant roles. Regulated industries, in general, would find their capital budgets subject to pressures other than market demand. Ultimately, all the capital requirements are evaluated at current asset prices and a capital budget is derived.

The technological characteristics of the capital budget create part of the demand for the other input factor. These include the general categories of labour and other expenses (henceforth to be referred to as "materials"). They comprise such items as maintenance, direct operating labour, rental of facilities, etc. In addition, the various components of the capital budget, as well as embedded capital, determine the value of capital costs. These include depreciation expenses, interest payments, taxes and, ultimately, the value of earnings applicable for dividend payments to equity holders. This is the residual, after payment to all factors, including debt capital, that ultimately compensates the owners of the firm. When calculated as a percentage of total invested capital, then it is known as the rate of return.

It is within this capital/other factor interaction that "quasi" partial productivity considerations make their first appearance. Quasi, because these are really measures of worker efficiency rather than true overall productivity measurements. They are industrial engineering measures such as "work units" which compare performance against established standards. They take no account of the negative contribution to overall productivity when capital is used to increase work units per unit of time. Naturally, the link between these measures and overall corporate performance is difficult to establish.

The other determinants of total expenses are only indirectly related to capital budgeting and are determined more as a result of overall business size and prosperity. These include all those luxury factors such as marketing, training, special studies, etc. - that is, the entire set of indirect, non-operating expenses.

Total revenues, including forecast demand at given prices and other, non-operating income, are combined with the total value of current input to determine the residual and, ultimately, the rate of return. If the rate of return is inadequate, in that it either fails to compensate existing capital at a fair rate or does not cover all capital expenditures without excessive external financing requirements then there occurs a budgetary recycling process where all or part of the plan is altered. Usually it is the latter, concentrating on the expense rather than capital budget items. Corrective action may include labour cuts, material cuts, output price changes and, as a last resort, capital budget cuts.

Significant by their absence are the aspects of simultaneity and some overall explicit recognition of productivity. The advantage of simultaneously calculating all the unknowns are obvious, but what are the advantages of including productivity? Simply that the implied technological relationship of a production function, as embodied in the explicitly recognized productivity number allows for a combination of inputs, given the output, that is in some sense optimum. This optimum provides an additional constraint to the general planning problem which serves to narrow the choice between the various input options to more manageable proportions.

Formally, the model postulates the existence of some cost function

$$C = g(w,m,r,Q,t) \tag{15}$$

where w = the price of labour
m = the price of materials (or intermediate expense items)
r = the periodic (say, annual) cost of using the capital stock. It includes:

δ = depreciation rate
ϕ = the rate of taxation
θ = the return to outstanding debt
π = the return to equity

Q = the volume of output produced
t = the technology indicator.

Following (Denny, Fuss & Everson (1979)) and (Denny, de Fontenay & Werner (1980)) we derive

$$-\frac{1}{C}\frac{\partial q}{\partial t} = \left(\frac{\partial g}{\partial Q}\frac{Q}{C}\right)\left(\frac{\partial Q}{\partial t}\frac{1}{Q}\right) - \Sigma\,\sigma_1\left(\frac{dX_i}{dt}\frac{1}{X_i}\right)$$

where X_i = L,M, and K and σ_i is the cost share of the i-th input.

If we assume that the cost elasticity, $\frac{\partial g}{\partial Q}\frac{Q}{C}$, is approximately unity over the period under consideration, and if we define the shift in the cost function due to technology the change in total factor productivity, $\dot{TFP}$ by X, then

$$\dot{TFP} = \frac{\partial Q}{\partial t}\frac{t}{Q} - \Sigma\,\sigma_i\left(\frac{dX_i}{dt}\frac{1}{X_i}\right). \tag{16}$$

Assuming the cost function to be approximated by a quadratic form, we may discretely approximate (16) as:

$$(17)\quad \overset{\Delta}{TFP} = (\ln Q_1 - \ln Q_0) - \Sigma\, \tfrac{1}{2}(\sigma_{i1} + \sigma_{i0})(\ln X_{i1} - \ln X_{i0})$$

We can now rearrange equation (17) so that it can be solved for any one of the X_{i1}, say K_1, then:

$$\ln K = \ln\left(\frac{Q_1}{Q_0}\right) + \sigma_L \ln L_0 + \sigma_M \ln M_0 + \sigma_K \ln K_0$$

$$(18)\qquad + (1-\sigma_K)\left[\ln\left(\frac{K_1}{L_1}\right)\right] - \sigma_M\left[\ln\left(\frac{M_1}{L_1}\right)\right] - \dot{TFP}$$

Equation (18) has several unknowns and is at present not solvable. Assuming the cost function g to be approximated by a translog we can derive equations for each of the cost shares σ_{i1}.

$$\sigma_{L1} = \alpha_L + \alpha_{LL} \ln w_1 + \alpha_{LM} \ln m_1 + \alpha_{LK} \ln r_1 + \alpha_{LQ} \ln Q_1 + \alpha_{Lt} t$$

$$\sigma_{M1} = \alpha_M + \alpha_{ML} \ln w_1 + \alpha_{MM} \ln m_1 + \alpha_{MK} \ln r_1 + \alpha_{MQ} \ln Q_1 + \alpha_{Mt} t$$

$$\sigma_{K1} = \alpha_K + \alpha_{KL} \ln w_1 + \alpha_{KM} \ln M_1 + \alpha_{KK} \ln r_1 + \alpha_{KQ} \ln Q + \alpha_{Kt} t$$

In the above system since $\Sigma\, \sigma_{i1} = 1$, we need only estimate any two and then solve for the third set of coefficients from the following conditions

$$\sum_i \alpha_i = 1 \;;\; \sum_i \alpha_{ij} = 0 \;; \qquad j = K, L, M, Q, \text{ and } t$$

For our model we assume, as management does in the planning process, that estimates are available for w_1, m_1 and t_1 and that r_1 is unknown. Therefore, in order to get estimates for the α_i and α_{ij}, we estimate econometrically the equation using available historical data. Then it follows from our earlier assumptions regarding w_1, m_1 and t that $\alpha_{i1} = h_i(r_1)$.

Further, from the definition:

$$\sigma_{i1} = \frac{q_{i1} X_i}{C}$$

we can find the ratios:

$$(19)\quad \frac{K_1}{L_1} = \frac{w_1}{r_1}\frac{\sigma_{K1}}{\sigma_{L1}} = \frac{w_1}{r_1}\frac{h_K(r_1)}{h_L(r_1)} \quad \text{and} \quad \frac{M_1}{L_1} = \frac{w_1}{m_1}\frac{\sigma_{M1}}{\sigma_{L1}} = \frac{w_1 h_M(r_1)}{m_1 h_L(r_1)}$$

Substituting in equation (18), we now have two unknowns, r_1 and K_1 and one equation, (18). Given that our aim is to integrate our model directly into the corporate planning routine, the economic cost of capital r must be related to the financial cost of capital, r^* where

$$(20)\qquad r^* = \delta + \lambda\theta + [(1-\lambda)/(1-\phi)]\pi$$

where λ is the proportion of total financial capital in the form of debt. The relation then can be postulated as:

$$(21) \qquad rK = r^{*}K^{B}$$

where K^{B} = the net original value of physical capital which, by definition equals the value of financial capital. In addition we also have, by definition:

$$A_1 = K^{B}{}_1 - K^{B}{}_0 + R_1 - (R_1 - R^{*}{}_1) = K^{B}{}_1 - (K^{B}{}_0 - R^{*}{}_1)$$

$$A_1 = q_1(K_1 - K_0) + R^{*}{}_1$$

where A_1 = the value of gross additions to the plant
$R^{*}{}_1$ = the value of retirements that are actually replaced
R_1 = the value of retirements
q_1 = the purchased price of capital (TPI)

We can now derive

$$(22) \quad K_1 = [r^{*}{}_1/(r_1 - q_1 r^{*}{}_1)](K^{B}{}_0 - q_1 K_0)$$

Equations (18) and (22) now form a system of two equations in the two unknowns r_1 and K_1. All the other unknowns of the general planning problem can now be derived from the solution to this system. Given a value for r_1, the share variables σ_{i1} assume values which, from (7), produce solutions for L_1 and M_1. This, along with the prices w_1, m_1 and r_1, puts a value on total cost which of course implies a total revenue requirement. Thus, we can see, that given the key constraints of demand forecasts ($\bar{Q}_1$), rate of return requirements (π_1), and desired productivity growth we have calculated the budgetary input requirements of the initial "top-down" budget. Together with the estimates w_1 and m_1, they yield the total cost:

$$C = r_1K_1 + w_1L_1 + m_1M_1 \quad .$$

Further, taking account of the accounting identity between total revenues and total costs, and the measure of output produced by the demand forecast, Q_1, we obtain the rate increase required to achieve the desired rate of return:

$$P_1 = C_1/Q_1$$

All the other details of a full-blown financial plan (including) depreciation expenses, taxes, interest payments, the various balance sheet items, source and uses statements and so on, can now be calculated.

5. Conclusion

The notion that productivity is an important part of business success, as stated at the outset, may not be a new concept, but to incorporate it explicitly into an overall corporate/budgetary plan is. In this paper we have demonstrated two ways of going about this integration. The first involves more of a static budgetary analysis in the form of NIPA and UNIPA. They take as given the financial/accounting information in any plan, and compute the relative impact of productivity, among other variables, on the growth in Net Income, which, after all, is the firm's ultimate measure of management success. While NIPA describes a given situation such as the plan, UNIPA views it from the perspective of variations with respect to a reference, such as the following period's actuals.

The other method of introducing productivity into the corporate/budgetary planning exercise involves a direct intervention in the process. TFP itself becomes a target variable and thus a parameter in the actual derivation of a complete guideline plan. Based on the desired levels of productivity, financial return and production (to meet the demand conbstraint), the planning model simultaneously calculates all the relevant variables of an entire plan which includes the income statement, balance sheet and funds flow information. While it does provide all the pertinent operating information the results of the model are not meant to replace the normal bottom-up planning process. Instead they offer a complete set of guidelines for senior management on the values of key operating indicators such as employee expenses, manhours, capital budgeting, etc. which, if not attained, will imply the untenability of management's key task targets, including financial return to investment, production level and productivity gains.

FOOTNOTE:

[1] The original work on the management use of TFP by a firm must be credited to the Electricite de France (EDF); its surplus analysis (Reimeringer, 1980) is the forerunner of all NIPA models. Certain multinational corporations, such as IBM or Xerox, are known to use TFP measures as general guidelines, and DRI is in the process of formalizing such an idea. In 1977, Teleglobe Canada and the British Columbia Telephone Company organized two symposia at which a number of Canadian telecommunications carriers came together to discuss the concept and measurement of TFP. Nevertheless, the active and systematic use of TFP as a management tool, introduced analytically in the management process, but for EDF, appears to have been pioneered by telecommunications carriers, with Teleglobe Canada and AT&T in the process of incorporating it in the formal budgeting and planning process and with Bell Canada developing similar internal uses. In addition, two other Canadian telecommunications carriers - British Columbia Telephone Company and Alberta Government Telephone - have on-going productivity studies. Finally, nine Canadian telecommunications carriers are participating with the Canadian Department of Communications in a major productivity project, which has as one of its goals the development of management uses of TFP analysis.

REFERENCES:

[1] Chaudry, M.A., and Burnside, M., Net Income and Productivity Analysis (NIPA) as a Planning Model.

[2] Chaudry, M.A., Net Income Productivity Analysis, paper given at the American Productivity Centre Conference, Houston (1980).

[3] Denny, M., de Fontenay, A., and Werner, M., Comparative Efficiency in Canadian Telecommunications: An Analysis of Methods and Uses,

Phase II: Productivity Employment and Technical Change in Canadian Telecommunications, Report to the Canadian Department of Communications (May 1980).

[4] Diewert, E., Exact And Superlative Index Numbers, Journal of Econometrics 4 (1976) 115-145.

[5] Kiss, F., and Joseph, J.R., Bell Canada Productivity: How Productivity Gains Increase Income and Absorb Inflation, proceedings of the 6th International Cost Engineering Congress, Mexico City, Vol. II (October 20-22, 1980).

[6] Reimeringer, J.N., La Productivite Globale des Facteurs a Electricite de France, Principe, Resultats et Utilisation, Electricite de France, Etudes Economiques Generales (November 1980).

[7] Werner, M., Productivity Based Planning Model for Teleglobe Canada, paper given at the Ninth International Teletraffic Congress, Spain, (October 1979).

[8] Werner, M., Teleglobe Canada Productivity Study, Teleglobe Canada, internal report (1977).

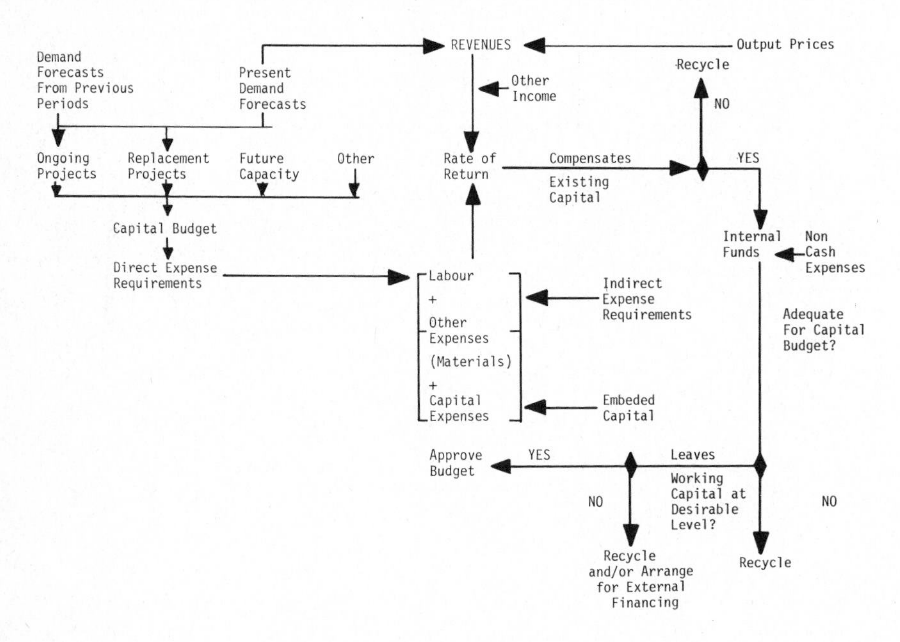

Figure 1

PART 2
DEMAND ANALYSIS

Modelling and Telecommunications Demand

Economic Analysis of Telecommunications:
Theory and Applications
L. Courville, A. de Fontenay and R. Dobell (eds.)

PROBLEMS AND ISSUES IN MODELING TELECOMMUNICATIONS DEMAND

L.D. Taylor

University of Arizona

1. A Critique of the Existing Econometric Literature on Telecommunications Demand[1]

The attributes of the telephone system that set telephone demand apart from the demand for most goods and services include (a) the distinction between access to the telephone network and use of the network once access has been acquired, (b) the presence of access and call externalities, (c) option demand as an important component of access demand, and (d) the importance of the opportunity cost of time. These attributes will be used in this section in reviewing the econometric literature on telecommunciations demand. This literature is large and diverse, and the discussion here is necessarily highly abbreviated. Readers interested in a detailed review and critique are referred to chapters 3 and 4 of my monograph.

As mentioned above, the distinction between access and use should provide the main point of departure in building models of telephone demand. As a practical matter, this means that the analysis should be approached in stages, with the first stage focusing on the demand for access. The most meaningful quantity to explain in this regard is the number of main-station telephones for residential subscribers and the number of main stations plus PBX extensions for business customers. Stage two of the analysis should then focus on the demand for use. However, the demand for use may itself need to be approached in stages, depending upon how use is priced. On the one hand, if a call is priced on a two-part tariff, in which the price for the initial period differs from the price of an overtime period (as is the case for toll calls in the U.S. and Canada), then the demand for use should involve two equations, one that explains the number of calls and one that explains average duration. On the other hand, if a call is not priced on a two-part tariff (as is the case in Sweden and the U.K.), a single equation that explains the number of conversation-minutes will suffice.

The access/use distinction is found in studies throughout the literature, but it is center stage in only three, namely, Alleman (1977), Pousette (1976), and Waverman (1974). Alleman (whose focus is the U.S.), restricts his analysis to a short-run elasticity and a "long-run" short-run elasticity, both of which are conditional on the number of main stations, and a long-run elasticity in which the number of main stations is allowed to vary in response to the change in income. Waverman's estimates of these three elasticities are 0.23, 0.32, and 1.25. Thus, we see that the indirect effect on local use of a change in income that arises through the equation for main stations is substantial.

The moral of the story is as follows: If the focus is entirely on estimating the effects on usage of changes in the price of usage, an access/use framework is not critical -- so long, that is, as usage is assumed to be conditional on the number of main stations -- since any impact of the price change on the number of main stations is probably small enough that it can be ignored. However, if the focus is on estimating the impact on usage of changes in income, an access/use framework is critical, because to ignore the indirect effect on usage that arises through the adjustment in main stations is likely to lead to a serious underestimate of the total effect of a change in income.[2]

Let me now turn to the access and call externalities. The access externality can in principle be taken into account by making the demand for access depend on system size. One way that this can be done is to relate the current number of main stations to the number of main stations in the preceding period. If there are no other dynamics, a positive access externality will be reflected in a coefficient on the preceding period's main stations that is greater than 1. The call externality is more problematic to measure empirically, although a case can be made that it too may be represented by system size.[3]

The evidence regarding the access and use externalities is thus to be found in the equations that include a measure of system size as a predictor. This occurs in a number of the state intrastate toll demand models and in the models of Feldman (1976), Davis et al. (1973), Pousette (1976), and Waverman (1974). In most of these models, system size is measured by the number of telephones less residence extensions, but in a few of the state models, the number of households or the population is used as a surrogate. Most of the equations in question refer to the demand for use. The only ones that involve the demand for access are the equation of Davis et al. for the total number of telephones (less residence extensions) and Waverman's equation for the number of main stations in Canada.

In general, however, the evidence concerning the two consumption externalities is inconclusive. The strongest suggestion that the externalities may be of some importance is given in Waverman's equation for local use in Sweden, which has an elasticity with respect to the number of telephone of 1.19.[4] However, the fact that the existing evidence is weak and mixed is hardly surprising since the empirical literature has not explicitly focused on the consumption externalities as factors to be taken into account. Many of the state toll demand models, for example, assume them away a priori by defining the dependent variable as the number of messages per main station (or price-deflated revenues per main station) and then not including main stations as a predictor. Moreover, it is not the case that the externalities have been considered and then dismissed as unimportant, for in general they have simply been ignored. Waverman, for example, does not see in his results for Sweden the suggestion that the externalities may be consequential, and Pousette, in his otherwise admirable study, ignores them from the start.

Let me now move on to option demand. The literature does not provide any empirical evidence, even inadvertently, regarding its importance. However, this too, is hardly surprising, for while option demand is an appealing concept, it is not easily given to measurement. In my monograph

I have suggested that option demand should be relatively more important in rural exchanges than in urban exchanges, and if this is so, the access-demand elasticity with respect to the price of access should be smaller in rural exchanges than in urban exchanges. This might be tested with the data set constructed from the 1970 U.S. Census used by Perl (1978). Also, it seems that option demand may be a factor in many subscribers' apparent preference for flat-rate pricing of local service over measured service. While it is not clear how this idea might be tested empirically, it has important implications for the pricing of access, and is therefore worthy of attention.

Let me now turn to other matters. A major deficiency in the empirical literature is the treatment of prices. Most telephone services are priced on a multi-part tariff of one form or another, and this has a number of important implications. With a multi-part tariff, one has to distinguish between the marginal price and the intra-marginal prices, and in some cases, the multi-part tariff changes the basic logic of the demand model. A toll call in the U.S. and Canada, for example, is priced on a two-part tariff, since the price per minute is less for overtime periods than for the initial period. Assuming that the research goal is to explain the number of conversation-minutes, the appropriate procedure (as discussed earlier in this section) is to decompose the number of conversation-minutes into the number of calls and the average duration of a call. The number of calls will then depend (besides on income, etc.) primarily on the price of the initial period, while the average duration of a call will depend primarily on the price of an overtime period.

The principles that underly these conclusions are discussed in detail in my monograph, but it is useful to summarize them here.[5] When a good is priced on a multi-part tariff, the different parts of the tariff affect a consumer's behavior in different ways. In equilibrium, the consumer equates marginal rates of substitution between pairs of goods to the ratios of their respective marginal prices. However, when the good involved is defined in several dimensions, then a tariff that is marginal to the decision in one dimension may be intra-marginal or extra-marginal to the decision in another dimension. Whith a toll call, there are two decisions to be made: whether to make the call, and how long to talk. The price of the initial period is the price that is most relevant to the decisions of whether to make the call, the time of day to make the call, and whether to direct dial, while the price of an overtime period is the price that is most relevant to the decision of how long to speak.

In general, these considerations are not reflected in the empirical literature. There are only two studies that seem genuinely cognizant of the problems that multi-part tariffs pose, namely, Deschamps (1974) and Pousette (1976). Deschamps discusses explicitly the complications created by multi-part tariffs, and his study stands alone in this regard. Pousette's actions, on the other hand, speak louder than his words, for although he does not mention the need to distinguish between the price of access and the price of use, his equations for new connections contain a price index that is a weighted average of the subscription fee and the call charge, while his equations for use contain only the call charge. Combining the access and use charges into a single index is not the ideal procedure, but it is certainly a step in the right direction.[6]

The failure to distinguish among the various components of a multi-part tariff usually leads to the use of an average price, and the worst situation is when the average price is obtained by dividing revenues by the quantity that is being explained. An average price for toll calls, for example, is frequently obtained by dividing toll revenues by the number of toll calls. Such an ex post procedure is to be avoided because it necessarily establishes a negative relationship between quantity demanded and price.[7] Perhaps the most serious lapse in this connection in the telephone demand literature is by Beauvais (1977) in his study of the demand for local calls. Beauvais defines an average price by dividing the monthly service charge by the number of local calls. However, this creates a potentially very serious bias in the estimate of the price elasticity because most of the variation in the price variable is caused by the variation across subscribers in the number of local calls. Other studies in which price is calculated as an ex post average price include Feldman (1976), Kwok, Lee, and Pearce (1975), Larsen and McCleary (1970), and Rash (1972).

The state toll demand models mentioned in footnote 2 all use a price index for intrastate MTS calls for the price variable. Laspeyres indices are used in about two-thirds of the models, and chain-weighted indices are used in the remainder. Never do the weights depend on current-period quantities demanded, so that the price variables used in these models do not suffer from the bias problem just referred to. However, they do have a problem in that the initial-period and overtime-period charges are combined in a single index. This is appropriate for the models in which the dependent variable is price-deflated revenues, in which case the dependent variable is a measure of conversation-minutes, but it is not appropriate for the models in which the dependent variable is the number of messages. In this case, the price of the initial period should be separated from the price of an overtime period.

Another shortcoming in the empirical literature is a focus on the number of calls, as opposed to conversation-minutes. As was noted, when the price of a toll call consists of an initial charge plus a charge that depends on the number of overtime periods, one must explain duration as well as the number of messages. If the price of a call were independent of duration, duration could be ignored, but in most instances, the price of a call does vary with duration, so that the possible dependence of duration on price must be taken into account. Only four studies in the literature do this, Feldman (1976), Gale (1974), Pousette (1976), and Waverman (1974). The studies of Feldman and Gale provide the most detailed analyses and, in both cases, duration is found to be negatively related to price.[8] Pousette and Waverman both explain the number of "pulses" (which is a physical measure of holding time), and duration per se is not singled out for analysis. However, Waverman also estimates an equation for the number of messages, and since the price elasticity in this equation is smaller than in the pulse equation, one can infer that there is also a nonzero price elasticity for duration.

The ignoring of duration is especially apparent in the U.S. state toll demand models. None of the state models have equations that focus directly on duration. About half of the state models whose elasticities are tabulated in Chapter 4 of my monograph have the number of messages as the dependent variable, while the other half have price-deflated revenues as the dependent variable. Since toll revenues can be decomposed into the product of the number of calls and the average revenue from a call, price-deflated revenues correspond, in principle, to the product of the number of calls and their average duration. The impact of a price change on average duration is thus reflected in the price elasticities that are estimated in the PDR equations.

Whether the dependent variable in the U.S. state models should be the number of messages or price-deflated revenues is a subject of some debate, and let me now offer a few comments. The state models have been developed mainly as planning tools and for isolating market reactions to tariff changes in rate filings before state Public Utility Commissions. In most cases, their primary use in rate filings has been to estimate the impact on toll revenues of a nonzero price elasticity of demand. For the models with the number of messages as the dependent variables, revenues (after repression[9]) are calculated by multiplying the estimated number of messages by a repriced average revenue per message. For the models with price-deflated revenues as the dependent variable, after-repression revenues are calculated by multiplying the estimate fo repression-adjusted real revenue by a price index which reflects the new rates.

Neither of these procedures is ideal in my opinion. The reason is that the number of messages and the average duration of a message respond in different ways to changes in the tariff structure. In the models with price-deflated revenues as the dependent variable, the price of a call is represented by a price index which combines the charge for the initial period with the charge for an overtime period, which means that the responses of messages and average duration are reflected in a single elasticity. In situations where the structure of the tariff schedule is to be changed, not allowing messages and duration to respond differently could lead to serious forecasting errors. However, the problem in the PDR models is not so much in the use of price-deflated revenues as the dependent variable, but rather in the use of a single price index. The use of two price indices, one for the initial-period charge and the other for the overtime-period charge, would largely overcome the problem. In contrast, the problem in the models with the number of messages as the dependent variable is that an equation is missing, namely, an equation for average duration. Either the models make no allowance at all for the effect of a rate change on duration or else it is assumed that the price elasticity for messages applies also to duration. Neither of these procedures is satisfactory. However, the solution (at least in priciple) is readily apparent: estimate an equation for average duration.

Let me now turn to some questions of dynamics. The models using the access/use distinction are dynamic by definition -- since use is predicated on system size -- and distributed-lag models are used extensively, especially in the analyses of U.S. intrastate toll demand. The tables in Chapters 3 and 4 of my monograph of existing estimates of price and income elasticities of demand provide a great deal of evidence that telephone demand is indeed a dynamic phenomenon, since when dynamic models are specified, the estimated long-run elasticities are nearly always considerably larger than the estimated short-run elasticities.

Perhaps the biggest problem with the treatment of dynamics in the empirical literature is that distributed-lag models are frequently forced to do too much. Dynamic adjustment can arise from two sources: the first reflects the dynamics inherent in the consumption of a service generated by a complementary durable good,[10] while the second reflects inertia that may exist in the short run.[11] A distributed-lag model can capture both types of dynamic processes, but the two processes cannot be separately identified in the same model. Much of the time, this cannot be avoided, since estimates of the stock of complementary durable goods either do not exist or else are of too poor quality to be used. But in the telephone industry, this is not the case, for the data on the stock of telephones are in general quite good. The access/use distinction can accordingly be modeled directly (thereby taking into account the dynamics associated with the consumption of a service generated by a durable good), and the distributed-lag model can focus exclusively on capturing short-run inertia.

On the other hand, one can question whether the Koyck model, which has been used in the vast majority of cases, may be too restrictive. The Koyck model postulates geometric decay in the distributed lag, and also constrains each independent variable to have the same lag structure. Much of the time, these restrictions are probably unrealistic, but are resorted to because multicollinearity precludes meaningful estimation of separate lag structures. However, other more flexible, distributed-lag models exist, and these should be analysed with the purpose of seeing whether price and income have different lag structures. The Almon polynomial-lag model would be a convenient (but not the only) model to use to this end. Transfer functions, which have been used in a couple of studies with promising results, also merit consideration, particularly when quarterly or monthly data are being analyzed.[12]

One thing that has not been adequately explored in the empirical literature is the time-series/cross-section nature of much of the telephone data base.[13] The data collected in AT & T's CMDS data base would allow for this, and so, too, would the data collected at the state level by the operating telephone companies. The major benefit from pooling is the increased variation in the independent variables. However, care must be taken that the structures being pooled are homogeneous -- i.e., that regression coefficients are constant across observational units, and similarly for the structure of the error term. Traditionally, the covariance model has been the model that has been employed in pooled time-series/cross-section analysis, but recent years have seen an increasing use of the variance-components model. Random coefficients models may also have a role to play, especially in situations where price or income elasticities are found to vary regionally.[14]

Since many of the econometric models of telephone demand are used in rate filings, the basic canons of econometrics need to be given a great deal of attention, for otherwise the results of the models, or even the models themselves, may be challenged on the grounds that proper econometric and statistical procedures have not been followed. Econometricians take many things for granted much of the time, especially when communicating with other econometricians; to do otherwise would be tedious, repetitive, and time-consuming. The validity of the t- and F-tests, for example, requires that the error term be normally distributed, but only rarely is normality tested for explicitly. Usually, the Classical Central Limit Theorem is relied upon to provide normality, but failing this, the econometricians know that the t- and F-tests are robust in the face of even quite substantial departures from normality. However, the people that ultimately have to be convinced in rate filings are not other econometricians, but rather people who may have little understanding or appreciation of econometric procedures. As a consequence, laxity and possible errors in procedure can be made to seem much more important than they in fact are. Hence, in using a model in a rate filing, formal procedures need to be followed with especial care, including (to return to the example) an explicit test for normality in the error term.[15]

Common sense suggests that access to the telephone system should be more of a necessity than local use and that local use should be more of a necessity than toll use. This would mean that the income elasticity for access would be smaller than the income elasticity for local use, which, in turn, would be smaller than the income elasticity for toll use. In general, we should expect the same relationships to hold among the price elasticities, although the reasoning is a bit more subtle. The effect on demand of a change in price, to recall, consists of two terms, an income effect and a substitution effect. If the substitution effect is held constant, the price elasticities will in general, increase pari passu with the income elasticities. However, common sense also suggests that the substitution effect will be very weak for access, and weaker for local use than for toll use. Thus, both the income effect and the substitution effect should make for progressively larger price elasticities as we move from the demand for access to the demand for local use, to the demand for short-haul toll, etc.

The empirical results tabulated in Chapters 3 and 4 of my monograph support these views. In general, the estimated elasticities for access are smaller than the elasticities for local use, which are smaller than the elasticities for toll demand. This is true for both the income and the price elasticities. Moreover, the empirical results also indicate that the elasticities for toll demand vary with distance, being smaller for short-haul than for long-haul calls. The elasticities in the U.S. intrastate models are, in general smaller than in the interstate models.

In Table 1, I have tabulated some point estimates from my monograph of price and income elasticities for the demand for access, local calls, and long distance calls and for the duration of long-distance calls. These estimates are my own interpretation of the existing empirical record. They are accordingly highly subjective, and are based on evidence from other countries as well as from the U.S. While the estimates tend to be near the midpoint of existing estimates, I have given some studies more weight than others. The estimates refer to steady-state, long-run elasticities, and are for residence and business demands combined, for the empirical record at this stage will not support an attempt to distinguish between residence and business customers. As a measure of the uncertainty to be associated with the estimates, I have appended a range to each estimate. These ranges are also subjective, and reflect my own views as to the intervals within which the true elasticities are likely to lie.

The greatest uncertainty attaches to the estimates of the income elasticities of the demand for use, particularly for local calls and interstate toll calls. I think that one can conclude that the income elasticity for toll calls is greater than 1, but how much greater is still an open question. Considerable uncertainty also surrounds the price elasticity for interstate toll, especially with respect to the critical value of 1 (in absolute value). There is a strong feeling within the telephone industry that the price elasticity for interstate toll is less than 1, and, indeed, is most likely in the neighborhood of 0.5. However,

Table 1

POINT AND INTERVAL ESTIMATES OF PRICE AND INCOME ELASTICITIES OF DEMAND FOR SELECTED TELEPHONE SERVICES

	ELASTICITY			
Type of Demand	Service-Connection Charge	Monthly Service Charge	Toll Price	Income
Access	-0.03 (±0.01)	-0.10 (±0.09)	--	0.50 (±0.10)
Local Calls	--	-0.20 (±0.05)	--	1.00 (±0.40)
Toll Calls (conversation-minutes)				
Intrastate	--	--	-0.65 (±0.15)	1.25 (±0.25)
Interstate	--	--	-0.75 (±0.20)	1.50 (±0.40)
International Calls	--	--	-0.90 (±0.30)	1.70 (±0.40)
Duration of Tool Calls	--	--	-0.15 (±0.05)	0.25 (±0.10)

Source: These estimates refer to long-run, steady state elasticities. The estimates for tool calls refer to conversation-minutes, rather than to just the number of messages. The estimates reflect my own interpretation of the empirical record (for both foreign countries and the United States), and are thus highly subjective. The numbers in parentheses are subjective standard errors.

my own view at this juncture, is that the long-run price elasticity for long-haul toll calls is as close to 1 as to 0.5.

The entries in the table provide only a partial listing of the categories of telephone demand. I have not included any elasticities for WATS and private line, vertical services, and coin stations, and, as mentioned, I have not attempted a residence/business breakdown. The empirical evidence in all of these areas is too weak for the tabulation of "best-guess" point estimates of elasticities. For WATS and private line, existing evidence suggests that own-price and income elasticities are more or less the same as for MTS, and there is solid evidence in the Feldman study (1976) that WATS, private line, and MTS are strong substitutes. Yet, interestingly, there is a suggestion in the study of Subissati (1973) that WATS and MTS (in Canada) may be complements. What Subissati finds is that current expenditures for MTS are positively related to the first difference in expenditures for WATS. Subissati suggests that this may reflect a stimulus to total toll calling induced by the presence of an additional way to make toll calls. If this is in fact the case, then WATS, private line, and MTS may actually be complements in the long run, but substitutes in the short run. In any event, the tradeoff between and among WATS, private line, and MTS is an important area for future research.

Let me now turn to some implications of the findings concerning the price and income elasticities of demand. To begin with, it must be emphasized that price elasticities exist; contrary to the views of many, they are not zero. On the other hand, it does appear that except possibly for very long-haul toll calls, telephone demand is inelastic. The demand for access, in particular, appears to be very price inelastic. Also, there is some evidence that the "transient" component of inward and outward movement is quite sensitive to the level of the service-connection charge.

The fact that the price elasticity for local use is small means that the telephone companies can look to the local market as a place to recoup the revenues that are almost certain to be lost in the private line, toll, and terminal-equipment markets as a result of competition. However, the key question in this connection is whether the price of access to residential customers is to be continued to be subsidized. In the past, the subsidization of residential access, primarily from "contributions" generated in the toll market, has been justified in terms of fostering universal service. However, the access externality is also an issue. If the access externality exists, then optimal social pricing requires that the price of access continue to be subsidized.[16] If it does not exist, then the price of access should be set equal to marginal cost.[17] Thus, we are once again reminded of the importance of establishing whether the access externality exists and estimating its quantitative magnitude.[18]

Let us suppose, for now, that the access externality is unimportant and that the telephone companies set out to increase the price of access to the level of marginal cost. What might be expected to happen? My own view is that there would probably be considerable resistance by residential customers. I base this observation on the empirical evidence offered by Perl (1978) showing the dependence of the price elasticity for access on the level of income and the level of the access price. Perl's results show the elasticity with respect to the access price to decrease with the level of income and to increase with the level of the monthly service charge. The access price elasticity is accordingly largest (although never absolutely large) for low-income households facing a high monthly service charge. On the other hand, the price elasticity is very small for high-income households facing a low monthly service charge. Thus, one can conclude that most of the access price elasticity arises from the income effect, rather than from the substitution effect. However, the substitution effect implies no loss in consumer welfare, whereas the income effect does. Consequently, as the price of access is moved toward marginal cost, I would expect consumer resistance, particularly on the part of low-income households living in areas where the monthly service charge is already fairly high.

Thus far in this review, the focus has been mostly on price elasticities. Income elasticities have been treated largely in passing. The reason for this is that for the most part, income elasticities are not controversial, but this is not the case with price elasticities, especially during a period of frequent (and sometimes large) rate increases. Income elasticities are very important to the telephone industry because they indicate how, holding prices, technology, and other nonincome determinants of demand constant, the industry will develop over time, which markets will require the most additional investment, and where additional revenues will accrue. In the markets where the aggregate income elasticity is greater than 1, the growth in revenues will be faster than the growth of the general economy, while in the markets where the aggregate income elasticity is smaller than 1, the growth in revenues will be slower than the growth of the general economy. Of course, nonincome factors do not remain constant, so that revenue growth in individual markets can be quite different than that implied by income elasticities alone. Nevertheless, in general, one should expect revenues to grow most rapidly in the markets with the highest income elasticities of demand.

As was noted earlier, the estimates of income elasticities are in general quite large, although there is a lot of variation in the estimates, especially in the ones for local use and interstate toll.[19] Table 1 suggests point estimates of at least 1 in all of the major telephone markets except for access. As noted, however, a great deal of uncertainty surrounds the estimate for local use, so that this elasticity can very well be less than 1. The income elasticity for toll calls, though, is almost certainly greater than 1, particularly in the long-haul interstate market. And there seems little question but that the elasticity for overseas calls is substantially in excess of 1. Indeed, for a number of years overseas revenues have been the fastest growing component of total Bell System revenues. Already these revenues are making an important contribution to Bell System profits, and, if present trends continue, there could be a day when they are the most important contributor.

2. Current Issues and Problems

In this sections, I shall present some concluding observations regarding the present state of demand analysis in the telephone industry and some suggestions as to where we might go from here. On the whole, the quality of analysis in the empirical literature is good. The models that have been analyzed are essentially state-of-the-art for applied demand analysis, and this is also true for the econometric techniques that have been used in estimation. The empirical literature contains a number of good studies and several really excellent ones. Included among the latter are Deschamps (1974), Gale (1974), Feldman (1976), Irish (1974), Larsen and McCleary (1970), Mahan (1980), Pavarini (1975, 1976, 1979), Perl (1978), Pousette (1976), Stuntebeck (1976), and Waverman (1974). The theoretical literature also contains some first-rate contributions, with the list being headed by Artle and Averous (1973), Littlechild (1975), Rohlfs (1974), and Squire (1973).[20]

Still, the quality of the empirical literature falls short of where it ought to be. The biggest problem is that the empirical and theoretical analyses of telephone demand have been like two ships passing in the night. The best empirical work has ignored the best theoretical work, and vice versa. Clearly, the theoreticians and the applied analysts need to join forces. Telephone demand modelers would also benefit from greater contact with the experience of demand analysts in other areas, particularly energy demand. Pooled time-series/cross-section models have been used extensively in analyzing energy demand, and the experience that has accumulated there is clearly relevant to the increased use of similar models for telephone demand. Energy demand analysts have also had extensive empirical experience in dealing with the problems caused by multi-part tariffs.[21]

It was noted earlier that existing estimates of price elasticities of demand provide solid support for the conclusion that telephone price elasticities are different from zero. Yet to state that price elasticities are different from zero is not to say what their values actually are. A great deal of uncertainty surrounds virtually all of the estimates, and one of the major tasks of future research is to reduce the zones of uncertainty. There is some evidence that the price elasticity for long-haul toll calls may be as large as -1, and since -1 is a decidedly critical value, it is particularly important that the uncertainty associated with the toll price elasticity be reduced. The efforts to do this, however, should distinguish more than has been the case in the past between the number of calls and the duration of calls, and price indices should be used that capture the essential characteristics of multi-part tariffs. There is also considerable uncertainty surrounding the existing estimates of income elasticities, especially the income elasticities for local calls and long-haul toll calls. However, the income elasticities appear to be at least 1 in all of the major markets except for access, and well in excess of 1 in the long-haul toll and international markets.

When we look at the empirical literature as a whole, toll demand has received a disproportionate share of attention in comparison with access and local-service demand. The demand for terminal equipment is also under-researched, and WATS and private line (particularly private line) have been practically ignored. International demand has received some attention, but not as much as its rapid growth warrants. Finally, residential demand has received considerably more attention than business demand. That toll demand has been the center of attention is readily understandable, for toll is relatively easy to model, and rate activity in recent years has tended to concentrate on the toll markets. However, in view of the current trends, in the telecommunications industry, the research focus needs to change, and much greater attention given to the demand for access and local use. As competition in the private-line, toll, and terminal equipment markets creates the pressure for additional revenues from access and local use, the industry and its regulators need better information on how customers -- especially residential customers --

will react to higher charges to access. A closely related question is how residential customers will react to the paced conversion to measured local service that is gaining increasing currency among the operating companies.[22] A major research effort is currently in progress at Bell Laboratories to examine these and related questions, but progress is slow and expensive.

Concerning toll demand, we have already mentioned the need to get a better fix on the long-haul toll price elasticity, and knowledge of point-to-point price elasticities is of obvious interest now that competition is a factor in many of the intercity markets. One of the biggest unanswered questions in the toll area concerns the tradeoffs among private line, WATS, and MTS for business customers, and the estimation of these tradeoffs needs to be approached in a well-articulated model of business demand. Some recent work by McFadden and Train (1979) provides some interesting and useful new suggestions in this direction.

FOOTNOTES:

[1] This section is taken mostly from Charpter 5 of Taylor (1980). An analytical treatment of the theory of telephone demand - in particular the attributes peculiar to telecommunications - is set out in Chapter 2 of that monograph.

[2] The state intrastate toll demand models in the U.S., of which between 30 and 35 existed in 1978, come to mind at this point. Most of the state models assume toll demand to be dependent on the number of main stations, which is of course as it should be, but the models do not include equations for the number of main stations. As noted in the text, there is no problem so long as the focus is on price changes and there exist good exogenous forecasts of the number of main stations. However, there will probably be occasions when the models will be used to forecast the impact of changes in income, and care must be taken to allow for feedbacks on the number of main stations. The explicit use of an access/use framework automatically reminds the model builder to do this.

[3] This is discussed in Chapter 2 of my monograph.

[4] Additional evidence concerning the externalities is found in the studies of Infosino (1976) and Wang (1976). Infosino finds that the number of local calls per line in a sample of residential customers in Los Angeles and San Francisco is positively related to the telephone density of the exchange, while Wang finds that the demand elasticity for yellow-page advertisements of a given size with respect to system size is greater than 1. This latter result implies that, with the size of an advertisement and price held constant, the demand for the space is stronger in a larger system.

[5] See also Taylor, Blattenberger, and Rennhack (1981).

[6] The studies of Waverman (1974), Perl (1978), and Larsen and McCleary (1970) should also be mentioned in this regard. Waverman distinguishes between the price of access and the price of use in his equations for Sweden, but not for Canada and the U.K., and Perl, in his analysis of access demand for the U.S., treats the service-

connection charge separately from the monthly service charge and also distinguishes between exchanges with measured local service and exchanges with flat-rate service. Finally, Larsen and McCleary, in their analysis of toll traffic between pairs of U.S. states, examine both the average charge per call and the average overtime charge per call. Unfortunately, though, Larsen and McCleary calculate both measures of price from ex post data.

[7] This is a long-standing problem in the analysis of electricity demand. See Taylor (1975).

[8] Gale's study is unique in the literature in that it is the only study that focuses directly and excusively on the dependence of duration on price.

[9] Repression in this context refers to the impact on revenues of a rate change when there is a nonzero price elasticity of demand.

[10] In the present context, the complementary durable good is the telephone system, while the service generated by the durable good is the use of the system.

[11] To dispel possible confusion, let me be more specific. The dynamics inherent in the access/use division represent the traditional distinction between the short run and the long run in the presence of a durable good. Suppose that the telephone system is in steady-state equilibrium (ignore the complications introduced by the access externality), and let this equilibrium be disturbed by an increase (say) in income. In the short run, there will be an adjustment in the number of calls that are made using the existing stock of telephones (and possibly also in average duration). In the long run, the stock of telephones may also adjust. However, it may be that in the short run (i.e., when the stock of telephones is fixed) there is a delay in adjusting the number of calls that are made to the higher income.

[12] For a discussion of transfer functions, see Box and Jenkins (1976). Cf. also Fask and Robinson (1977).

[13] There are only six studies in the literature that estimate models using pooled time-series/cross-section data. Kearns (1978) and Reitman (1977) pool time-series data across states in estimating models of the demand for residence extensions (Kearns, Reitman) and the demand for total vertical services (Reitman). Stuntebeck (1976) and Wert (1976) both use pooled data in analyzing the daytime/nighttime composition of toll traffic. Finally, Deschamps (1974) uses a pooled time-series/cross-section data set in his analysis of toll demand in Belgium, and Rea and Lage (1978) do the same in their analyses of international telephone, telex, and telegraph demand.

[14] Of the studies just listed, Deschamps and Rea and Lage are the only ones to use a variance-components framework. Kearns, Stuntebeck, and Wert use a covariance model, but Reitman pools directly without either the covariance or variance-components adjustment. The consequences of pooling directly are especially severe in Reitman's

study, for a model is used in which the lagged value of the dependent variable is included as a predictor. For models of this type, a covariance or variance-components framework is necessary; otherwise, the individual state effects will be reflected in the lagged value of the dependent variable, and the estimate of its coefficient can be severely biased.

[15] I have singled out normality of the error term because it is not usually considered a trouble point. The conventional list of statistical problems includes autocorrelation, heteroscedasticity, and multicollinearity. Autocorrelation is nearly always a problem with aggregate time-series data, and extreme care should be exercised in checking for autocorrelation in situations where t-ratios are around 2. Heteroscedasticity, on the other hand, is more likely to be a problem with cross-section data than with time-series data. In general, there is probably more laxity in checking for heteroscedasticity than in checking for autocorrelation. Finally, multicollinearity is nearly always present in some degree, no matter what the source of the data, although it is typically strongest with time-series data.

[16] By optimal social pricing in this context, I mean a policy that has the objective of maximizing the sum of consumers' and producers' surplus. See Zajac (1979).

[17] Because of rate-of-return (or other regulatory) constraints, optimal social pricing might require that the price for access deviate from marginal cost in a way that depends upon the price elasticity of the demand for access. This is usually referred to as Ramsey-pricing after F.P. Ramsey (1927). See Baumol and Bradford (1970) and Zajac (1979).

[18] If the access externality is quantitatively significant and if the telephone companies cannot subsidize access from the local-use market because of competition, how the subsidy is to be financed obviously becomes an important social question.

[19] See Tables 4 and 5 in Chapter 3 of my monograph.

[20] I view the studies cited here as constituting a "bare-bones" reading list for anyone wishing to become familiar with the literature on telephone demand. Two older studies are also highly recommended, namely, Kraepelien (1958) and Leunbach (1958), and also the soon-to-be-published monograph of Brandon and Brandon (1981).

[21] See, for example, Taylor, Blattenberger, and Rennhack (1981) and Acton, Mitchell, and Mowill (1976).

[22] For discussion of measured local service from a Bell System perspective, see the recent articles in the Public Utilities Fortnightly by Garfinkel and Linhart (1979, 1980) and Cosgrove and Linhart (1979).

REFERENCES:

[1] Acton, J.P., Mitchell, B.M., and Mowill, R.S., Residential Demand

for Electricity in Los Angeles: An Econometric Study of Disaggregated Data, Rand Corporation (R-1899-NSF), Santa Monica, California (September 1976).

[2] Alleman, J.H., The Pricing of Local Telephone Service, U.S. Department of Commerce, Office of Telecommunications, OT 77-14 (April 1977).

[3] Artle, R., and Averous, C., The Telephone System as a Public Good: Static and Dynamic Aspects, Bell Journal of Economics and Management Science 4 (Spring 1973) 89-100.

[4] Baumol, W.J., and Bradford, D.F., Optimal Departures from Marginal Cost Pricing, American Economic Review 60 (June 1970) 265-283.

[5] Beauvais, E.C., The Demand for Residential Telephone Service Under Non-Metered Tariffs: Implications for Alternative Pricing Policies, paper presented at the Western Economic Association Meetings, Anaheim, California (June 1977).

[6] Becker, Gary S., A Theory of the Allocation of Time, Economic Journal (September 1965).

[7] Box, C.E.P., and Jenkins, G.M., Time Series Analysis Forecasting and Control, revised edition (Holden-Day, Inc., San Francisco, 1976).

[8] Brandon, B.B. ed., The Effect of the Demographics of Individual Households on Their Telephone Usage in Chicago, (Ballinger Publishing Co., Cambridge, 1981).

[9] Cosgrove, J.G., and Linhart, P.B., Customer Choices Under Local Measured Telephone Service, Public Utilities Fortnightly, (August 1979).

[10] Davis, B.E., Caccappolo, C.J., and Chaudry, M.A., An Econometric Planning Model for American Telephone and Telegraph Company, Bell Journal of Economics and Management Science 4 (Spring 1973) 29-55.

[11] Deschamps, P.J., The Demand for Telephone Calls in Belgium, 1961-1969, paper presented at the Birmingham International Conference in Telecommunications Economics, Birmingham, England, (May 1974).

[12] Fask, A., and Robinson, P.B., The Analysis of Telephone Demand by Dynamic Regression, unpublished paper, AT&T (September 1977).

[13] Feldman, J., A Preliminary Cross Sectional Analysis of Services, unpublished paper, AT&T (February 1976).

[14] Gale, W.A., Duration of Interstate Calls, March 1969, unpublished Bell Laboratories Memorandum (December 1971).

[15] Garfinkel, L. and Linhart, P.B., The Transition to Local Measured Telephone Service, Public Utilities Fortnightly (August 1979).

[16] Gronau, R., The Value of Time in Passenger Transportation: The Demand for Air Travel, National Bureau of Economics Research, Inc.

(New York 1970).

[17] Infosino, W.J., Estimating Flat Rate Residence Average Local Calling Rate from Demographic Variables, unpublished Bell Laboratories memorandum (August 1976).

[18] Irish, W.F., A Market Analysis of the Demand for Telephone Sets by Class of Service, North Carolina Utilities Commission (December 1974).

[19] Kearns, T.J., Modeling the Demand for Residence Extension, unpublished Bell Laboratories memorandum (February 1978).

[20] Kreapelien, H.Y., The Influence of Telephone Rates on Local Traffic, Ericsson Technics, No. 2, published by Telefonaktiebolaget (Stockholm: L.M. Ericsson 1958).

[21] Krutilla, J.V., Observation Reconsidered, American Economic Review 57 (September 1967) 777-786.

[22] Kwok, P.K., Lee, P.C., and Pearce, J.C., Econometric Models of the Demand for Bell Originated Message Toll, Bell Canada, CCNS Network Market Planning, Ottawa, Ontario (January 1975).

[23] Leunbach, G., Factors Influencing the Demand for Telephone Service, paper presented at the Second International Tele-traffic Congress, The Hague (July 7-11, 1958).

[24] Larsen, W.A., and McCleary, S.J., Exploratory Attempts to Model Point-to-Point Cross-Sectional Interstate Telephone Demand, unpublished Bell Laboratories memorandum (July 1970).

[25] Linder, S.B., The Harried Leisure Class (Columbia University Press, New York, 1970).

[26] Littlechild, S.C., Two-Part Tariffs and Consumption Externalities, Bell Journal of Economics 5 (Autumn 1975) 661-670.

[27] Mahan, G.P., The Demand for Residential Telephone Service, East Lansing, Michigan, Michigan State University Public Utilities Papers (1979).

[28] McFadden, D., and Train, K., Specification Manual for Business Telecommunications Demand Model, unpublished paper, AT&T (March 1979).

[29] Meyer, J.R., Wilson, R.W., Baughcum, M.A., Burton, E., and Caouette, L., The Economics of Competition in the Telecommunications Industry (Charles River Associates Inc., Boston, Massachusetts, August 1979).

[30] Pavarini, C., Identifying Normal and Price-Stimulated Usage Variations of Groups of Customers Part II: Individual Customer Comparison, unpublished Bell Laboratories memorandum (October 1975).

[31] Pavarini, C., Residence and Business Flat Rate Local Telephone Usage, unpublished Bell Laboratories memorandum (July 1976a).

[32] Pavarini, C., The Effect of Flat-to-Measured Rate Conversions on the Demand for Local Telephone Usage, unpublished Bell Laboratories memorandum (September 1976b).

[33] Pavarini, C., The Effect of Flat-to-Measured Rate Conversions on Local Telephone Usage, paper prepared for the Mountain Bell Economics Seminar, Keystone, Colorado, August 27-30, 1978, published in Pricing in Regulated Industries, Vol. II, J.T. Wenders, ed. Denver: Mountain States Telephone and Telegraph Co. (1979).

[34] Perl, L.J., Economic and Demographic Determinants of Residential Demand for Basic Telephone Service, National Economic Research Associates, Inc. (March 1978).

[35] Pousette, T., The Demand for Telephones and Telephone Services in Sweden, presented at the European Meetings of the Econometric Society, Helsinki, Finland (August 1976).

[36] Ramsey, F.P., A Contribution to the Theory of Taxation, Economic Journal 37 (March 1927) 47-61.

[37] Rash, I.M., Residence Sector Demand for Message Toll Service, Bell Canada Working Paper, Ottawa, Ontario (April 1972).

[38] Rea, J.D., and Lage, G.M., Estimates of Demand Elasticities for International Telecommunications Services, Journal of Industrial Economics 26 (June 1978) 363-381.

[39] Reitman, C.F., The Measurement and Analysis of Demand Elasticities of Residential Vertical Telephone Service, unpublished Ph.D. Dissertation, New School for Social Research (May 1977).

[40] Rohlfs, J., A Theory of Independent Demand for a Communications Service, Bell Journal of Economics and Management Science 5 (Spring 1974) 16-37.

[41] Squire, L., Some Aspects of Optimal Pricing for Telecommunications, Bell Journal of Economics and Management Science 4 (Autumn 1973) 515-525.

[42] Stuntebeck, S., The Relation of Day/Night Price Differentials and Message Volume Distributions for Intrastate Toll Calls, unpublished Bell Laboratories Memorandum (October 1976).

[43] Subisatti, E., An Econometric Model for WATS and Business Toll Services, Computer Communications and Network Services - Bell Canada, Ottawa, Ontario (August 1973).

[44] Taylor, L.D., The Demand for Electricity: A Survey, Bell Journal of Economics 6 (Spring 1975) 74-110.

[45] Taylor, L.D., Telecommunications Demand: A Survey and Critique, (Ballinger Publishing Co., Cambridge, 1980).

[46] Taylor, L.D., Blattenberger, G., and Rennhack, R., The Residential

Demand for Energy in the United States, Report to Electric Power Research Institute on RP1098, Data Resources, Inc., Lexington, Massachusetts (1981).

[47] Wang, R.C. Demand for Yellow Pages Advertising Space, unpublished paper, Mountain Bell (October 1976).

[48] Waverman, L., Demand for Telephone Services in Great Britain, Canada, and Sweden, paper presented at the Birmingham International Conferance in Telecommunications Economics, Birmingham, England (May 1974).

[49] Wert, G.M., The Effects of Day/Night Price Ratios on Residence Weekday Interstate MTS Traffic, unpublished Bell Laboratories memorandum (June 1976).

[50] Zajac, E.E., Fairness or Efficiency: An Introduction to Public Utility Pricing (Ballinger Publishing Co., Cambridge 1978).

Economic Analysis of Telecommunications:
Theory and Applications
L. Courville, A. de Fontenay and R. Dobell (eds.)

B.C./ALBERTA LONG DISTANCE CALLING

A. de Fontenay and J.T. Marshall Lee

Government of Canada and British Columbia Telephone Company

1. Introduction and Outline

In the last ten years there have been a large number of telephone demand studies on both theoretical and empirical levels. From these studies two observations are in order: first, analysts appear to be very divided with respect to the significance of network externality, and second, most studies simply invoke the constant price and income elasticities assumption by estimating a double-log equation without first investigating the validity of the assumption. In light of these two observations, the purpose of this study is (1) to critically discuss the presence of network externality and other related issues, (2) to investigate the validity of the commonly employed constant elasticity assumption, and finally (3) to explore the use of pooled time series and mileage band data in obtaining a better and more general demand equation for toll where price and income elasticities may vary. For this purpose, a common data base (B.C. to Alberta Monday to Friday day DDD residence call minutes from 1973 - 1979 Q4), is set up to evaluate alternative specifications and competing hypotheses as well as to estimate the proposed demand equation.

Taylor (1980) has presented a thorough and enlightening review of the empirical demand literature. Starting from Taylor, the second part of this paper reviews the subsequent literature in Canada. Particular attention is given to the estimation of toll elasticities. Issues relating to dynamic adjustment processes normalization and degree of disaggregation are also considered. It is also argued in this section that the assumption of constant price elasticity is questionable. In most empirical studies where calls are disaggregated by mileage bands, the common finding is that price elasticities in absolute value increase with distance. Since the price of a call usually increases with distance, the above finding implies that price elasticities actually increase with prices. The translog demand function is later introduced to cope with this observation.

In section 3, the issue of consumption and network externalities is discussed. In section 4, the translog demand function is introduced to cope with the empirical observation cited earlier that price elasticities in absolute value increase with distance. The estimated results of the translog demand function are presented in this section 5. Pooled time-series mileage bands data are used in the estimation. The pooling of data is based on the presumption that consumer's welfare is independent of distance, distance affecting solely the price and opportunity cost of the call. Finally, in section 6, some concluding remarks are offered.

2. Modelling Message Toll Demand: A Constructive Review of the Literature

2.1 The Double Log (DL) Model

In this section the Canadian literature subsequent to Taylor (1980) in the area of the empirical analysis of demand for message toll services is reviewed. This literature is divided between work done by carriers (Bell Canada, 1976; Dreessen, 1977; and 1979; Bell Canada, 1980; Piekaar, 1980) and work done at various universities, often under the Department of Communications' sponsorship, (Bernstein et al., 1977; Corbo et al., 1978; Fuss and Waverman, 1978; Corbo et al., 1979; Breslaw and Smith, 1980; Breslaw, 1980; Bernstein, 1980; Breslaw and Smith, 1981 (a) and (b) and Fuss and Waverman, 1981).

Two factors characterize this literature, namely its almost universal adoption of double-log (DL) types of demand models and the sharp differences in the treatment of the network externality and in the magnitudes of estimated elasticities. To illustrate these differences, it suffices to note that elasticity estimates are ranging from a low of -.18 in Bell Canada (1980)'s estimate for mileage bands of 100 miles and under to highs in the neighbourhood of -1.3 obtained at Concordia University (see for instance Breslaw and Smith, 1980) or at the University of Toronto (see for instance Fuss and Waverman, 1981).

In spite of the general acceptance of the DL model, analysts differ in the particular specifications they have adopted. These may involve various aggregations, the time structure of the model, the variables used, etc. For instance, industry studies have been by far the most disaggregated, both in terms of the message toll services outputs considered and in terms of time. In fact, the Intra B.C. model and Bell Canada (1976) are based on monthly series. These differences in specification will be reviewed, as will be the concept of network externality, and conclusions will be drawn regarding the desired specification of a demand model.

While the strength of the DL model lies in its very simplicity, its justification has been rather scanty. It has been derived through the Box-Cox transformation in Corbo et al. (1979) (see also Fuss and Waverman, 1981), and Breslaw and Smith (1981, (a) and (b)) have gone to great length to set it within a utility maximization framework. Alternative forms have been explored, hence, Bernstein et al. (1977) considers the Rotterdam model while Corbo et al. (1979), investigates the application of flexible functional forms such as the translog and the generalized Leontief. To date, those efforts have not been successful, and the DL model appears more entrenched than ever.

In this study, a new specification for message toll demand is proposed. The aim is to cope with the empirically observed relationship between the message toll price elasticity and distance which will be hypothesized to be the result of an ex ante dependence between the elasticity and the price. Unlike Breslaw and Smith (1981), the relationship between the utility maximization process and the specification of the demand curve shall not be investigated as such a process would, under most common specifications, imply additional constraints which have not been considered. Instead, we will introduce the direct translog demand function (not to be confused with the demand function which can be derived as a first order condition from a translog utility function; Corbo et al., (1979). Simply, this function is a second order approximation to an arbitrary demand curve in which the variables are expressed in terms of the logarithms. Hence, it is of the form of a translog function, and a straightforward generalization of the DL model.

2.2 The Aggregation Problem

Outside of the industry studies, a common flaw among the Canadian literature appears to be related to the data base used. All these studies, with the exception of part of Bernstein et al. (1977) and the whole of Bernstein (1980), are based on the same data base, Olley's Bell Canada Productivity Study data base (Bell Canada, 1969, 1973 and 1980).

Certain observations regarding the applicability of the productivity data base to demand analysis are in order. First, these series are deflated settled revenues; only accidentally would settled revenues correspond to the output measures which correspond to the quantity demanded. The relevant output measure in a study of subscribers' demand is the output which corresponds to calls originated by subscribers (even though the relevant measures would be more complex whenever Taylor's call externalities are introduces) and the revenue measure to which it corresponds is the originating revenue. The industry studies are free from that flaw since they use deflated originating revenues (Bell Canada), or actual quantities (B.C. Tel).

Another major flaw of most studies is their level of aggregation. As noted earlier, among those studies based on Bell Canada's productivity data base, only Breslaw (1980) considers the disaggregation between classes of message toll services while Fuss and Waverman's (1981) study, in one of their models, considers a disaggregation by mileage bands. Bell Canada's study is restricted to Intra-Bell message toll services, which is disaggregated into two categories depending upon whether or not the distance is greater than 100 miles. B.C. Tel's analysis is by far the most disaggregated since not only the Intra-B.C. Tel category alone is considered, but it is divided further into seven mileage bands. The aggregation problem is extremely serious in view of the simple observation that in almost all models, estimates of demand elasticity increases with the mileage, an observation already made by Taylor (1980) and Fuss and Waverman (1981).

The aggregation problem appears at two levels: those of the time and output characteristics. Industry studies are based on monthly or quarterly series, over relatively short time periods, typically 6-7 years. All the other analyses but for one are based on yearly series and cover a much longer time span, typically over twenty five years. The rationale for the rather short period used in industry studies is typically based not only on data availability but also on the presumption that the one-minute minimum duration was introduced in the early seventies and that the access market reached saturation around that period, the net effect being a structural change in the beginning of the seventies.

A message toll call can be indexed in terms of its class (intra, adjacent, etc.), its originating customer (business, resident), its type (DDD, SOH, PP), its time of day (TOD), (day, evening, night) - day of week (week day, week-end and holidays) characteristics, its duration and its distance. Only the B.C. Tel's study takes into account the complete disaggregation in terms of types of call, TOD and distance. Even though one would expect that message and duration are determined simultaneously, one would expect that the absence of an access charge, in the form of a higher price for the first minute, should minimize the desirability to go beyond the message - minute as the output unit. The extent of the disaggregation, in the Intra B.C. model, is such as to raise some estimation problems (Piekaar, 1980). Bell Canada restricted its attention to DDD in terms of two mileage bands but regardless of TOD (also excluding holidays), duration and type of customer, and to person-to-person with the same restrictions, but without mileage bands differentiation. This led Bell Canada to use deflated revenue output measures. Only Fuss and Waverman (1981), in one of their attempts, consider duration explicitly. Finally, Bernstein (1980) uses messages as his output measure.

2.3 Review of the Demand Specification

In this section, we shall review the main characteristics of the various demand studies cited earlier to attempt to draw some conclusions with respect to the specification of a demand model.

2.3.1 Money Illusion

Taylor (1980)'s specification of the demand for telephone services allows for the possibility of money illusion. Only the earlier models such as Bell Canada (1976) and Dreessen (1977, 1978) have maintained this format. All other studies reject a priori any money illusion. This will also be the approach adopted here.

2.3.2 Cross-price Elasticities

The only attempts at measuring cross-price elasticities are found in the work of Concordia University and in Dreessen (1977, 1978). Cross-price elasticities were estimated unsuccessfully in Bernstein et al. (1977) and Corbo et al. (1978, 1979). The problem follows from the lack of variability of relative prices, the degree of multicollinearity being very high, and from the aggregate nature of the local services series. Similarly, in the Intra-B.C. model, Piekaar (1980) abandoned Dreessen's (1977, 1978) previous attempts to introduce, in some mileage bands, the price of local services as an explanatory variable. If the price of local services is seen as an access charge, it should be deducted from the income variable and not be introduced as a price variable (Bernstein, 1980). Toll calls over other mileage bands are not substitutes, and their price level would act only through the income constraint. In other words, cross-price elasticities between mileage bands should not be a problem to worry about. Cross price elasticities between types of call and between different periods of the day or days of the week are still outstanding problems.

2.3.3 The Dynamic Structure of Demand

The most general linear demand model, in terms of its dynamic structure would be the transfer function:

(1) $r(B)q_t = B^s\, s(B)p_t + B^u u(B)Y_t + B^v v(B)X_t + e_t$, and $\phi(B)e_t = \theta(B)a_t$

where $r(B)$, $s(B)$, $u(B)$ and $v(B)$ are proper rational functions in B which is itself the lag operator, i.e. such that $Bz_t = Z_{t-1}$, s, u and v are non-negative scalars which indicate the dead time, $\emptyset(B)$ and $\theta(B)$ are proper polynomials in B and a_t is $N(o,b^2)$, q_t, p_t, Y_t and x_t denote the quantity demanded, the price of the service, the income and other exogeneous variables, after proper deflation and transformation, as required. This general model, without transformation of the variables and with $r(B) = (1-r_1B)$, $s(B) = s_0$, $u(B) = u_0$, $v(B) = 0$ and $\emptyset(B) = \theta(B) = 1$ is the Houthaker-Taylor flow adjustment model.

As noted earlier, the standard application in modelling the demand for telephone services is the DL model, which implies that the variables are expressed as logarithms. Then we have the habit formation model; this model was used by Piekaar (1980). If in addition to the habit formation hypothesis, one also assumes that $\emptyset(B) = (1-\emptyset_1B)$ while $\theta(B) = 1$, i.e. if one introduces a correction for autocorrelation, then we obtain the model adopted in Corbo et al. (1978) and Fuss and Waverman (1981). The most extensive study of $\theta(B)$ can be found in Corbo et al. (1979), in which even $\emptyset(B) = (1-\emptyset_1B-\emptyset_2B^2-\emptyset_3B^3)$ was investigated. However as all of these tests in Corbo et al. (1979) are applied to regressions which also contained cross-price elasticities and as it is unlikely that we can disentangle those elasticities from one another, the utility of these tests is limited. On the other hand, the attempt to go beyond a first degree polynomial in the specification of $\emptyset(B)$ is welcomed since its higher degree polynomials introduce the possibility of complex roots corresponding to cyclical movements.

Economic theory has nothing to tell us as to the proper dynamic structure of the model, and one must turn toward time series analysis. Box and Jenkins (1970) present a methodology to identify a transfer function; however, their methodology cannot be applied, at this stage, to our problem because the series are too short. Furthermore it has also been noted that different models may yield very similar summary statistics (Granger and Newbold, 1978). In this context, it seems wise to follow Box and Jenkins' parsimony principle, i.e. to select the simplest of the models which can reasonably be entertained. The testing, as indicated above, will remain rather ad hoc as long as we do not have longer time series.

2.3.4 The Seasonality of Demand

As noted earlier when referring to the construction of price indexes, seasonality affects the demand for message toll services. Again the time series framework presented in the previous section can accommodate this new dimension without any problem. The problem, however, is that the requirement on data is much greater, hence, the practical application of standard time series procedure will have to wait for a few years, when, barring major structural changes, we will have sufficiently long series!

The seasonality question affects only the industry studies. There, the Intra-B.C. model is based on seasonally adjusted data, the seasonal adjustment procedure being the Bureau of the Census X-11, while Bell Canada (1980) utilizes seasonal dummies for the intercept. Cleveland (1972) has shown that the X-11 program can be approximated by a seasonal multiplicative autoregressive-integrated-moving average (ARIMA) model. It can also be shown that the use of dummy variables can be analyzed from the point of view of ARIMA models with common roots which were studied by Abraham and Box (1979). Courchesne, Fontenay and Poirier (1980) have developed a general analytical framework within which the use of the X-11 and the use of dummy variables both can be evaluated within the general ARIMA specification. They show that it is an empirical matter which of the dummy variables approach and the use of variables adjusted by the X-11 dominates. Hence, once again, as with the previous sections, we cannot derive a general rule.

Finally, even though it cannot be said exactly how many degrees of freedom are lost through a prior adjustment by the X-11 seasonal adjustment program, if one uses its ARIMA approximation as given in Cleveland (1972) or Cleveland and Tiao (1976), one can obtain a reasonable estimate. This correction was not done in the Intra B.C. model which treats the seasonally adjusted data as raw series.

2.4 Fuss and Waverman (1981) Demand for Toll Calls by Distance and Length of Call

As noted earlier, most approaches adopted in the demand for message toll services are tailored to the data base available to the author. A particularly interesting example is one of Fuss and Waverman (1981)'s models which they tailored around a 1977 Quebec interrogatory which made public the number of calls by duration and by mileage band for Bell Canada. Even thought the econometric estimation, by the authors' own account, was unsuccessful, it is worthwhile to present their model.

Bell Canada's message toll tariffs are two-part tariffs, the first minute being more expensive than subsequent minutes which are all always priced at the same rate, and they attempt to tailor their model to that feature. The demand is differentiated in terms of mileage and in terms of duration; while mileage is used to index the demand curves and won't be considered further, the duration is used to specify different forms depending on the number of minutes.

For calls of one minute duration or less, it is assumed that the quantity, i.e. the number of calls, x_1, is a function of the price of one minute calls, p_1, and the expenditures on all toll calls, E:

(2) $x_1 = D^1 (p_1, E)$

For calls of j minutes duration, j > 1, the price of the jth minute (which is independent of j for j > 1), p_2, the expenditure on all toll calls and the access charge for each of the calls lasting j minutes, which will be denoted by $E^*_j = (P1-P2)x_j$, are assumed to be arguments of the demand function, such that

(3) $x_j = D^j (P_2, E^*_j, E) \quad j = 2,3,...6$

These models were estimated, using the following specification of the demand functional form:

(4) $S_{i,j,t} = \alpha_{i,j} + \beta_i \ln P_{i,j,t} + \gamma_i \ln E^*_{i,j,t} + \delta_i \ln E_t$
where $S_{i,j}$ is the share of toll call expenditures in mileage band i of duration j, $P_{i,j}$ is the price of the last minute for a call in mileage band i lasting j minutes, i.e. if j = 1, $P_{i,j} = P_{i,1}$ and if

$j \neq 1,\ P_{i,j} = P_{i2},$

$$(5) \quad E^*_{i,j} = (P_1 - P_2)\ x_{i,j}$$

No mention is made by the authors, in their estimation, of the fact that $\sum_{i+j} S_{i,j,t} = 1$ implies the constraints

$$(6) \quad \sum_{i,j} \alpha_{i,j} = 1, \qquad \sum_i \beta_{ii} = \sum_i \gamma_i = \sum_i \delta_i = 0$$

There are numerous other reasons why one would expect the problems they faced in their estimation of this model. First of all, the disaggregation of the data is misleading; even though they consider seventeen distinct mileage bands, the data are nevertheless aggregated over (i) types of customers, i.e. business and residential, (ii) types of call, i.e. DDD, SOH, and P-P, (iii) carriers involved, i.e. intra-Bell, or adjacent, or TCTS, and finally (iv) time of day - day of week. Experience with the Intra-B.C. model leads us to believe that (ii) is crucial, while other things equal, the fact that the call is intra, adjacent, TCTS or US will generally affect its rate. In addition, the methodology to derive the price indexes is not presented.

Second, the specification says nothing about the determinant of the total expenditures on toll calls, E, and it is not fully consistent with the conceptual argumentation for the analytical model. In the latter, it is stated, that the substitutability between calls is fundamental to the argumentation, yet, in the application, each duration is taken by itself independently of the others.

"...The demand for one minute calls can decrease when the price beyond the first block falls, since a longer call is a substitute for one minute call."

Finally, it is hard to accept the hypothesis that the demand for each duration is inversely related to price; it does not seem to be far fetched to imagine the demand for 1 minute calls, and even, possibly, 2 minute calls to be upward sloping.

2.5 Bell Canada's Intra Model

In considering the modelling of the demand for message toll services, one must repeatedly refer to Bell Canada's path-breaking study, as we have done already and as we will have to do further on in the paper. In this section the intent is to complement our comments by a review of Bell Canada's methodology. As illustrated in Table 1, the main features of Bell Canada (1980) are the externality variables which we reject on the grounds that the estimate of its coefficient cannot be justified on economic ground, (this is discussed in detail in the next section) and the level of aggregation across time of day, rate groups and subscribers. The level of aggregation across mileage bands found in Bell's model does not appear justifiable in view of the observed differences in estimated elasticities at lower aggregation levels. The level of aggregation across subscribers raises just as many problems. For the business subscribers, telephone services are one input in their production process, and in their decision process, they will select a level of demand in terms of the price of that service, that of other inputs such as labour, capital and the level of demand. On the other hand the residential customer should consider the income constraint, the price of the service and the price of other goods and services he may demand. It follows that the price of the service is the only variable which is common to both decision processes. The ambiguity in the choice of exogenous variables in Bell's model follows from this problem.

Table 1

The Bell Canada Intra Model

$\ln Q_t$	Deflated revenue for business-residential/DDD/ Day-Eve-Night disaggregated into two mileage bands
=	
a	
+	
$b \ln P_t$	Laspeyres chained price index
+	
$c \ln Y_t$	Average quarterly number of employed persons 15 years and over, Quebec & Ontario(1)
+	
$(b+c) \ln PD_t$	GNE deflator(2)
+	
$d \ln EXT_t$	Sum of business and residential main telephones in service(3) replaced in 1981
+	
e SEAS	Seasonal dummy variables
+	
f DAYS	Correction for the number of weekdays in the quarter

NOTES:

(1) Bell Canada considers the aggregate over business and residence; as such there does not exist a "proper" income variable.

(2) The GNE deflator is appropriate only for business customers. Even then, it is not proper here as it accounts for price fluctuations across the whole of Canada, i.e. it reflects price changes which do not intervene in the decision process of an Ontario or Quebec business subscriber.

(3) In addition to the observations made in this paper, it should be noted that business main telephones, while technically well defined by any one carrier, is not an unambiguous concept from the demand for message toll services.

However these features are not unique to the Bell model. What is more fundamental in that model is the thoroughness of the statistical testing and the implied philosophy. A clear set of criteria are set to evaluate all equations; these cover (i) the use of a residual plot, and (ii) of a normal probability plot, (iii) the F statistic at 1% significance level, (iv) both the R^2 and the R^2 together with the standard error of regression, (v) the t-test at 5% level, (vi) the D.W. statistics and Box-Jenkins' procedure (vii) the Anderson-Darling test of normality, (viii) the Goldfeld-Quandt method to test for heteroscedasticity at the 5% level, (ix) Chow's tests for stability, (x) Klein's procedure to evaluate multicollinearity, (xi) the General Linear hypothesis, and (xii) the appropriateness of deflating by a general price indicator. This is complemented by an ex-post forecast analysis of the results.

The philosophy is outlined at the very beginning of Bell Canada (1980):

> "Because econometric models produce statistically optimal estimates only when the models satisfy certain statistical assumptions, failure of the models to satisfy all of these assumptions can lead to erroneous conclusions". (p. 1)

While this position is useful, it is meaningless by itself since it cannot be used to evaluate spurious relations. There are two possible approachs to go around that problem. The first approach is statistical and again suffers from the shortcomings of remaining within the statistical discourse, however it would help us cope with the problem. That solution consists of testing for Granger Causality. The second approach consists of first deriving the bounds to meaningful results from economic analysis. It is hinted at for instance (pp. 5 and 8) when it is stated that "the explanatory variables included in the models were selected on the basis of market characteristics and conventional economic principles..." and "the estimated coefficients of the economic variables must have a plausible sign based on economic principles". However at no time does the study consider the economic implication of the model beyond those few generalities. It follows that, even though the statistical testing is meritorious, it is also vacuous. Certainly, economic priciples cannot be used in the form of a statistical test such as the t-test. To wit, even though it is known that the price elasticity should be negative, it is also known that there exists situations in which it will be positive. All this means is that econometrics is by necessity as much an art as a science and that its statistical components cannot meaningfully be used mechanically.

3. Network Externalities

Taylor (1980) considers two forms of externalities when he notes that:

"A completed telephone call requires the participation of a second party, and the utility of this party is accordingly affected... an externality is thereby created... the externality is a call (or use) externality".

and that:

"Connection of a new subscriber confers a benefit on existing subscribers because the number of telephone that can be reached is increased. In this case, the externality is an access (or system) externality".

The first form of externality is fundamental to any form of communication and in particular to telephony. It is an externality since, presumably, the second party would be willing to pay a price to receive (or, maybe, not to receive) that call. As, in practice, the cost can usually be shared by taking turns, and in the case of a connection which is charged in terms of usage, this form of externality should not be relevant as long as both parties are facing the same price. For this study, the comparability of income between Alberta and British Columbia and the fact that the rates are the same in both directions should minimize the impact of the form of externality.[3]

Now if one additional subscriber joins the network, it is contended that every subscriber will benefit since every subscriber can now reach that new subscriber. This form of externality should have an enormous if not explosive impact on the network since the nth subscriber increases the number of possible originating calls by 2 (n-1), that is the number of possible connections by (n-1). This form of externality, however, depends crucially upon the existence of a potential call externality since it is necessary that the other subscribers increase their usage of the network or at least increase their option demand to gain access to the network to create an externality. In this context, it is wise to distinguish between two types of expansion in the number of subscribers depending upon whether (i) the population does not change but the penetration rate increases, i.e. the increase in n, the number of subscribers, is solely due to individuals who did not have the telephone previously and who are now getting it, or (ii) the population increases at the same rate as the number of subscribers, leaving the penetration rate constant. Whereas type (i) expansion might characterize networks such as the French one, it seems clear that, at the very least through the seventies, it is type (ii) expansion which characterizes the Canadian situation.[4] While the access externality could be expected to be significant given type (i)[5], this should not be true of type (ii) expansion.

It is recognized that the issue is empirical but it is suggested that the econometric approach will not provide the proper test to sustain or refute our contention. First of all, we begin by suggesting that the proportion of inward movements or net gains caused by such effects as "keeping up with the Jones's" is, for all practical purpose, null. We also suggest that new subscribers do not consider, in their decision to join the network, whether others are subscribing. Rather we assume that they take it for granted that almost everyone is a subscriber. In other words we suggest that access is not affected, in a Canadian context, by the so-called network externality. Our hypothesis could be investigated through a proper survey of new subscribers to establish their rationale for joining the network. To look at the access externality impact on usage, we suggest that a proxy might consist in considering calls per station in two communities with the same penetration rate, relatively similar income and socio-demographic characteristics except for their populations. (We are also making the assumption that usage behaviour doesn't vary much across Canada). Ideally, these two communities should be in the same rate group to avoid a price differential effect. However, it is suggested that as long as it is not too widely different communities, in terms of culture, say Edmonton and Quebec, which are selected, even differences in rate group, should not affect significantly calling patterns. For instance, it is suggested that on a main station basis Victoria would not differ all that much from Vancouver. This hypothesis is based on the contention that the subset of subscribers any one subscriber is likely to call with non-zero probability is extremely small compared to the set itself. Hence a change in the set in the form of a new subscriber could not affect the calling pattern of but a very small subset of subscribers with a non-zero probability; the possibility to reach millions of subscribers by accessing the network is of no real relevance to me as a new subscriber since I am concerned only with the few I am likely to ever reach. Furthermore, even though some old subscribers will reach me with a non-zero probability, it is contended that this will be dominated by substitution in their calling patterns. The argument presented here could be re-phrased in terms of time allocation by consumers who maximize their utility function; in a near saturated market the constraint in using the telephone is time, while in a market with low penetration, the opportunity cost to an increased penetration is likely to be high.

It could be contended that what one measures through such an externality is a change of behaviour, the new subscriber being different from the old. Once again, however, this would have to be established by direct comparison since it appears extremely unlikely to be very significant. In the case of Bell Canada (1980), if this were the interpretation given, one would have to accept that the new subscribers consume in the order of 133 to 145% more message toll services than established customers within Bell territories. If an independent investigation were to establish that this were true, the interpretation of the estimated coefficient of the price variable as an elasticity measure would still be invalid; the regression coefficient is obtained from the reduced form of the aggregation of two distinct subscriber populations, the old and the new subscribers, while the composition of that population is changed.

Finally, it should be noted that, in any case, the number of subscribers that can be reached will vary significantly and simultaneously with the mileage band and the exchange the call originates from. Unless one takes a point to point approach, or considers the demand originating at the exchange level, it is clear that the number of subscribers cannot be specified and that it will always differ from the total number of subscribers. Hence, even though the number of subscribers that can be reached could be roughly approximated by the total number of subscribers as a first approximation if the intent is only to deflate the output variables, its interpretation as an externality variable is invalid if it is used as an explanatory variable.

4. A Message Toll Services Demand Model

It was pointed out earlier that recent Canadian results confirm the result that the own price elasticity of demand is not constant with respect to distance. In addition to Fuss and Waverman (1981)'s conclusion to that effect, we can also cite Breslaw's (1980) disaggregation of Bell Canada's message toll services. However, the most convincing evidences are provided by the Bell Canada (1980) model and by the Intra-B.C. model.

In Bell Canada (1980), the DDD business and residential (0-100 miles) service and the (101+ miles) service yield very similar results on the whole, except for the income elasticity which appears to decrease as the distance increases: from .38 to .24, and the own price elasticity which almost doubles: from -.18 to -.32. As such, one has to be careful when interpreting the income elasticity variable adopted by Bell Canada which is only a proxy for income; in fact it is hard to understand why such a variable (the average number of employed persons in Quebec and Ontario) was adopted by Bell Canada when, consistently, in recent rate applications, Bell Canada has also been presenting in the form of an exhibit to support their application the graph of the year-to-year percent increase in toll messages shown next to that of the year-to-year percent increase in the GNE measured in constant dollars (Bell Canada, Exhibit B-81-220). In addition, at least one of these elasticity estimates is not significantly different from zero.

The Intra-B.C. model price elasticities, (given in Table A.1) once again confirm that, especially with respect to DDD-type calls, the elasticity increases rather systematically with distance. In addition, the income elasticity also has a tendency to decrease with distance which is generally consistnet with that observed in Bell Canada (1980).

We suggest that such evidence is sufficient to raise serious questions as to the applicability of the DL model to the analysis of demand for message toll services. We will further use this result to hypothesize such an observation implies that the own price elasticity and the income elasticity of message toll services should, at least, be allowed to vary systematically with price and income.

A telephone conversation enables two parties to communicate with one another. There are two features which are unique to modern telecommunications: first, the communication is nearly instantaneous, second the quality of the communication is approximately independent of distance. On theoretical grounds one cannot, however, assume that distance is not an argument in the utility function of the consumer since the opportunity cost to the consumer, to the extent it can be defined, may be expected to increase very rapidly with distance ("may" since it won't as long as the alternative is the mail while it will in all other cases since transportation will be involved). Since there is no readily available close substitute, any form of transportation being time consuming and, with the exception of the mail, very costly, we shall nevertheless assume that distance is not an argument on the consumer's utility function.[6] Evidently our assumption also assumes that all the functions fulfilled by the telephone are not distant specific; this is clearly not the case when we consider the complete set of distances since such services as emergency services are relevant solely over very short distances. Since we restrict our attention to message toll services, it is felt that it is a very weak assumption. Given such an hypothesis, it is possible to pool observations across mileage bands in the same demand model. To be consistent with existing observations, this model must be such as to account for the higher elasticity associated with calls over longer distances. To account for this higher elasticity, while at the same time recognizing the rejection of distance as the determinant of this elasticity variation, we not that, empirically, distance is positively correlated with the price of a message-minute. Hence, we assume that the observed correlation in elasticity with distance is the result of a utility function which is such as to associate a higher elasticity with a higher price. Evidently, we do not contend that the price level is the cause of the elasticity, rather that the observed correlation is the result of the form of the utility function.

As the data which we are using in this paper are not point-to-point, it is not possible to clearly identify the population within B.C. which can make the calls and the population within Alberta which and be reached. Furthermore, it is likely that there will be considerable variations within as well as between mileage bands since, for the smallest mileage bands, B.C. Tel subscribers living in Vancouver or Victoria are excluded. In fact, the two longest mileage bands (F and G) include the toll calls from Vancouver and/or Victoria to Calgary and/or Edmonton respectively. Even if the number of message-minutes are indexed in terms of the B.C. Tel subscriber population and the Alberta subscriber population, a B.C. Tel subscriber selected at random is more likely to make a call in mileage bands F and G. This will be accounted for through dummy variables which let the levels of the demand curve vary with the mileage bands.

Furthermore, to account for variation in demand characteristics between various point-to-point combinations, variations which could be due to demographic or geographic characteristics, we begin by specifying the demand function as a general function such that

$$(7)\quad q_{i,t} = F(P_{i,t}, P_t, Y_t, D_i),\ i = 1, 2, \ldots, 7$$

where $q_{i,t}$ is the number of message-minutes in mileage band i, period t, by subscribers in B.C. Tel, to subscribers in Alberta, $P_{i,t}$ is the price index of a message-minute in mileage band i, period t, P_t is the general price level in period t, Y_t is the per capita personal disposable income in period t, and D_i is a vector of dummy variables, (δ_j), such that

$$\delta_{i,1} = 0 \text{ for all } i, \text{ and}$$

$$\delta_{i,j} = \begin{matrix} 1 \\ 0 \end{matrix} \text{ for } \begin{matrix} i=j \\ i \neq j \end{matrix} \qquad j = 2, 3, \ldots, 7$$

It can further be assumed that there exists a flexible functional form, defined by Diewert (1973), which "contains precisely the number of parameters needed to provide a second order approximation to an arbitrary twich differentiable ... function satisfying the appropriate regularity conditions...", where F is that arbitrary function. A possible approach consists of considering it as a second-order Taylor's series approximation (Blackorby, Primont and Russell, 1978).

Since accumulated experience indicates that the DL model gives good fit, it is reasonable to look for a flexible form which is closely related to the DL model. As the DL model can be seen as a first-order Taylor's series approximation in the log to any appropriate demand function, the logical extension would be the second order approximation in the log.

Assuming no money illusions, i.e. zero homogeneity with respect to all prices, the demand function can be written

$$(8)\quad q_{i,t} = f(P_{i,t}, Y_t, D_i)$$

where $P_{i,t}$ and Y_t correspond respectively to $(P_{i,t}/P_t)$ and (Y_t/P_t).

Following Jorgenson and Nishimizn (1978), the translog demand function will be

$$(9)\quad \sum_{i=1}^{7} \alpha_{i,j}\, \delta_{i,j} + \alpha_p \ln p_{i,t} + \alpha_y \ln y_t$$

$$+ \sum_{j=1}^{7} \beta_{p,j}\, \delta_{i,j} \ln p_{i,t} + \sum_{j=1}^{7} \beta_{y,j}\, \delta_{i,j} \ln y_t$$

$$+ \beta_{p,y} \ln p_{i,t} \ln y_t$$

$$+ \tfrac{1}{2} \beta_{pp} (\ln p_{i,t})^2 + \tfrac{1}{2} \beta_{yy} (\ln y_t)^2$$

where $\alpha_{o,o}$ and $\sum_{j=1}^{7} \beta_{j,j} \delta_{jj}^{2}$ have been omitted since δ_{jj}^{2} cannot be differentiated from δ_j , and where $\delta_{i,j}$ is as previously specified.

This form of the translog reduces to the DL, applied independently a mileage band at a time, whenever

$\beta_{p,Y} = \beta_{p,p} = \beta_{Y,Y} = 0^7$ since then

$$(10) \quad \ln q_{i,t} = \gamma_{o,i} + \gamma_{p,i} \ln p_{i,t} + \gamma_{y,i} \ln y_t$$

where

$$\gamma_{o,i} = \alpha_{o,i}$$

$$\gamma_{p,i} = \alpha_p + \beta_{p,i}$$

$$\gamma_{Y,i} = \alpha_Y + \beta_{Y,i}$$

Now the elasticities vary systematically with the price and the income:

$$(11) \quad \frac{d \ln q_{i,t}}{d \ln p_{i,t}} = (\alpha_p + \beta_{p,i}) + \beta_{p,y} \ln y_t + \beta_{p,p} \ln p_{i,t}$$

$$\frac{d \ln q_{i,t}}{d \ln y_t} = (\alpha_y + \beta_{y,i}) + \beta_{p,y} \ln p_{i,t} + \beta_{y,y} \ln y_t$$

The number of observations on the demand function has now been multiplied by seven, while the same was done to the number of parameters to which three new parameters are added. Part of our intention in developing this model was based on the desire to retain as many degrees of freedom. In this context it seems plausible to further assume that the geographic and demographic characteristics of demand only affect the level of demand without having any impact on the elasticities.

The demand function becomes

$$(12) \quad \ln q_{i,t} = \sum_{j=1}^{7} \alpha_{i,j} \delta_{i,j} + \alpha_p \ln p_{i,t} + \alpha_y \ln y_t + \beta_{p,y} \ln p_{i,t} \ln y_t + \tfrac{1}{2} \beta_{p,p} (\ln p_{i,t})^2 + \tfrac{1}{2} \beta_{y,y} (\ln y_t)^2$$

which yields as own-price elasticity

$$(13) \quad \frac{d \ln q_{i,t}}{d \ln p_{i,t}} = \alpha_p + \beta_{p,y} \ln y_t + \beta_{p,p} \ln p_{i,t}$$

and, as own-income elasticity

(14) $$\frac{d \ln q_{i,t}}{d \ln y_t} = \alpha_y + \beta_{p,y} \ln p_{i,t} + \beta_{y,y} \ln y_t$$

To the extent that the observed differences in estimated elasticities are indeed the result of differences in the price level prevailing in various groupings, in Bell Canada (1980) and Piekaar (1980), then the following hypothesis can be entertained:

$$\beta_{p,p} > 0$$

i.e. that the own-price elasticity decreases as the price level decreases. This hypothesis is interesting since it enables us to model and test an empirical observation that might be interpreted by some as a structural change, namely the fact that patterns of usage of message toll services over longer distances appear to have evolved in recent years. An alternative justification which might be offered is that younger subscribers have an inherently different demand function; such an hypothesis could only be tested if (i) one could define a younger subscriber, (ii) and one could relate chronologically a sample of "younger subscribers" and one of "non-younger subscribers".

It can be expected that, as the price of the service decreases, the quantity demanded, while it may increase, will tend toward some maximum rather than increase to infinity. This expectation can be based on the observation that, at the local level, even though additional calls are free, the number of calls for any subscriber is finite. It can conceptually be justified in terms of the opportunity cost of a call to a subscriber, the opportunity cost of a local call being measured in terms of other uses of his leisure time, in a Becker-type analysis.[8] One way to describe the service to the subscriber is to suggest that, as the price goes down, the subscriber considers the service progressively less as a luxury and progressively more as a necessity. As there are no close complements or substitutes, it would seem reasonable that, in the price effect, the income effect dominates. Then it would seem reasonable to assume that, if the own-price elasticity decreases as the price decreases, it is mostly because successively lower price levels create successive income effects which have decreasing impacts of the subscriber's own-price elasticity. In other words we may entertain the hypothesis that

$$\beta_{p,y} < 0$$

All of this leads us to suggest that on Engel's curve it is appropriate to describe the demand for message toll services, hence that

$$\beta_{y,y} > 0$$

This hypothesis is consistent with Bell (1980).

Strictly speaking, these hypotheses, together with the independence of the income and price elasticities with respect to the mileage band and the DL model can be tested since they are all nested in the general translog form. However the lack of price variability within any mileage band together with the shortness of the time series make it unlikely that the test would be meaningful.

5. Empirical Results

5.1 Data

In this section, the estimated results of the proposed translog demand model are presented. As indicated in Section I, the market segment being studied is B.C. to Alberta residence weekday daytime dialed direct traffic. Based on 1979 data, this market segment represents approximately 9.3% and 10.3% of total B.C. to Alberta call minutes and revenue respectively. The data used consist of a sample of calls and corresponding dollars that are compiled on a monthly basis. These calls are regrouped into 7 mileage bands which are listed in Column A of Table 2 with corresponding rate steps in the B.C. to Alberta tariff listed in Column B. Quarterly observations for 1973 - 1979 are used in the estimation. This period covers one rate change and is chosen because the conversion to DDD was not completed until late 1972. Prior to 1973, some operator handled calls were wrongly classified as DDD equivalent. The choice of 1973 as the start year avoids this problem.

In Tables 3 and 4, the major characteristics of this segment are presented. Table 4 contains a brief description for each mileage band of the major routes, originating and terminating points. Table 3 shows the shares of each mileage band in total revenue and call minutes for this market segment. As indicated, mileage bands F and G together represents approximately 60% of the traffic. This is to be expected since these two bands include the two largest centres in each province; Vancouver and Victoria in B.C. and Edmonton and Calgary in Alberta.

As discussed in section 4, the dependent variable is minutes of calling per residence main station. The independent variables are own price, per capita income and a vector of dummy variables to account for seasonality and discontinuous shocks such as strikes. The own price variable is measured as revenue per minute of call. This definition is employed for data reasons as the present sampling file does not sample data by rate steps so that actual price per minute of call as given in the rate table can be used.[9] The correspondence between this price and the actual price as given in the rate table is presented in Table 5. Part A shows the prices for the initial and each additional minute of call for customer dialed calls and part B shows the implicit price employed in this study. For each part, the price before and after the August 1975 rate change are presented. As can be inferred from this table, there are at least two problems associated with the use of implicit prices. First, any change in the mix of calls (for example, changes in the relative shares of steps 106, 107 and 108) will show up as a price change even when there is no rate change. And second, multi-part tariffs for some segments are also ignored. In addition, there is also the problem of Christmas and New Year days that fall on a week-day. For those days, there are discounts for DDD calls but our sampling procedure does not allow for that.[10] These are important problems that should be borne in mind when interpreting the estimation results.

Table 2

Definition of Mileage Bands

A		B	
Data Base		Tariff	
Mileage Band	Mileage	Rate Step	Mileage
A	0 - 20	102	0 - 20
B	21 - 80	103	21 - 36
		104	37 - 56
		105	57 - 80
C	81 - 180	106	81 - 110
		107	111 - 144
		108	145 - 180
D	181 - 290	109	181 - 228
		110	229 - 290
E	291 - 400	111	291 - 400
F	401 - 500	112	over 400
G	500 +		

Note:

(1) Prior to August 1975, rate step 102 was subdivided into 2 steps - 1 to 8 and 9 to 20.

Table 3

Mileage Band

B.C. to Alberta Call Minutes and Revenue by Mileage Bands

	A	B	C	D	E	F	G
MINUTES	3518	14802	56112	83172	110160	196258	155210
SHARE	0.6	2.4	9.1	13.4	17.8	31.7	25.1
REVENUE	402	4283	24431	50054	74523	142084	112034
SHARE	0.1	1.1	6.0	12.3	18.3	34.8	27.5

Notes:

(1) Total is approximately 9.3% of total B.C. to Alberta call min 10.3% of total B.C. to Alberta revenue.
(2) Based on 1979 sample.

Table 4

B.C. - Alberta Long Distance Calling Market Characteristics

Mileage Band	Rate Steps	Major Characteristics
A	102	All calls originate from B.C./Alberta border with one route (Dawson Creek to Bonanza) accounting for 45% of total traffic.
B	103	All calls originate from B.C./Alberta border, with one route (Sparwood to Blairmore) accounting for 42% of traffic. In fact, approximately 55% of total traffic originates from Sparwood.
B	104	All calls originate from B.C./Alberta border with no dominating route. However, the five routes (out of 47) with the most traffic account for over 40% of total traffic.
B	105	All calls originate from B.C./Alberta border with one route (Dawson Creek to Grande Prairie) accounting for 53% of total traffic.
C	106	All calls originate from B.C./Alberta border with five routes out of 159 accounting for over 60% of traffic.
C	107	Most calls from South-eastern B.C. with one route (Cranbrook to Calgary) accounting for over 36% of traffic and the five routes (out of 213) with the most traffic accounting for over 60% of total traffic.
C	108	Most traffic originates from B.C./Alberta border and South-eastern B.C. Out of 330 routes, five accounted for over 40% of total revenue.
D	109	Most traffic originates from South East of B.C. Out of 673 routes, five accounted for 38% of total revenue. These five routes all terminate at Calgary.
D	110	Most traffic from the Okanagan Valley with five out of 1307 routes accounting for 46% of revenue. These five routes all terminate at Calgary.
E	111	Calls appear to come from all over B.C. with no single route accounting for more than 10% of traffic. However, five out of 3413 accounted for 26% of total revenue. The five routes are from Dawson Creek, Fort St. John, Kelowna and Prince George and all terminating at Edmonton.
F & G	112	There are two mileage bands in this rate step. The major routes are Vancouver and Victoria to Calgary in mileage band F and Vancouver and Victoria to Edmonton in mileage band G. These four routes accounted for over

60% of traffic in this rate step.

Notes:

(1) Based on December 1979 data.
(2) Per cent figures are approximate.
(3) Include all types of calls.

The income variable used is per capita personal disposable income in B.C. in real terms. It has the same value for all mileage bands at any given point in time and has not been adjusted for expenditure on telephone services. The effect of the latter is insignificant as the proportion of household expenditure on telephone services is likely to be very small. The lack of a household income measure for each mileage band separately, on the other hand, is more serious. Personal disposable income in B.C. is dominated by incomes in Vancouver and Victoria which may not reflect income movements in other regions of B.C.

The other explanatory variables are all dummy variables to account for seasonality and discontinuous shocks such as strikes in the economy. There are three seasonal dummies and three strike dummies. The value of each strike dummy is determined by the ratio of number of weekdays affected by the strike to the total number of weekdays in the quarter.

Table 5

Comparison of Actual and Implicit Price: $ per minute

	A: Actual			B: Implicit Price		
Rate Steps	Before Rate Change		After Rate Change	Mileage Bands	Before Rate Change	After Rate Change
102	1st min	.17	.15	A	0.11	0.12
	add min	.08	.10			
103	1st min	.21	>.22			
	add min	.15				
104	1st min	.25	>.28	B	0.23	0.29
	add min	.21				
105	1st min	.30	>.34			
	add min	.27				
106	1st min	.35	>.40			
	add min	.33				
107	1st min	.40	>.46	C	0.38	0.44
	add min	.39				
108	each min	.45	.52			
109	each min	.50	.58	D	0.53	0.62
110	each min	.55	.64			
111	each min	.60	.70	E	0.59	0.69
112	each min	.65	.75	F	0.64	0.75
				G	0.64	0.75

Notes:

(1) Average of two rate steps. Prior to August 1975 rate step 102 was subdivided into two rate steps.
(2) Based on 1974 Q2 Data
(3) Based on 1978 Q2 Data
(4) There is a minimum charge of .20¢ per call.

All dollars values are converted to real terms by deflating by the Vancouver CPI. Precise definitions of these and other variables are given in the Appendix, Table A.2.

5.2 Translog Demand Specification

In this section, estimates from the proposed Tranlog demand function are presented. As discussed earlier, the strength of the model is to allow mileage band data to be pooled together; this should introduce wider variations in the explanatory variables and hence partially alleviate the problem of multicollinearity. The number of observations is 196 (7 mileage bands with 28 observations each). The estimated equation is (12) with the following additional variables: three seasonal dummies and three strike dummies (PS, OKTS and BCTS) i.e.

$$\ln q_{i,t} = \sum_{j=1}^{7} \alpha_{i,j}\, \delta_{i,j} + \alpha_p \ln p_{i,t} + \alpha_y \ln y_t + \beta_{p,y} \ln p_{i,t} \ln y_t + \tfrac{1}{2} \beta_{p,p} (\ln p_{i,t})^2 + \tfrac{1}{2} \beta_{y,y} (\ln y_t)^2 + a_1\, S1_t + a_2\, S2_t + a_3\, S4_t + a_4\, OKTS_t + a_5\, BCTS_t + a_6\, PS_t + u_{i,t} \quad ; i=1,2,\ldots,7 \tag{15}$$

where the dummy variables are defined as before.

This equation was estimated using OLS and the results are presented in Table 6. The use of OLS was justified for the following reasons. First, the results in terms of signs and magnitudes of the estimated coefficients are not significantly different when an alternative error structure is assumed.[11] And second, OLS is unbiased in any event. The results indicate that all price and income coefficients have the expected signs and four of the five coefficients are significant at the 5% and one at the 10% level. Because the translog model is an extension of the double-log model, an F test was also performed to test the composite hypothesis $\beta_{p,y} = \beta_{p,p} = \beta_{y,y} = 0$ to find out whether the higher order terms add any explanatory power to the model.[12] The test statistics decisively rejected the null hypothesis at the 1% level.

Table 6

Trans Log Model = OLS

	B Values	T for H:B=0	PROB |+| > 0
Constant	2.06	8.73	0.00
lnp	-1.85	-5.36	0.00
$(\ln p)^2$	-0.15	-1.69	0.09
(lnP)(lny)	0.88	1.97	0.05
lny	2.73	3.45	0.00
$(\ln y)^2$	-4.16	-2.14	0.03
BCTS	-0.04	-0.76	0.45
OKTS	-0.06	-1.43	0.15
PS	0.50	5.72	0.00
S1	-0.06	-2.73	0.01
S2	-0.09	-4.11	0.00
S3	-0.02	-0.88	0.38
δ1	-6.09	-27.28	0.00
δ2	-3.55	-23.57	0.00
δ3	-1.73	-17.99	0.00
δ4	-0.97	-21.36	0.00
δ5	-0.49	-16.49	0.00
δ6	0.21	8.42	0.00

Degree of freedom for t statistics = 178

F test for $(\beta_{p,y} = \beta_{y,y} = \beta_{p,p} = 0) = 26.00$

PROB + > 0 indicates the probability of getting a larger absolute t if B = 0.

In Table 7 the estimated price and income elasticities with their corresponding standard errors are presented for each mileage band. The elasticity estimates are calculated using equations (13) and (14) and the corresponding standard errors provided to indicate the precision of the elasticity estimates. As the elasticities vary systematically with income and price they will change through time; hence they are evaluated at three data points (the mean and the beginning and end of the study period). For rate decisions and planning purposes the most relevant elasticity estimates are those at the end of the period.[13] The results confirm the entertained hypothesis. The price elasticity estimates suggest demand for calls in this market segment is highly price elastic. Both price and income elasticity estimates appear to decline over time, from very elastic to moderately elastic or inelastic. In one case, the shortest mileage band, the income elasticity even turns negative.[14]

The general trend of declining income elasticities indicates that telephone calls in this segment are becoming a necessity as households' incomes grow. The declining trend in price elasticities on the other hand indicates consumers tend to be less sensitive to price changes at lower prices. The reason for this is the income effects which offset part of the negative price effects. In fact, examination of the size of the price and income coefficients (Table 6) suggests that income effects appear to dominate over price effects. This is indicated by the relative size of the coefficients $\beta_{p,p}$, $\beta_{p,y}$ and $\beta_{y,y}$. Take price elasticities for example. Both $\beta_{p,p}$ and $\beta_{p,y}$ have the right sizes but the size of $\beta_{p,y}$ (.88) gives income much greater impact on elasticity than that of price through $\beta_{p,p}$(-.15). Combining both the price and income elasticity estimates, the results suggest that as time goes on, if income does rise and new technology lowers cost and price, the service will become more and more of a necessity. These results are not unexpected as we indicated in Section 4.

Table 7

Price and Income Elasticities Trans Log Model: OLS

	A	B	C	D	E	F	G
Price Elasticities							
73.Q1	-1.33	-1.22	-1.65	-1.77	-1.80	-1.83	-1.83
	(0.22)	(0.17)	(0.14)	(0.20)	(0.22)	(0.22)	(0.20)
MEAN	-1.12	-1.37	-1.50	-1.60	-1.63	-1.65	-1.65
	(0.20)	(0.14)	(0.14)	(0.20)	(0.20)	(0.20)	(0.20)
79Q4	-0.94	-1.22	-1.34	-1.43	-1.46	-1.48	-1.48
	(0.20)	(0.14)	(0.14)	(0.20)	(0.20)	(0.20)	(0.20)
Income Elasticities							
73Q1	2.18	2.73	3.12	3.46	3.55	3.66	3.65
	(0.72)	(0.54)	(0.47)	(0.47)	(0.48)	(0.50)	(0.50)
MEAN	0.82	1.55	1.93	2.23	2.32	2.39	2.39
	(0.57)	(0.30)	(0.26)	(0.36)	(0.39)	(0.41)	(0.41)
79Q4	-0.33	0.48	0.83	1.10	1.20	1.26	1.25
	(0.69)	(0.54)	(0.56)	(0.61)	(0.63)	(0.66)	(0.66)

*Standard errors are given in parentheses.

6. Conclusion

In this study, several issues relating to modelling telephone demand are critically discussed, and a model more general than the double log demand model for toll was proposed. It was used to specify an alternative specification which was empirically tested. The proposed model makes it possible to cope with the empirical observation that price elasticities in absolute value increase with distance. The estimated price and income elasticities exhibit the expected patterns: own price elasticity decreases with price and increases with income, and income elasticity increase with price and decreases with income.

The results also indicate that the demand for calls is price elastic and that long distance calls are becoming a necessity as household income grows. However, because of weakness of the data - such as the lack of an appropriate income variable that reflects the household income of the major calling groups within each band - and the implicit price-aggregation problem, more work is needed to develop a better data base. Further subdivision of the traffic data by mileage bands as well as other characteristics and estimation of income series using census and income tax information would be a useful starting point.

FOOTNOTES:

[1] In many instances this follows from the use the demand models are put to. Typically they are designed as one input toward the construction of an overall production model. Since most authors have utilized their demand model to justify their hypothesis regarding profit maximization on some of the firm's outputs, namely toll services, and since the econometric analysis of production has not been able to tackle more than three outputs - three inputs production or cost functions these authors have been forced to restrict themselves to an aggregate message toll demand curve.

[2] The evidence is not as clear in the context of Bell Canada (1980) as that model is far too aggregated to draw this kind of conclusion.

[3] Whenever this is not the case, then this form of externality becomes very relevant. Two examples would be the difference in rates in a call between Canada and the USA (Europe) depending whether it originates in Canada or in the USA (Europe) and the GTE USP experiment in Illinois. In the latter, the usage charge of a call originating from a multiparty line is zero, which is not the case for most calls originating from individual lines. That situation does not apply to B.C.-Alberta calls. As long as there is a complete uniformity in rates, then the incentive to shift calls is minimized, even though differences in income, social or demographic characteristics could still be expected to play some role.

[4] It is contended here that the Canadian market is fundamentally saturated, yet it is recognized that saturation is a vague concept. Thus Bell Canada gives the number of residential main stations per person 15 and over (B-81-206), a series which still exhibits growth, even if it is at a slightly decreasing rate. It also shows the total number of business telephones per person employed (B-81-212) which also exhibits continued growth. Both observations, however, while useful from a marketing point of view, cannot be used to reject the hypothesis that saturation generally characterizes the

Canadian market. Hence a more appropriate measure, for the residence market, would be to express main stations in terms of households, the size of which has been changing with the age composition, etc. Similarly, if in the context of business services, one considers the trend toward service industry, and if one excludes new services such as data transmission,...it is likely that, once again, there is near saturation in any one industry.

[5] For a counter argument, see Curien and Vilmin (1981).

[6] This is distinct from assuming that calls should be uniformly distributed, independently of distance. The latter is not assumed in this paper.

[7] We owe this point to Jon Breslaw.

[8] This condition may change as the network is put to new usage such as data transmission, etc.

[9] There are other definitions of price such as a chained price index that can be employed. A chained price index may be obtained from repricing a given volume with new rates each time they come into effect. Unfortunately, such a price index is not available at present.

[10] Two attempts were made in the estimation to account for this problem. First, for the quarters that were affected, the average prices for the other two months were used instead, and second a dummy variable was inserted into the equation. Unfortunately, the results are not as expected with some of the coefficients having the wrong signs and are generally less significant than those reported later on in Tables 6 and A.4.

[11] Two alternatives have been considered: a variance components model and a first-order autoregressive model with contemporaneous correlation. These models are described in Drummond and Gallant (1979) who implemented the estimation procedure for SAS. Attempts to estimate the first model were unsuccessful because of insufficient cross-sectional observations. Attempts to estimate the second model produced results quite similar to that of OLS. These results are reported in Tables A.3 and A.4. Briefly, all the price and income coefficients have the expected signs but the coefficients are now less significant. There is however one significant difference; the income squared coefficient is now much smaller in magnitude.

[12] Because the translog and the double log model are not nested, an attempt was also made to test for the truth of each model with the other as the alternative hypothesis, the Davidson-MacKinnon (1981) J test for non-nesting hypothesis was applied. The double log model used is equation (10) rewritten in its "stack form" so that the number of observations is the same in each model. Unfortunately, the results reject both hypotheses.

[13] In fact for revenue forecasting following a proposed rate change, it would become necessary to forecast or extrapolate the elasticities.

[14] This is probably due to data problems such as personal disposable income for B.C. which is dominated by incomes in Vancouver and Victoria and is not representative of income of callers in mileage band A.

REFERENCES:

[1] Abrahan, Bovas and George E.P. Box, Linear Models, Time Series and Outlines 3: Stochastic Difference Equation Models, Department of Statistics Technical Report No. 430, University of Wisconsin, Madison, 1975.

[2] Bell Canada P(CRTC)23 Dec. 76-500, Tab. K.

[3] Bell Canada, Productivity Measures, Exhibit No. B242, CTC, 1969.

[4] Bell Canada, Memorandum on Bell Canada Productivity, Exhibit No. B-73-62, 1973.

[5] Bell Canada, Econometric Models of Demand for Selected Bell Canada Services, Attachment 1 in Bell (CRTC)03 Apr. 80-809, 1980.

[6] B.C. Tel, Some Comments on Jeffrey I. Bernstein's A Corporate Econometric Model of British Columbia Telephone Company, Mimeo, B.C. Telephone Company, Burnaby, 1980.

[7] Bernstein, Jeffrey I. et al, A Study of the Productive Factors and Financial Characteristics of Telephone Carriers, DGCE Working Paper 52, Department of Communications, Ottawa, 1977.

[8] Bernstein, Jeffrey I., A Corporate Econometric Model of the British Columbia Telephone Company, Public Utilities Forecasting, ed. O. Anderson, (North Holland, Amsterdam, 1980).

[9] Blackorby, C., Primont, D. and Russell, R., Duality, Separability and Functional Structure: Theory and Economic Applications, (North Holland, New York, 1978).

[10] Breslaw, Jon A., Simulations of Bell Canada Under Various Rate Scenarios, DGCE Working Paper No. 161, Department of Communications, Ottawa, 1980.

[11] Breslaw, Jon A. and Smith, J.B., Efficiency, Equity and Regulation: An Econometric Model of Bell Canada, DGCE Working Paper No. 145, Department of Communications, Ottawa, 1980.

[12] Breslaw, Jon A. and Smith, J.B., Efficiency, Equity and Regulation: A Model of Bell Canada, presented at this Conference, 1981.

[13] Breslaw, Jon A. and Smith, J.B., Efficiency, Equity and Regulation: An econometric Model of Bell Canada, Working Paper No. 81-01, Department of Economics and Institute of Applied Economic Research, Concordia University, Montreal, 1981.

[14] Christensen, L.R., Jorgenson, D. and Lau, L.J., Transcendental

Logarithmic Production Frontiers, Review of Economic and Statistics, LV-1 (1973) 28-45.

[15] Cleveland, William P., Analysis and Forecasting of Seasonal Time Series, Ph.D. Dissertation, University of Wisconsin, Madison, 1972.

[16] Cleveland, William P. and Diao, George C., Decomposition of a Seasonal Time Series: A Model for the Census X-11 Program, Journal of the American Statistical Association. Vol 71 (1976) 581-587.

[17] Corbo, V., Breslaw, J.A. and Vrljicak, J.M., A Simulation Model of Bell Canada, DGCE Working Paper No. 73, Department of Communications, Ottawa, 1978.

[18] Corbo, V., Breslaw, J.A., Dufour, J.M. and Vrljicak, J.M., A Simulation Model of Bell Canada: Phase II, DGCE Working Paper, Department of Communications, Ottawa, 1979.

[19] Courchesne, C., de Fontenay, A. and Poirier, J., An Empirical Study of Seasonality in Econometric Modelling, in Time Series Analysis. Anderson, O.D. and M.R. Perryman, eds. (North Holland, Amsterdam, 1981).

[20] Diewert, W.E., Functional Forms for Profit and Transformation Functions, Journal of Economic Theory, 6 (1971) 284-316.

[21] Dreessen, Erwin A.J., The Demand for Intra-B.C. Toll Calling - A Preliminary Report, Costs, Prices and Economics' Working Paper, B.C. Telephone Company, Burnaby, 1977.

[22] Dreessen, Erwin A.J., Elasticity is ....., Costs, Prices and Economics' Working Paper, B.C. Telephone Company, Burnaby, 1978.

[23] Dreessen, Erwin A.J., "REMARKS" to the Demand Estimation Session, this Conference, 1981.

[24] Drummond, D.J. and Gallant, A.R., TSCSREG: A SAS Procedure for the Analysis of Time-Series Cross-Section Data, SAS Technical Report S-106, Raleigh: SAS Institute Inc, 1979.

[25] Fuss, M. and Waverman, L., Multi-product Multi-input Cost Functions for a Regulated Utility: The Case of Telecommunications in Canada, Paper presented at the NBER Conference on Public Regulation, Washington, D.C. (December, 1977).

[26] Fuss, M. and Waverman, L., The Regulation of Telecommunications in Canada, forthcoming report to the Economic Council of Canada, 1981.

[27] Granger, Clive W.J. and Newbold, P., Forecasting Economic Time Series (Academic Press, New York, 1977).

[28] Jorgenson, D.W. and Nishimizn, K., U.S. and Japanese Economic Growth, 1952-74: An International Comparison, Economic Journal 88 (1978) 707-726.

[29] Kmenta, J., Elements of Econometrics, New York (MacMillan, 1971).

[30] Maddala, G.S., Econometrics, New York (McGraw-Hill, 1977).

[31] Piekaar, Ed., The Intra B.C. Demand Model: Further Developments, mimeo, B.C. Telephone Company, Burnaby, 1980.

[32] Rea, John D., and Lage, G.M., Estimates of Demand Elasticities for International Telecommunications Services, Journal of Industrial Economics XXVI, (1978) 363-381.

[33] SAS Institute, SAS User's Guide: 1979 Edition, SAS Institute (Raleigh, North Carolina, 1979).

[34] Taylor, Lester D., Telecommunications Demand: A Survey and Critique, Ballinger, 1980.

APPENDIX

Table A.1

INTRA BC Model = Long-Run Price Elasticities

Segment					Mileage Band A	B	C	D	F	G	H
#01	RES	Mon-Fri	DDD	Day	- .16	- .36	- .54	- .88	-1.60	-2.02	-2.31
#02	RES	Mon-Thu	DDD	Eve	- .11	- .78	- .88	-1.14	-1.34	-1.41	-1.33
#03	RES	Fri	DDD	Eve	- .27	- .80	- .43	- .68	-1.33	- .94	- .75
#04	RES	Sat	DDD	Day	- .49	- .18	- .28	- .42	- .74	- .83	- .70
#05	RES	Sat	DDD	Eve	- .32	- .51	- .27	- .50	-1.33	-1.02	- .67
#06	RES	Sun	DDD	D+E	- .39	- .82	- .67	- .91	-1.75	-1.81	-1.45
#07	RES	Mon-Sun	DDD	LNI	-1.08	-1.45	-2.08	-2.48	-1.71	-1.77	-1.72
#08	RES	Mon-Fri	SOH	Day	- .69	- .39	- .16	- .33	- .81	-1.49	-1.46
#09	RES	Mon-Fri	SOH	Eve	- .81	- .44	- .23	- .45	- .78	- .64	- .72
#10	BUS	Mon-Fri	DDD	Day	- .19	- .27	- .73	- .97	-1.09	-1.09	-1.93
#11	BUS	Mon-Fri	SOH	Day	- .50	- .18	- .24	- .28	+ .18	+ .39	+ .36
#12	BUS	Mon-Fri	P-P	Day	-1.49	-1.22	-1.00	-1.05	- .68	- .27	-1.08

INTRA BC Model = Long-Run Income Elasticities

Segment					Mileage Band A	B	C	D	F	G	H
#01	RES	Mon-Fri	DDD	Day	1.69	1.09	2.38	2.51	2.16	1.80	.72
#02	RES	Mon-Thu	DDD	Eve	2.62	1.30	2.00	1.46	.78	.80	.38
#03	RES	Fri	DDD	Eve	3.22	1.16	2.46	1.75	.43	.15	.68
#04	RES	Sat	DDD	Day	1.22	1.22	2.11	2.37	1.17	1.03	1.37
#05	RES	Sat	DDD	Eve	1.94	1.08	1.83	1.52	.35	- .04	.67

					A	B	C	D	F	G	H
#06	RES	Sun	DDD	D+E	3.25	1.03	2.82	2.66	.84	.17	.15
#07	RES	Mon-Sun	DDD	LNI	2.84	.87	2.15	1.15	1.22	.65	-.19
#08	RES	Mon-Fri	SOH	Day	1.72	2.48	2.74	2.95	2.60	1.53	2.75
#09	RES	Mon-Fri	SOH	Eve	1.16	1.60	1.73	1.98	1.06	1.28	2.81
#10	BUS	Mon-Fri	DDD	Day	1.82	.28	1.99	2.47	1.69	1.44	.69
#11	BUS	Mon-Fri	SOH	Day	- .22	-1.19	.35	.63	- .08	.90	.71
#12	BUS	Mon-Fri	P-P	Day	7.59	2.49	2.93	2.94	.95	.68	3.82

Source: Piekaar (1980)

Note: The mileage for the mileage bands are (0-20, 21-80, 81-180, 181-290, 291-400, 401-500 and 500+ miles) respectively.

Table A.2

Definition and Description of Variables

I Quarterly current dollar personal disposable income in B.C. Estimated as three month sums of monthly series. Monthly figures are obtained by applying monthly wages and salaries in B.C. series from CANSIM to annual income figures. Annual current dollar income figures are obtained from the B.C. Economics Accounts as the differences between total personal expenditure and current transfers to government.

CPI Quarterly Vancouver consumer price index (1971=1.00), calculated as 3 month averages of the monthly series obtained from CANSIM.

N Quarterly population of B.C. Estimated as monthly averages. Monthly figures are derived by applying monthly pattern of population over age 15 for men and women to quarterly population of B.C. All series are obtained from CANSIM and all figures are in thousands of persons.

T Numbers of residence main stations in service at B.C. Telephone. Three month averages of monthly data.

Q B.C. to Alberta Monday to Friday day-time DDD call minutes. Figures are in thousands and are obtained from toll sample data.

P Revenue per call minute (M). Revenue figures are obtained from toll sample data.

q = Q/T

p = P/CPI

Y = Y/N

OKTS* = OK Tel Strike dummy (1973Q3-1974Q1)

BCTS* = B.C. Tel Strike dummy (1977Q4-1978Q1)

PS* = Postal Strike dummy (1975Q4 and 1978Q4)

S1-S3 = Seasonal dummies.

* All strike dummies are calculated as the ratio of the number of week-days affected by the strike to the total number of week-days in the quarter.

Table A.3

Trans Log Model

	B Values	T for H:B=0	PROB + > 0
CONSTANT	2.02	7.12	0.00
lnp	-1.88	-4.55	0.00
(lnp)	-0.17	-1.71	0.09
(lnp)(lny)	0.65	1.41	0.16
lny	2.33	2.49	0.01
(lny)	-2.11	-0.73	0.46
BCTS	-0.01	-0.07	0.94
OKTS	-0.05	-0.89	0.37
PS	0.44	3.36	0.00
S1	-0.06	-1.93	0.05
S2	-0.09	-2.63	0.01
S3	-0.01	-0.20	0.84
δ1	-6.03	-26.71	0.00
δ2	-3.53	-20.63	0.00
δ3	-1.72	-15.38	0.00
δ4	-0.96	-19.45	0.00
δ5	-0.48	-20.57	0.00
δ6	0.21	22.12	0.00

Degree of freedom for t statistics = 178

PROB + > 0 indicates the probability of getting a larger absolute t if B = 0.

Table A.4

Price and Income Elasticities Trans Log Model: GLS

	A	B	C	D	E	F	G
Price Elasticities							
73.Q1	-1.23	-1.45	-1.59	-1.72	-1.76	-1.80	-1.80
MEAN	-1.04	-1.32	-1.47	-1.58	-1.62	-1.65	-1.65
79Q4	-0.87	-1.18	-1.32	-1.42	-1.46	-1.48	-1.48
Income Elasticities							
73Q1	1.58	1.99	2.28	2.53	2.60	2.68	2.67
MEAN	0.84	0.79	1.05	1.25	1.32	1.37	1.36
79Q4	-0.19	0.79	1.05	1.25	1.32	1.37	1.36

PART 2
DEMAND ANALYSIS

New Theoretical Developments

Economic Analysis of Telecommunications:
Theory and Applications
L. Courville, A. de Fontenay and R. Dobell (eds.)

TELEPHONE DEMAND AND USAGE: A GLOBAL RESIDENTIAL MODEL

N. Curien and E. Vilmin

Direction generale des Telecommunications, P.T.T. France

1. Introduction

In the framework of the theory of telephone demand ([1] to [4]), a microeconomic integrated model is proposed in section 1, taking into account both residential demand for access and demand for usage. The classical utility maximization under budget constraint is carried out and some assumptions on the utility function (as introduced in previous work [6]) allow to explicitly derive the thresehold income for access and the level of use, without using the surplus approximation. By aggregating, modelization is then developped in both directions: access demand (section 2) and traffic demand per main-station (section 3). The causality of individual decisions, as described in section 1 by the microeconomic generating model, first leads to identifying and classifying variables affecting access and traffic, as income, tariffs, supply externality, habit of use. The theory provides conjectures about the relationship between the two aspects of demand as, for instance, the negative impact on usage produced by the growth of the penetration rate. Econometric estimations are presented and discussed in details and they give good support to theoritical predictions. Price elasticities are the subject of a particular attention.

2. Microeconomic Decision Model

a) Utility function

Working within the classical framework of microeconomic consumption theory and in terms of an individual consumer i, let:

- x = telephone consumption over a given period (expressed in terms of unit pulses

- G = the real telephone penetration rate (applications not included), reflecting the level of supply

- T_i = telephone subscription seniority ($T_i = 0$ if i is a non-subscriber)

- X = the quantity of all non-telephone goods consumed over a given period, viewed as combined in a single unit.

We shall assume that the utility in x and X of consumer i can be represented by a factorable function of the type:

(1) $U_i(x,X) = Xu_i(x,G,T_i)$

in which the telephone element $u_i(x,G,T_i)$ is:

- positive, increasing, concave in x ($u_i > 0$, $\frac{\delta u_i}{\delta_x} > 0$, $\frac{\delta^2 u_i}{\delta_x 2} < 0$)

- increasing in G: $\frac{\delta u_i}{\delta} > 0$

- such that $u_i(0, ., .) = 0$, $u_i(, ., .) \neq \infty$

A utility function such as (1) reflects a number of aspects:

- the classical properties of preferences over the consumption space [x>0, X>0]: continuity, growth, convexity (see, for example, [5])

- the marginality and isolability of the telephone good "x" in relation to the total of all goods "X" (in this connection, see [6] and [7])

- the positive externality of telephone supply, represented by the penetration rate G (see [4])

- the influence of the subscription seniority T_i on the satisfaction associated with the consumption of telephone good "x"

The dependence of telephone utility u_i on T_i is not traditionally recognized, and requires some justification: it allows us, first of all, to take into account a possible shift in utility when the consumer passes from the non-subscriber state ($T_i = 0$) to the subscriber state ($T_i > 0$); secondly, it is supported by empirical data showing (see section 3 b) that the dynamics of consumption evolution vary on the basis of the value of T_i, with recent subscribers showing a higher consumption trend than longer-term subscribers.

b) Decisions on Telephone Subscription and Consumption

In order to have access to consumption x at marginal price p (value of the unit pulse rate), the consumer must pay a fixed charge A: in the case of continuing consumption ($T_i > 0$), this charge corresponds to the telephone rental rate; at the time of the initial decision to subscribe ($T_i = 0$), it includes the regular rental rate plus a rent equivalent to the installation charge.

The period of the initial decision to subscribe ($T_i = 0$) and later consumption periods ($T_i > 0$) can be described by means of the same plan of analysis, with the user in each case making two choices:

- the decision to become a subscriber ($T_i = 0$) or to remain a subscriber ($T_i > 0$)

- determination of this level of consumption if he decides to become a subscriber ($T_i = 0$) or to remain a subscriber ($T_i > 0$).

Formulation is as follows: user i, with income R_i, faced with general price level P, calculates first the utility optimization program conditional on access:

(2) $\text{Max } Xu_i(x, G, T_i)$

$$px + PX = R_i - A$$

and then the program conditional on non-access:

(3) $\text{Max } Xu_i\ (x, G, T_i)$

$$x = 0$$

$$px + PX = R_i$$

and, from the two relative optima thus obtained, chooses the one which offers him the greatest utility. The solution is readily calculated

If tariffs in relation to income $\frac{p}{R_i}, \frac{A}{R_i}$ are on the plane $(\frac{p}{R_i}, \frac{A}{R_i})$, located above the arc of parametric equations

(4)
$$\frac{P}{R_i} = -\frac{v_i'(\xi)}{v_i(0)} \qquad 0 \leq \xi \leq \infty$$

$$\frac{A}{R_i} = 1 - \frac{1}{v_i(0)}\left[v_1(\xi) - \xi\, v_i'(\xi)\right]$$

where

(5) $$v_i(\xi) = \frac{1}{u_i(\xi, G, T_i)},$$

then the consumer decides not to subscribe ($T_i = 0$) or to cease subscribing ($T_i > 0$)

If the pair $(\frac{p}{R_i}, \frac{A}{R_i})$ is located below arc (4), then the consumer becomes a subscriber ($T_i = 0$) or remains a subscriber ($T_i > 0$), and the level of consumption is the sole solution to the implicit equation in x:

(6) $$\frac{px}{R_i - A} = \frac{-xv_i'(x)}{v_i(x) - xv_i'(x)},$$

or:

(7) $\bar{x}_i\ (p, A, R_i, G, T_i)$

Equation (6), in which the total good "X" is eliminated, reflects the property of isolability of the telephone good "x" and can be interpreted in behavioural terms as the adjustment of the budget actually paid for telephone consumption to disposable net income, or $\frac{px}{R_i - A}$, to a desired relative budget, only depending on preferences.

It can readily be shown from (6) that consumption $\bar{x}_i$ is a decreasing function of $\frac{p}{R_i - A}$, that is, an increasing function of R_i and a decreasing function of p and A.

Elimination of the parameter ξ between equations (4) of the arc separating the access and non-access zones makes it possible to define a threshold income function:

(8) R_{si} (p, A, G, T_i)

This function, which is homogeneous of degree 1 in tariffs p and A, represents the minimum income required by individual i, characterized by utility (1), to decide to become a subscriber (T_i = 0) or to remain a subscriber (T_i > 0), when the tariffs are p and A, and the penetration rate is G; it can be shown that R_{si}, which is an inverse indicator of telephone affinity, increases when p and A increase, and decreases when G increases.

Note that the concept of threshold income, endogenous here to the model, as in [4], has already been employed exogenously in [8] and [9] to model the possession of durable goods.

Now, by the process of aggregation, we shall develop, first, a model reflecting the demand for access, from expression (8) of the threshold income R_{si}, and, secondly, a model reflecting usage, from expression (7) of equilibrium consumption x_i.

3. Access Demand Model

a) Theoretic formulation

In this section, we are interested solely in the access or new-subscription model, for which T_i = 0, and not the symmetrical cancellation model, in which the nonnull variable T_i can play an explanatory role (influence of seniority on the probability of cancellation). In order to simplify the expression we shall therefore omit the variable T_i = 0 and write (8), for example, in the form:

(8') R_{si} (p, A, G)

In order to proceed with aggregation of individuals i, we formulate as in [8] two "natural" distribution hypotheses:

- the distribution of incomes R_i in i is lognormal, with mean m and standard deviations s
- the distribution in i of utilities u_i leads to a lognormal distribution of threshold incomes R_0 with mean μ and standard deviation σ. μ and σ, like R_{si}, are according to (8') functions of tariffs p and A and of the

penetration rate G; in particular, $\mu(p, A, G)$ increases in p and A and decreases in G. It will also be noted that because of the independence intrinsic in microeconomic models between utility function and income, distributions R_i and R_{si} are independent.

Given this property and hypotheses i) and ii), the rate of demand, that is, the penetration rate of subscribers and applicants G^{S+A}, is written successively as follows (see [6]).

$$G^{S+A} = \text{Prob}\ (R_i > R_{si})$$

$$G^{S+A} = \text{Prob}\ (\text{Log}R_i - \text{Log}R_{si} > 0)$$

Now, since the distribution $(\text{Log}R - \text{Log}\ R_s)_i$ is normal, with mean m - μ and standard deviation $\sqrt{s^2 + {}^2}$, we can deduce that:

$$(9) \quad G^{S+A} = N_{0,1}\left(\frac{m - \mu}{\sqrt{s^2 + \sigma^2}}\right)$$

where $N_{0,1}$ designates the standardized normal distribution function.

We assume the parameters of dispersion s^2 and σ^2 to be constant; according to (9), the cumulative demand G^{S+A} varies then under the effect of the income shift (evolution of m), under the effect of tariffs p and A and under the effect of penetration rate G, all variables of the function μ(p, A, G.

Since μ is a decreasing function of G, G^{S+A} is an increasing function of G which has a saturation threshold S, attained when $G^{S+A} = G = S$ (nullity of applications), and defined by the implicit equation:

$$(10) \quad S = N_{0,1}\left(\frac{m-\mu(p,A,S)}{\sqrt{s^2 + \sigma^2}}\right)$$

The saturation rate S is a function of tariffs p and A; it represents, for a given set of tariffs, the equilibrium size of the telephone market in the sense of [2]: if the supply offered, for example, forced a penetration rate $G_0 > S$, then certain households i, having income R_i below the necessary threshold income R_{si} (p, A, G), would decide to cease subscribing, thus reducing the rate towards its equilibrium value S.

In the present model, unlike the classical logistic model, the saturation level is determined endogenously; however, as in the logistic model, we find an evolution according to a "S" curve, generated here by a normal integral rather than a logistic function.

If we designate the number of households by M, the demand D expressed over a given period can be written:

$$D = MG^{S+A}$$

or, by calculating the temporal derivative G^{S+A} from (9):

$$(11) \quad D = \frac{1}{\sqrt{2\Pi}} \exp\left[-\frac{(m-\mu(p.A.G))^2}{s^2 + \sigma^2}\right] \times \frac{1}{\sqrt{s^2+\sigma^2}}\left[\dot{m} - \frac{\delta\mu}{\delta p}\dot{p} - \frac{\delta\mu}{\delta A}\dot{A} - \frac{\delta\mu}{\delta G}\dot{G}\right]$$

The effects taken into account can be read naturally from this equation:

- the Gaussian as a general factor corresponds to the "S" curve demand trend

- this trend is modulated by a series of additive effects:

  - the increase in mean income ($\dot{m}$)

  - the evolution of usage charges ($- \frac{\delta\mu}{\delta p} \dot{p}$)

  - the evolution of access charges ($- \frac{\delta\mu}{\delta A} \dot{A}$)

  - the network externality ($- \frac{\delta\mu}{\delta G} \dot{G}$; $\frac{\delta\mu}{\delta G} < 0$)

Equation (11) describes the dynamics of the demand, but does not provide any information on the composition of this demand in terms of income distribution; it can, in fact, be shown that the telephone clientele includes progressively larger numbers of subscribers with progressively lower incomes. A visual illustration of this point is provided in the following graph, which shows:

- at the top, first, the (normal) distribution of the logarithms of the incomes in the population as a whole (n), and, secondly, this same distribution among telephone subscribers at different successive points in time [1], [2], [3] (we have omitted the income shift $\dot{m}$)

- at the bottom, again for successive points in time [1], [2], [3], the distribution of demand as a function of income G^{S+A} (R), which is by definition simply the distribution function (normal integral) of the logarithm of the threshold incomes (Prob ($R_s < R$); this curve moves towards the left with time, because of the shift in its mean μ(p, A, G), under the effect of the evolution of the tariffs and of the penetration rate; with constant tariffs, it tends towards a limit curve [ℓ] corresponding to the saturation of the penetration rate (G = S).

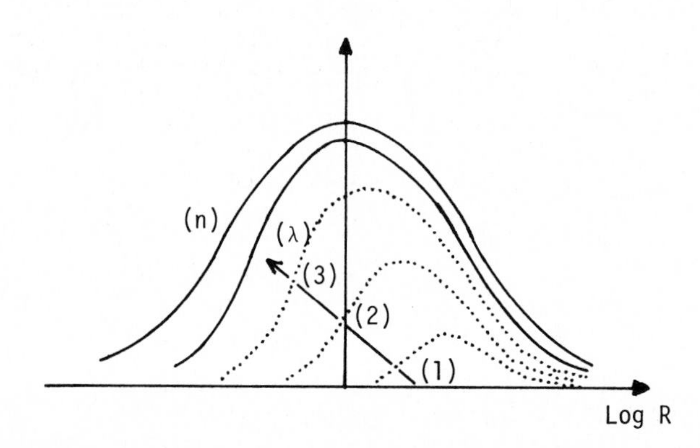

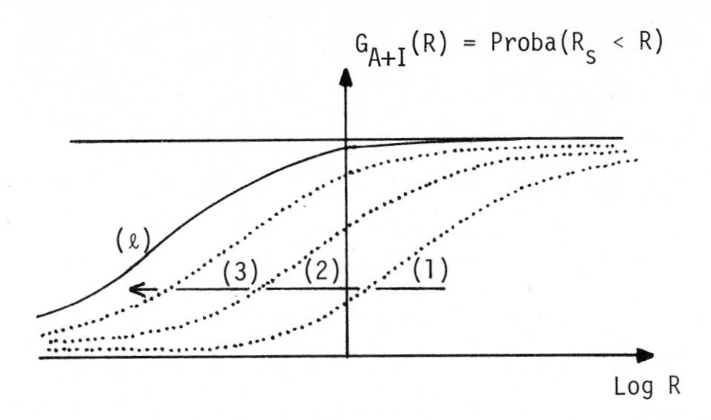

Each "upper" curve of income distribution among subscribers [1], [2], [3], is obtained roughly by "multiplying" the related "lower" curve by the normal income distribution curve within the population (n). It will be observed that the distribution of incomes among subscribers moves progressively towards the left, or towards the lowest incomes; this distribution tends towards a limit [λ] associated with [ℓ] and corresponding to the saturation of the market.

This result will be useful in producing the consumption model (see section 3) since it implies that the inclusion of new subscribers having lower average incomes and hence, according to (7), consuming less, tends to reduce the average usage per subscriber.

b) Econometric specification

Since parameters m and s have been provided by classical surveys of household income, econometric specification of the central equation (11) is based on a modeling of μ as a function of p, A, G and on an estimate of σ. Now, μ and σ are available for observation as the mean and standard deviation of the distribution of the demand in relation to income, or G^{S+A} (R) (since $G^{S+A}(R)$ = Prob (R_s < R)). The distribution G^{S+A} (R) is known for the period 1971-1978 from surveys on the "economic situation" and "household living conditions" performed by INSEE (Institut National de la Statistique et des Etudes Economiques) with the co-operation of the Direction Generale des Telecommunications.. A significant adjustment confirms the hypothesis of lognormality of the distribution of threshold incomes, provided households are divided into socio-professional categories (SPC), supports the invariability of the standard deviation σ, and provides, for μ (the mean of the logarithm of the threshold incomes) the following model, differentiated for each SPC n:

$$(12) \quad \mu_n = a_n + b_n \sqrt{G_n} \, \text{Log}\, G_n + \underset{(3.0)}{0.17} \, \text{Log TX} + \underset{(2.6)}{0.76} \, \text{Log TB}$$

where:

- G_n is the penetration rate in SPC n
- TX is the installation charge (in constant francs)
- TB is the unit pulse rate (in constant francs)
- coefficients a_n and b_n are given by the following table:

Table 1

Coefficients of the mean of the logarithm of treshold incomes according to different socio-economic categories

SPC n	Constant a_n (Student t)	Coefficient of $\sqrt{G_n}$ Log G_n b_n (Student t)
Farmers and farm workers	2.50 (22.1)	-0.10 (17.7)
Employers, industry and commerce	2.55 (24.8)	-0.09 (11.1)
Liberal professions and managers	2.82 (20.6)	-0.08 (7.0)
White-collar workers	2.52 (23.8)	-0.07 (11.9)
Employees and other members of the labour force	2.64 (23.8)	-0.09 (12.0)
Labourers and service personnel	2.63 (21.8)	-0.10 (13.4)
Retired persons	2.09 (19.6)	-0.05 (5.1)

Standard deviation of the regression: 2.8 R^2 = 0.9952
N° observations: 56

The explanatory tariff variables actually used in the econometric model, TX and TB, are related to the theoretical variables of the mathematical model, A and p, by the relations:

(13) $A = aTX + K.TB$

$p = TB$

where a is the discount rate and K the number of unit pulse rates in the rental charge; in fact, the marginal price of traffic p is equal to the base rate TB, and the fixed charge A is equal, at the time of the decision to subscribe, to the rental charge K.TB plus the rent equivalent to the installation charge aTX.

If e_t is the elasticity of the threshold income with respect to the dummy tariff variable t, the transformation formulae (13) and the results of adjustment (12) enable us to write

$$(14) \quad e_{TX} = \frac{aTX}{A} e_A = 0{,}17$$

$$e_{TB} = e_p + \frac{K.TB}{A} e_A \simeq 0{,}76$$

Since the theory indicates that the threshold income is a homogeneous function of degree 1 of tariffs p and A (as a consequence of equations (4)), this property should be reflected by the complementary to 1 of elasticities e_p and e_A; verification is acceptable since from (14) we deduce:

$$(15) \quad e_p + e_A = e_{TX} + e_{TB} = 0.93 \quad 1$$

Moreover, the condition of positivity of e_p in system (14) enables us to establish numerically (with TX = 500Fr, KTB = 600Fr/year) that the annual household actualization rate should not be lower than 30%, which reflects probable consumption behaviour in terms of durable goods.

The model which we have just presented is used for medium- and long-term forecasting (see [10]). Used for simulation during the period 1960-1978, it gives satisfactory results and provides an acceptable explanation in particular for the "explosion" in demand for new main lines between 1974 and 1976.

4. Usage Model

a) Theoretical formulation

We are interested in this section in modeling the evolution of average usage per residential line. This average use at date t, or us(t), is obtained by aggregation on the N(t) lines in service P(t) at this date of the individual consumptions $\bar{x}_i(p,A,R_i,G,T_i)$ produced by the microeconomic model (see section 1), according to the equation:

$$(16) \quad us(t) - \frac{1}{N(t)} \sum_{i \varepsilon P(t)} \bar{x}_i(p,A,R_i,G,T_i)$$

This equation can easily be rewritten, in summing i, by regrouping individuals on the basis of their date of subscription $\theta < t$; since all individuals within the same "seniority group" "θ" have the same subscription seniority $T_i = t = \theta$, this gives us.

$$(17) \quad us(t) = \frac{1}{N(t)} \int_{-\infty}^{t} \left[\sum_{i \varepsilon "\theta"} x_i(p,A,R_i,G,t-\theta) \right] d\theta$$

If, (t,θ) designates the number of lines in seniority group "θ" not cancelled at date t and $us(t,\theta)$ the average usage per line at date t within the same seniority group, we have, by definition

$$(18) \quad N(t) = \int_{-\infty}^{t} N(t,\theta)\, d\theta$$

$$(19) \quad us(t,\theta) = \frac{1}{N(t,\theta)} \sum_{i \varepsilon \theta} \bar{x}_i(p,A,R_i,G,t-\theta)$$

and equation (17) becomes:

$$(20) \quad us(t) = \frac{1}{N(t)} \int_{-\infty}^{t} N(t,\theta) us(t,\theta)\, d\theta$$

By simple differential calculus from (18) and (20), the rate of growth of usage per line, or $\frac{\dot{u}s(t)}{us(t)}$ can be expressed in additive form:

$$(21) \quad \frac{\dot{u}s(t)}{us(t)} = NA(t) + R(t) + E(t)$$

with

$$(22) \quad NA(t) = -\frac{N(t,t)}{N(t)} \left[1 - \frac{us(t,t)}{us(t)}\right]$$

$$(23) \quad R(t) = \int_{-\infty}^{t} \frac{-\frac{\delta N}{\delta t}(t,\theta)}{N(t,\theta)} \left[1 - \frac{us(t,\theta)}{us(t)}\right] \frac{N(t,\theta)}{N(t)}\, d\theta$$

$$(24) \quad E(t) = \int_{-\infty}^{t} \frac{\frac{\delta us}{\delta t}(t,\theta)}{us(t,\theta)} \frac{us(t,\theta)}{us(t)} \frac{N(t,\theta)}{N(t)}\, d\theta$$

NA(t) represents the effect of the introduction of new subscribers to the clientele: in fact, this term is equal to the rate of appearance of new subscribers at date t, $\frac{N(t,t)}{N(t)}$, weighted by the relative deviation, $\frac{us(t,t)}{us(t)}-1$. of the usage of new subscribers, us(t,t), from the average usage of long-term subscribers us(t). Now, we have seen in section 2 a) that new subscribers have lower average incomes than long-term subscribers, and as a result have lower average consumption $\bar{x}_i$, where $us(t,t) < us(t)$, $NA(t) < 0$: the "new subscribers" effect is negative. More specifically, this negative effect is explained not only by the inclusion of a larger proportion of low-income subscribers but also by the fact that, within any given income group, new subscribers consume less than long-term subscribers: in fact, given certain hypotheses as to the regularity of the telephone utility function u_i, we can show that if two individuals have the same income $R_i = R_j$ and if their threshold incomes are such that $R_{si} < R_{sj}$ - that is, if i subscribes first - then $\bar{x}_i > \bar{x}_j$, that is, i consumes more.

R(t) represents the effect of cancellations; those customers subscribing on or about date θ, and constituting at date t a proportion of the clientele $\frac{N(t,\theta)}{N(t)}\,d\theta$, contribute to this term through their cancellation rate at date t, $-\frac{\frac{\delta N}{\delta t}(t,\theta)}{N(t,\theta)}$[14], weighted by the relative deviation from the mean of their usage level $1 - \frac{us(t,\theta)}{us(t)}$; the longest-established cancelling lines, because their use is above arerage, thus have a negative effect on the rate of variation of the average use per line and vice versa.

E(t) represents the usage progression effect: in fact, each seniority group "θ" contributes to this term by the rate of evolution of its use per line
$\frac{\frac{\delta us}{\delta t}(t,\theta)}{us(t,\theta)}$, weighted by its usage "index" $\frac{us(t,\theta)}{us(t)}$. A seniority group with average usage thus affects the rate of variation of usage per line

$\frac{\dot{u}s(t)}{us(t)}$ through its own growth rate; this effect is amplified (or reduced) for an older (or newer) seniority group whose level of use us(t,θ) is higher (or lower) than the us(t).

In order to simplify the model, we shall assume that the probability of cancellation of a line i is largely independent of its seniority T_i, so that on or about a given date t all seniority groups "θ" have the same rate of cancellation μ(t) (which may however be sensitive to the tariff or economic situation at date t), that is:

$$(25)\qquad \frac{-\frac{\delta N}{\delta t}(t,\theta)}{N(t,\theta)} = \mu(t),\ \forall\theta$$

Introducing (25) in (23) and using (20), we therefore deduce

$$(26)\qquad R(t) = 0$$

Cancellations thus have, according to this hypothesis, a resultant null effect on average usage growth, since cancellations within recent low-consumption seniority groups exactly balance those within longer-established higher-consumption seniority groups.

Using this same hypothesis, we can also write, working from (18):

$$(27)\qquad N(t,t) = N(t) + \mu(t)\,N(t),$$

whence the "new subscribers" effect:

$$(28)\qquad NA(t) = -\left[\frac{N(t,t)}{N(t)} + \mu(t)\right]\left[1 - i(t)\right]$$

where:

$$(29)\qquad i(t) = \frac{us(t,t)}{us(t)}$$

the new subscriber usage index.

Except for cancellations, $-(1 - i(t))$ thus appears in (28) as the elasticity of usage $us(t)$ with respect to the size of the clientele $N(t)$. In addition, we can deduce theoretically from the demand model that the index $i(t)$ (or elasticity $|1-i(t)|$) is decreasing (or increasing) and tends towards a limit. In fact, since the income distribution among subscribers shifts towards the lower incomes when the penetration rate increases (see section 2 a), the average income, and hence the average usage of the most recent subscribers, continues to decline relative to these same averages calculated for all subscribers, with the gap widening to a positive limit associated with the saturation S of the clientele. We shall see in section 3 b) that this property of stabilisation of the index $i(t)$ can be readily verified by empirical means.

Finally, the expression of the growth term $E(t)$ can be simplified by breaking down, as in (19), the usage trend for each seniority group "θ",

that is, $\frac{\delta us}{\delta t}(t,\theta)/us(t,\theta)$, into two components, one, $g(p,A,G)$, related to

the tariff variables p, A and the penetration rate variable G, and the other, $f(t-\theta)$, related to the seniority variable $T_i = t-\theta$; whence the expression:

$$(30) \quad \frac{\frac{\delta us}{\delta t}(t,\theta)}{us(t,\theta)} = g(p,A,G) + f(t-\theta)$$

then, by replacing (30) in (24) and using (20), we have:

$$(31) \quad E(t) = g(p,A,G) + \int_{-\infty}^{t} f(t,\theta)\,\frac{us(t,\theta)}{us(t)}\,\frac{N(t,\theta)}{N(t)}\,d\theta$$

It will in fact be demonstrated in section 3 b) that the most recent seniority groups ($t - \theta < 6$ years) have, in addition to the basic trend $g(p,A,G)$, a positive additional trend coefficient $f(t-\theta)$, which partially offsets the lower consumption of new subscribers in comparison to longer-term subscribers (see [11]). The coefficient $f(t-\theta)$, considered in relation to each subscriber, indicates a rise in the consumption trend over a period of approximately 6 years from the date of subscription; this phenomenon of acceleration in use relates back, in the microeconomic model of section 1, to the evolution of the utility function u_i under the influence of the seniority variable T_i; this evolution should therefore be interpreted, not as a process of learning to use the "telephone tool", which would be a short-term effect, but rather as a progressive extension of the role assigned to the telephone in the performance of the communication tasks of the subscriber household.

According to (21), (26), (28) and (31), the evolution of average consumption per line is finally governed by the following differential equation:

(32) $$\frac{\dot{us}(t)}{us(t)} = -\left[\frac{\dot{N}(t)}{N(t)} + \mu(t)\right]\left[1-i(t)\right] + g(p,A,G) + \int_{\infty}^{t} f(t-\theta)\frac{us(t,\theta)}{us(t)}\frac{N(t,\theta)}{N(t)}d\theta$$

where the terms $us(t,\theta)$, $N(t-\theta)$, appearing in the integral term, are themselves generated by dynamic equations (25) and (30).

The central equation (32) clearly summarizes the three additive effects contributing to the evolution of consumption per residential line:

- effect of clientele demography, or new-subscribers effect NA(t)
- basic usage trend common to all subscribers, g(p,A,G), varying with tariffs and penetration rate
- differentiation of usage trends by seniority as a result of acceleration in usage (integral term in $f(t-\theta)$).

Econometrics will show in fact (see section 3 b) that the dynamics in usage results essentially from the first two effects (the first acting negatively, the second positively), while the third effect, which is weaker, acts as a corrective factor.

b) Econometric Specification

Econometric specification of equation (32) is thus based on the adjustments for the cancellation rate $\mu(t)$, the usage index of new subscribers i(t), the modeling of the trend function g(p,A,G) and the seniority coefficient $f(t-\theta)$. All these adjustments are done for each region; the methods used and the results obtained are as follows:

- Cancellation Rate $\mu(t)$

By averaging the relatively stable values observed in the past, we have obtained a constant value by region as shown in the table of the country which varies between 1.1 and 3.0.

- New Subscribers Usage Index i(t)

The pattern followed by this index at the national level was first reconstructed by calculating the fictitious evolution which line usage would have followed if the various "seniority groups" of longer-term subscribers had maintained a constant level usage (that of 1976). The movement of the curve thus obtained indicates how usage per line varies as a result of changes in clientele alone, with the trend effect eliminated; its slope represents exactly the elasticity usage line elasticity with respect to the growth of the clientele. We see that this elasticity has varied strongly in the past: for instance, it is equivalent to an average of -0.22 for the decade 1961-71 and an average of -0.35 for the period 1971-75.

The new subscribers usage index which, according to the basic equation of usage growth (32), is equal to the elasticity with respect to total clientele plus 1, has thus diminished progressively in the past to its present value, near 0.7.

However, despite this decline in the past, we can assume that the index is constant for purposes of forecasting: observation, in accordance with theoretical prediction (see section 3 a), shows in fact that within each SPC (Socio-Professional Category) the index stabilizes clearly towards an asymptotic value. We can therefore simulate its future evolution for all households by weighting the indices for each SPC by the variable proportion which this SPC represents in the subscriber shift and by its level of consumption in relation to the average household level. It will then be observed that the global index differs only slightly from the 0.7 value, which we shall therefore maintain as a constant in the model.

Table 2

Evolution and Projection of New Subscribers Usage Index by SPC

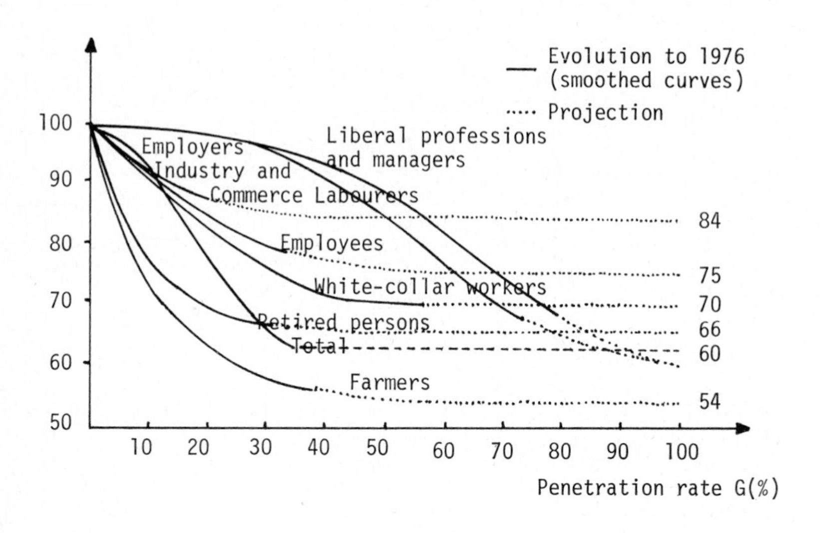

- Usage Trend Function

The method of adjustment consists of isolating within each region n a constant group of subscribers selected from an active sample of telephone lines - the PANEL (see [12]) - and of observing the evolution of consumption of this sub-sample of fixed size; in practice, we have followed the bimonthly use per line $us_n(t)$ of subscribers included in the clientele prior to 1974 and have developed a relation of the type:

(33) $\text{Log } us_n(t) = a_n t + b \text{ Log TB} + c \text{ Log IPI} + cst$

where t is time, TB the unit pulse rate, IPI the industrial production index.

The empirical model (33) shows a number of differences from the theoretical mathematical model in which the trend usage is generated by variables p, A and G, through function g(p,a,G):

- the penetration rate G is replaced by time t, which is strongly correlated with G
- the unit pulse rate TB includes both the charge for calls p = TB and the rental charge A, which corresponds to the fact that p and A varied in parallel with and proportionally to TB over the estimation period (A = K.TB)
- the IPI index has been introduced to take into account the effects of economic conditions.

The basic trend a_n is approximately 10%, with slight regional differences. The consumption trend varies between 7.5 and 10.6 according to the regions.

These values would be relatively instable if,a posteriori, the congruence of the prediction model with reality over the period 1968-77 did not argue for their relative stability in time. However, in making predictions for 1985, it appears reasonable to postulate a slight easing of the trends, corresponding to an alignment with the trends observed in other European countries. A study now underway is designed to take into account such an evolution in trends by introducing explicitly into the model (33) the penetration rate G (thus bringing the econometric model closer to the mathematical model); in fact, the saturation of G, by creating at the same time a saturation of the associated externality, should eventually bring about a slowdown in the growth of consumption. Such an approach is compatible with current observations, the regular growth noted in usage being explained by the continuing regular growth of the clientele itself in the vicinity of the logistic inflection.

The elasticity b with respect to the unit pulse rate has not been found to be significantly different from zero, but this result is unreliable because of the extreme variability of the usage series of the low number of tariff changes occurring during the estimation period. In addition, the new modeling approach being used (explicit introduction of the penetration rate G) appears to give a significant and positive value for b. If this is the case, the microeconomic model then makes it possible to predict that this value b corresponds essentially to an elasticity with respect to the tariff price p and not to the fixed charge for rental A (which is currently econometrically undecidable, both tariffs having evolved proportionally over the adjustment period). In fact, since the individual consumptions $\bar{x}_i$ are entirely a function of $(R_i-A)/p$ (see section 1 b), it is easy to show that their elasticities with respect to p are much stronger than their elasticities with respect to A, providing the rental charge A is negligible in relation to income R_i. Until we have empirical confirmation, this theoretical result will be useful for purposes of forecasting since changes in the unit pulse rate and the rental charge have been separated since June 1979.

The data produced by the PANEL relate to a period which is too short to provide direct evidence of elasticity with respect to the IPI. This elasticity has therefore been estimated indirectly from total consumption statistics; it is in the vicinity of 0.1.

- Coefficients of Seniority $f(t-\theta)$

An attempt has been made to adjust the coefficients of seniority or coefficients of acceleration in usage $f(t-\theta)$, components which add to the basic usage trend for recent subscriber groups. These coefficients, which are low in absolute value, are difficult to estimate directly with any accuracy from the observed trend for each seniority group since this trend is composed essentially of the basic component ($a \simeq 10\%$), the tarif component (b Log TB) and the economic situation component (c Log IPI). This is why, in order to eliminate these "strong" components, we have attempted instead to obtain $f(t-\theta)$ as the difference between the usage trends of two distinct subscriber populations: to do this, we have looked, first, at those subscribers who joined the clientele before 1971 and, secondly, at those who entered the clientele at a given later date, t_0, between 1971 and 1974.

Let $i(t,t_0)$ be the index which, at each date t subsequent to 1974, relate the usage level of the second population to that of the first. The rate of evolution of this index in terms of the current date t is a measurement of the rate at which those subscribing at t_0 are catching up with those present within the clientele before 1971; this catch-up rate is exactly identical to the coefficient of accelerating use $f(t-t_0)$, provided that the duration of the period of accelerating use does not exceed 4 years. (If this were not in fact the case, the 1971 subscribers would still be showing accelerating use during 1974 and the catch-up rate would be equal to the acceleration in use of the t_0 subscribers, or $f(t-t_0)$ less a complex corrective term expressing the acceleration use amond subscribers from the period prior to 1971).

Figure 2 below shows several values of t_0, the evolution of the index $i(t,t_0)$ as a function of t.

Figure 2

Evolution of the Usage Index $i(t,t_0)$ of Households by First Year of Subscription t_0 (100 = subscribers before 1971)

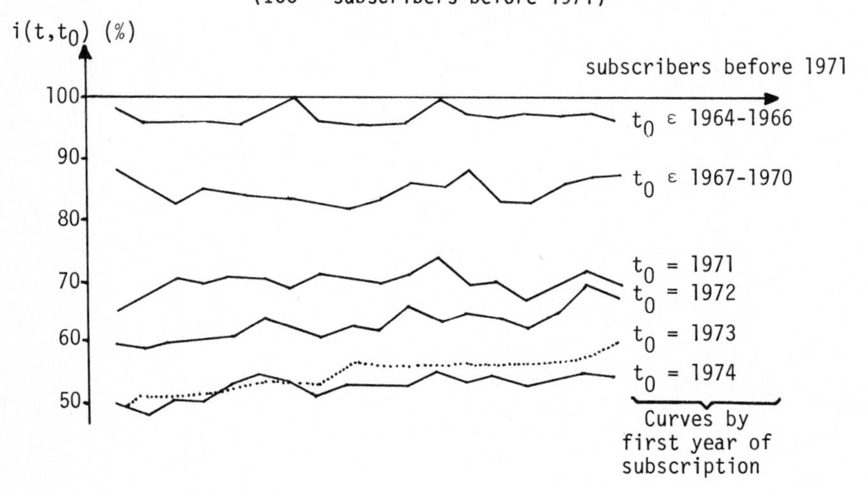

It will be noted that:

- first, there is a certain long-term narrowing of the gap between those subscribing at a given date (1971) and the longer-established subscribers who were within the clientele prior to this same date; in fact, those subscribing in 1971 with an initial use index of 0.6 reach an index of 0.7 by 1977

- secondly, for t_0 subsequent to 1971, the index $i(t,t_0)$ increases most rapidly as a function of t for the most recent subscribers (t_0 = 1973 or 1974).

It appears then that there is a slow acceleration in use for households having subscribed over a total period of 5 to 7 years, with a maximum between 2 and 3 years.

Since the hypothesis on which the estimate is based, that the duration of the period of increased usage should not exceed 4 years, has not been rigorously verified, a simple calculation of the over-estimation involved shows that the relative error committed as a result in evaluating the coefficients is in any case less than 10%.

The residential usage model which we have just presented, applied in conjuction with a professionel model (see [11]), reflects past growth by backcasting correctly back to 1966, beyond its estimation period (1974-79) as in figure 3.

Figure 3

Actual vs. Predicted Evolution of Usage Rate

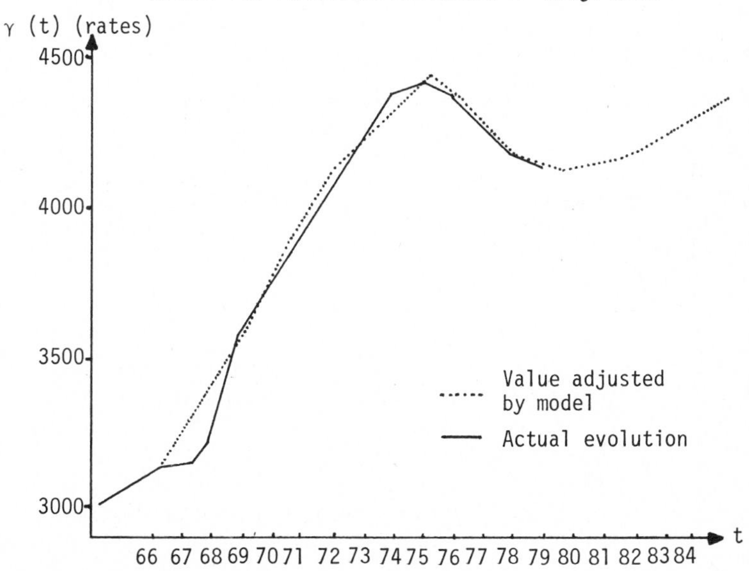

It predicts, for the period 1980-85, a resumption in the growth of use per line at an average rate of 1.2%, due essentially to two factors.

One is the maintenance of the sustained effect of the trends, even assuming a slight weakening of these trends corresponding to an alignment to the average trends observed in other European countries.

The second is a stabilization of the growth of the relative proportion of households in the clientele shift near saturation; in fact, the strong growth of this relative proportion during the period 1975-80 produced a very important negative effect on average consumption, since household usage per line is much lower than business usage.

REFERENCES:

[1] Artle, R. and Averous, C., The Telephone system as a public good: static and dynamic aspects, The Bell Journal of Economics and Management Science, 4 (1973).

[2] Rohlfs, J., A theory of interdependent demand for a communications service, The Bell Journal of Economics and Management Science, 5 (1974).

[3] Squire, L., Some aspects of optimal pricing for telecommunications, The Bell Journal of Economics and Management Science, 4 (1973).

[4] Taylor, L., Telecommunications demand: a survey and critique (Ballinger Company, Cambridge, Massachusetts, 1980).

[5] Gamot, G., Theorie microeconomique (Microeconomic theory), Course offered at the Ecole Nationale Superieure des Telecommunications.

[6] Curien, N., La consommation telephonique: une approche microeconomique (Telephone consumption: a microeconomic approach), 3rd International Conference on Analysis, Prediction and Planning in the Public Services (June 1980).

[7] Curien, N., Le telephone et les biens a acces: modeles de demande et de consommation (The telephone and access goods: demand and consumption models), Note, Direction Generale des Telecommunications (September 1980).

[8] Rault, C., Etude econometrique de la possession d'un ensemble de biens durables de consommation (Econometric study of the possession of a set of durable consumer goods), Annales de l'INSEE, 1.

[9] Von Rabeneau and Stahl, Dynamic aspect of public goods: a further analysis of the telephone system, The Bell Journal of Economics and Management Science, 5 (Autumn 1974).

[10] Gensollen, M. and Vilmin, E., Prevision de la demande telephonique (Prediction of telephone demand), 3rd International Conference on Analysis, Prediction and Planning in the Public Services (June 1980).

[11] Curien, N., and Dang Nguyen, G., Prevision du produit du trafic telephonique (Prediction of the product of telephone traffic), 3rd

International Conference on Analysis, Prediction and Planning in the Public Services (June 1980).

[12] Berthon, D., Chabrol, J.L., Leroy, F. and Werkoff, M., Le Panel des abonnes au telephone (The panel of telephone subscribers), 3rd International Conference on Analysis, Prediction and Planning in the Public Services (June 1980).

PART 2
DEMAND ANALYSIS

Local Measured Service

Economic Analysis of Telecommunications:
Theory and Applications
L. Courville, A. de Fontenay and R. Dobell (eds.)

THE ESTIMATION OF USAGE REPRESSION UNDER LOCAL MEASURED SERVICE EMPIRICAL EVIDENCE FROM THE GTE EXPERIMENT

G.F. Wilkinson

G.T.E.

1. Introduction

One of the major changes forthcoming in the pricing of local telephone service in the United States is the conversion to some form of usage sensitive pricing. Presently, most consumers within the U.S. pay a single flat rate charge each month for which they can make an unlimited number of calls within a specified local calling area. Currently, the Bell System and the major independent telephone companies are pursuing plans to convert, at least portions of their serving territory, to measured service.[1]

In order to learn about the impact of converting to measured service both upon the consumer and company operations, the GTE system initiated a measured service experiment in three exchanges in central Illinois in 1975. Data have been gathered since May, 1975 on consumer telephone usage, and these data have been used as the basis for a number of papers on measured service. In particular, Jensik (1979) and Park (1981) had similar research objectives and each used time-series data from the same experimental exchanges.[2] This paper uses models of the type used by Jensik to elaborate further on the changes in consumer usage under measured service.

2. The Measured Service Experiment and the Tariffs

The GTE measured service experiment is being conducted in the Jacksonville, Clinton, and Tuscola exchanges in Illinois.[3] These exchange were chosen as the study area since they are reasonably representative of a large portion of GTE serving territories. Since May, 1975, data have been gathered on calls and minutes of usage by class of service (residence one-party, residence multi-party, business one-party, key trunks, and PBX trunks). From May, 1975 until August, 1977, flat rate tariffs were in effect in these exchanges.

In September, 1977, residence one-party, business one-party, and key and PBX trunks were converted to measured tariffs. In the Jacksonville Exchange, the initial rates were structured to include a fixed charge each month (approximately 40% of the former flat rate) and a charge for each call and minute of usage. In the Clinton and Tuscola exchanges the initial measured service tariff included only a charge per minute and a fixed monthly charge. These usage tariffs were designed to charge the same amount for a four minute call in all of the experimental exchanges (which is approximately the average length of a residence call). In 1979, usage rates were increased and the tariff structures in the three exchanges were made the same. Table 1 illustrates the changes in the residence one-party rates throughout the experiment. Business one-party, key, and PBX usage rates have been and continue to be the same as the residence usage rates although the fixed monthly charges for these classes of service are higher.

Since measured service was implemented, there has been a 20% discount on all usage between 5 and 11 p.m. daily and all day Sunday, and a 50% discount on all usage between 11 p.m. and 8 a.m. No distance sensitivity was included in the tariffs.

Table 1

Tariffs for the Measured Service Experimental Exchanges

Residence One-Party

	Flat Rate (5/75 - 9/77)	Measured Rate (9/77 - 6/79)		Measured Rate (6/79 - Present)	
		Monthly Charge	Usage Charge	Monthly Charge	Usage Charge
J'Ville	$7.95	$3.15	2¢/call 1¢/min.	$3.15	2.5¢/call 1¢/min.
Clinton	6.20	2.50	1.5¢/min.	2.50	2.5¢/call 1¢/min.
Tuscola	5.90	2.50	1.5¢/min.	2.50	2.5¢/call 1¢/min.

Time-Of-Day Pricing Periods

	Monday through Saturday	Sunday
8AM - 5PM	Full Rate Period	20% Discount
5PM - 11PM	20% Discount	20% Discount
11PM - 8AM	50% Discount	50% Discount

3. Study Methodology

The intent of this study was to estimate changes in usage characteristics as a result of measured service tariffs for the three experimental exchanges. The primary focus was on residence one-party service although models for business one-party service were also developed and used for comparisons. For research of this type, the ideal circumstance would have been to compare the telephone usage of control groups having very similar demographic and economic characteristics which did not experience the changes in local service tariffs with the usage of the exchanges which had been converted to measured service. However, because no equivalent control groups were established for the experimental exchanges (primarily due to the high cost of data gathering and processing), it was necessary to infer the effects of the tariff changes by estimating deviations from the historical trend which had been established prior to the change in rates. The particular data series which were analyzed in this manner were the monthly calls and minutes per mainstation.

The time-series analysis techniques developed by Box and Jenkins[4] were chosen to identify the changes in usage characteristics. These Box-Jenkins models are particularly well suited for this type of analysis because of the seasonal characteristics of the data. For these time series, transfer function models using intervention terms were used to estimate the changes. Ideally, transfer functions using economic variables to explain the changes in the behavior of the series would provide better indicators for understanding the series. However, economic time series data which adequately describe the dynamics of consumer telephone usage in these exchanges have not been found and thus it was necessary to rely on intervention analysis.

4. Model Structure

The models which have been developed for usage per mainstation employ very simple intervention structures. In building intervention models, in contrast to the identification of the transfer function structure based upon cross correlation functions, it is necessary to hypothesize the nature of the change attributable to a tariff (such as a permanent change in number of calls per main per month due to a price change, i.e. a step function) and then test whether the actual behaviour of the time series supports this hypothesis. The models used in this analysis can be viewed as having an intervention component which describes step changes in the series due to tariff changes or pulses which account for unusual events, a noise component which describes all other systematic behaviour of the series, and a random error.

In order to briefly explain the model structures, let us first examine the noise component of the models. This noise component can be viewed as being the univariate model of the time series having already accounted for the effects of the interventions. Like all univariate models, the noise component describes the behaviour of the series based upon its past history using autoregressive or moving average operators (or both). Prior to identifying the types of operators which are needed and estimating their values, it may be necessary to difference and transform the series to make it stationary.

The noise structures for most of the models used in this study are of the form:

$$\nabla^1\nabla^{12} \ln Y_t=(1-B)(1-B^{12})a_t$$

In this model, the usage per mainstation series (Y_t) was transformed by natural logarithms (ln) and has been differenced using first order and seasonal operators ($\nabla^1\nabla^{12}$). This transformed and differenced series which is stationary is described by a first order and seasonal moving average process $(1-B)$ $(1-B^{12})$. The forecast errors are the a_t's (Two operators are used in Box-Jenkins nomenclature for writing these models in a compact form: they are the backshift operator B for shifting backward in time (i.e. $(B)Z_t = Z_{t-1}$) and a differencing operator which indicates a new series has been created by differencing the original series according to the degree of the operator (i.e. $\nabla^1 = Z_t - Z_{t-1}$). See Box and Jenkins (1970) for a full explanation of these operators and the structure of the models).

The intervention component of the models are terms which describe changes in the level of the series (steps) or unusual events (pulses). Like dummy variables in standard regression equations, these interventions are 1's when the intervention is active and 0 everywhere else. The form of the intervention structure for the models used in this study are simply Y_t = $_t$, where is an estimated parameter which describes the transfer function between the intervention term ($_t$) and the usage series (Y_t).

5. Empirical Results

Initially univariate models were developed for the residence calls per mainstation and minutes per mainstation series for the aggregate of the three exchanges. (See Figure 1 and 2.) The models were fitted with 65 monthly observations from May, 1975 to September, 1980. These univariate models can be very helpful in understanding the behaviour of the series and developing hypotheses for the interventions.

The largest residuals from these univariate models were associated with the change to measured service or unusual weather conditions which were known to affect calling patterns. Although no large residuals occurred following rate change in June, 1979, an intervention term for the rate case was tested for significance. Using the residuals from these univariate models as guides, transfer function models were estimated using the same time series data. The interventions for rate changes were modelled as step functions and the interventions for the months with excessively bad weather were modelled as pulses in a single intervention variable. In the formulas, the intervention terms are designated as ξ_1 for a step function beginning in September, 1977 when measured service was introduced, ξ_2 for pulses at January, 1977 and March, 1978 for unusually severe winter weather, and ξ_3 as a step function beginning in June, 1979 for the increases in the measured rates.

The transfer function model for calls per residence mainstation was the following:

$$\nabla^1\nabla^{12}\ln Y_t = -.16\xi_{1t} + .17\xi_{2t} - .08\xi_{3t} + (1-.83B)\,(1-.75B^{12})a_t$$

$$\pm.03 \qquad \pm.03 \qquad \pm.03 \qquad \pm.08 \qquad \pm.07$$

This model indicates that there has been a decline of about 15% (calculated as the complement of e to the exponent of the coefficient) in calls per mainstation following the introduction of measured service. The severe weather during January, 1977 and March, 1978 accounted for an average increase of 19% during these months. The decline of about 8% on the third intervention is difficult to intrepret. It is tempting to attribute this decline in calling entirely to the June, 1979 rate increase; however, I think this would be incorrect. While the previous interventions do not appear to have other variables significantly affecting the estimate, an alternate hypothesis that the decline in calling is due to the recent recession could be formulated for this third intervention term. In order to test this intervention, the same model was estimated with only 55 data points, and the ξ_3 term was insignificant (-.05 ±.03). Further evidence that the recession was a contributing factor in this intervention coefficient were large negative residuals in the Spring and Summer months of 1980, over 9 months after the new rates were in effect (in the original 65 observation model). Thus it is probably more reasonable to attribute this 8% decline in calling to the combination of rate increase and recession effects. Since this coefficient is within two and three standard errors, it would be premature to draw strong conclusions about its effect upon the time-series.

The model for business one-party calls per mainstation indicated that businesses reacted differently than residences in their response to measured services. (See Figure 3 for a graph of this series). The model for this series was the following:

$$\nabla^1\nabla^{12}\ln Y_t = \begin{array}{ccccc} -.04\xi_{1t} & +.08\xi_{2t} & -.11\xi_{3t} & +\ (1-.88B) & (1-.70B^{12})a_t \\ \pm.03 & \pm.05 & \pm.03 & \pm.06 & \pm.08 \end{array}$$

Thus, there was not a significant reduction in business calls per mainstation, due to measured service or an increase in calling due to the unusual weather for January, 1978 and March, 1978. However, the coefficient on the ξ_3 intervention is significant. This is probably due to a combination of rate effects and the recession.

Looking at the minutes per main series, the residence model exhibits the same structure as the calls per main model. The model was the following:

$$\nabla^1\nabla^{12}\ln Y_t = \begin{array}{ccccc} -.26\xi_{1t} & +.14\xi_{2t} & -.05\xi_{3t} & +\ (1-.78\xi) & (1-.79B^{12})a_t \\ \pm.04 & \pm.04 & \pm.04 & \pm.08 & \pm.06 \end{array}$$

The model indicates that residence minutes per main declined by about 23% due to the implementation of measured service. The weather increased minutes per mainstation by about 15%, and the third intervention (again related to rate increase recession) indicated a decline in minutes per main of about 5% (this coefficient is insignificantly different from zero at two standard errors).

The business minute per main per month model (see Figure 4 for a graph of this series) was the following:

$$\nabla^1\nabla^{12}\ln Y_t = \begin{array}{ccccc} -.05\xi_{1t} & +.06\xi_{2t} & -.05\xi_{3t} & +\ (1-.88B) & (1-.72B^{12})a_t \\ \pm.03 & \pm.05 & \pm.03 & \pm.06 & \pm.07 \end{array}$$

This model is very similar to the business calls per main model with the exception of the coefficient on the ξ_3 intervention term which, unlike the calls model, is insignificant. This indicates that business calls per mainstation have been reduced to a somewhat greater degree than minutes per main as a consequence of the rate case/recession influence.

The previous models were developed using aggregate data for the three experimental exchanges. In order to examine whether residence usage was similar among the experimental exchanges separate residence minutes per mainstation models were developed.

Jacksonville: $\nabla^1\nabla^{12}\ln Y_t = -.24\xi_{1t} + .15\xi_{2t} - .04\xi_{3t} + (1-.80B)\ (1-.78B^{12})a_t$

$\pm.03 \quad \pm.05 \quad \pm.03 \quad \pm.08 \quad \pm.07$

Clinton: $\nabla^1\nabla^{12}\ln Y_t = -.25\xi_{1t} + .15\xi_{2t} - .06\xi_{3t} + (1-.86B)\ (1-.81B^{12})a_t$

$\pm.04 \quad \pm.07 \quad \pm.04 \quad \pm.06 \quad \pm.06$

Tuscola: $\nabla^1\nabla^{12}\ln Y_t = -.26\xi_{1t} + .21\xi_{2t} - .07\xi_{3t} + (1-.80B)\ (1-81B^{12})a_t$

$\pm.05 \quad \pm.07 \quad \pm.04 \quad \pm.09 \quad \pm.06$

It is evident from these models that the reaction to measured service, winter storms, and the rate case/recession effects was similar among the exchanges. All coefficients for any exchange are within one standard error of the corresponding coefficients for the others.

The next models identified customer reaction to changes in the tariff structure by discount period. Recall from the tariff table that a 20% discount was applied to usage charge between 5 and 11 p.m. each day and all day Sunday, and a 50% discount was applied to usage between 11 p.m. and 8 a.m. These discount periods are designated as P1 for no discount, P2 for a 20% discount, and P3 for a 50% discount. The following models were developed for residence minutes per mainstation within each discount period:

P1

$$\nabla^1\nabla^{12}\ln Y_t = -.28\xi_{1t} + .15\xi_{2t} - .06\xi_{3t} + (1-.62B)\ (1-68B^{12})a_t$$

$\pm.06 \quad \pm.05 \quad \pm.07 \quad \pm.12 \quad \pm.13$

P2

$$\nabla^1\nabla^{12}\ln Y_t = -.27\xi_{1t} + .14\xi_{2t} - .03\xi_{3t} + (1-.73B)\ (1-.66B^{12})a_t$$

$\pm.04 \quad \pm.04 \quad \pm.06 \quad \pm.12 \quad \pm.13$

P3

$$\nabla^{12}\ln Y_t = -.14\xi_{1t} + .23\xi_{2t} - .07\xi_{3t} +$$

$\pm.05 \quad \pm.06 \quad \pm.08 \quad \pm.11$

The models for the P1 and P2 pricing periods show that the decline in minutes per mainstation (about 24%) due to the introduction of measured service was virtually the same, in spite of the 20% discount on usage in P2. The noise model for P3 usage was structured a little differently than the previous noise models using an autoregressive operator and only seasonal differencing. It indicates that a 13% decline in minutes per mainstation occurred due to the introduction of measured service. Models for business minutes per mainstation were also estimated by pricing period. Although the coefficients were very similar between the pricing period models, such a large percentage of the traffic was concentrated in the P1 period, that the P1 model was the only model which was meaningful. The business P1 model had the same coefficients as the overall model.

The final phase of this research identified a model which was somewhat better than the step function for explaining the effects of the introduction of measured service. Although it might be expected that the effects would gradually increase over a few months, the contrary occurred. A model which had a lower residual variance compared with the step function had a structure for the intervention which hypothesized an over-reaction to measured service. In this formulation, the model indicates that there was approximately a 29% reduction in minutes per mainstation during the first month of conversion, but from the second month onward, there has been only a 21% reduction in usage.

$$\nabla^1\nabla^{12}\ln Y_t = \underset{\pm.05\ \pm.05}{(-.36\ +.13)}\xi_{1t} \underset{\pm.04}{+.14}\xi_{2t} \underset{\pm.03}{-.05}\xi_{3t} + (1-\underset{\pm.07}{.82}B)(1-\underset{\pm.06}{.76}B^{12})a_t$$

Unfortunately, this particular model had a very high correlation between the ω_0 and ω_1 terms on the intervention for measured service. This high correlation indicates that there are probably a number of combinations of parameter estimates very close to these maximum likelihood estimates. Therefore, these results are suspect.

6. Conclusion

For these particular exchanges and tariffs, the introduction of measured service resulted in a decline in calls per residence mainstation of about 15% and for minutes per residence main, about 23%. Usage reductions (both calls and minutes per main) due to measured service for business one-party were not significant. (Although the models have not been presented here, Key and PBX usage also did not decline significantly due to the introduction of measured service). For residence, the evening discount did not significantly influence the overall impact on usage of the conversion to measured service. In fact, the coefficient indicated that approximately the same percentage repression occurred during P2 as during the daytime period. It appears that usage has declined further due to a combination of the June, 1979 rate and the current economic recession. It was not possible to isolate each effect independently in this analysis, although further work is being done in this area. Another interesting result is that severe weather significantly affects residence usage, in this case, increasing calls per mainstation by about 15%.

Overall, it seems that intervention modelling can be a useful tool for identifying the effects of price changes upon usage. The technique produces reasonable parameter estimates of usage repression. It is especially appropriate for telephone usage data because of its seasonal nature.

FOOTNOTES:

[1] See for example Garfinkel, L., and Linhart, P.B., "The Transition to Local Measured Service," Public Utilities Fortnightly, 104, Aug 1979 and Schmidt, L.W., "Local Measured Service: A Telephone Industry

Perspective," in J.A. Baude, et. al. eds. Perspective on Local Measured Service, Kansas City: Telecommunications Industry Organizing Committee, 1979.

[2] Jensik, John M., "Dynamics of Consumer Usage", in J.A. Baude, et al. eds. Perspectives on Local Measured Service, Kansas City: Telecommunications Industry Workshop Organizing Committee, 1979; and Park, R.F. and Wetsel, Bruce M. and Mitchell, B.M., "Charging for Local Telephone Calls: Price Elasticity Estimates from the GTE Illinois Experiment", The Rand Corporation, R-2635-NSF, 1982.

[3] For a more extensive description of the GTE Measured Service Experiment see G. Cohen, "Usage Sensitive Pricing", Fifth Annual Rate Symposium on Problems of Regulated Industries, Columbus, Mo. 1979.

[4] For background on this technique see Box, G.E.P. and Jenkins, G.M., Time Series Analysis: Forecasting and Control, San Francisco: Holden Day, 1976. A more technical illustration of intervention modeling of telephone calls can be found in Jenkins, G.M. and Wilkinson, G.F., The Estimation of a Change in Price Structure in the U.S. Telephone Industry, in Jenkins, G.M. and McLeod, G., eds., Case Studies in Time Series - Volume 1, GJP Publications, St. Helier and Lancaster, 1982.

REFERENCES:

[1] Baude, J.A., Cohen, G., Garfinkel, L., Krehmeyer, M., Ogg, J., eds., Perspectives on Local Measured Service, Telecommunications Industry Workshop: Kansas City, Mo. (1979).

[2] Box, G.E.P. and Jenkins, G.M., Time Series Analysis: Forecasting and Control, (2nd ed.), Holden Day: San Francisco (1976).

[3] Box, G.E.P. and Tiao, G.C., Intervention Analysis with Applications to Economic and Environmental Problems, Journal of the American Statistical Association, Vol. 70, 349 (March 1975).

[4] Cohen, G., Usage Sensitive Pricing, Fifth Annual Rate Symposium on Problems of Regulated Industries, Columbia, Mo. (February 1979).

[5] Doherty, A.N., Econometric Estimation of Local Telephone Price Elasticities, Assessing New Pricing Concepts in Public Utilities, Edited by H. Trebing, Michigan State University: East Lansing, Mi., (1978).

[6] Jenkins, G.M., Practical Experiences with Modelling and Forecasting Time Series, GJP Publications: St. Helier (1979).

[7] Park, R.E., Wetzel, Bruce, M., and Mitchell, B.M., Charging for Local Telephone Calls: Price Elasticity Estimates from the GTE Illinois Experiment, The Rand Corporation, R-2638-NSF (1982).

[8] Pavarini, C., The Effect of Flat-to-Measured Rate Conversions on Local Telephone Usage, Mountain Bell Economics Seminar, Keystone, Colorado (August, 1978).

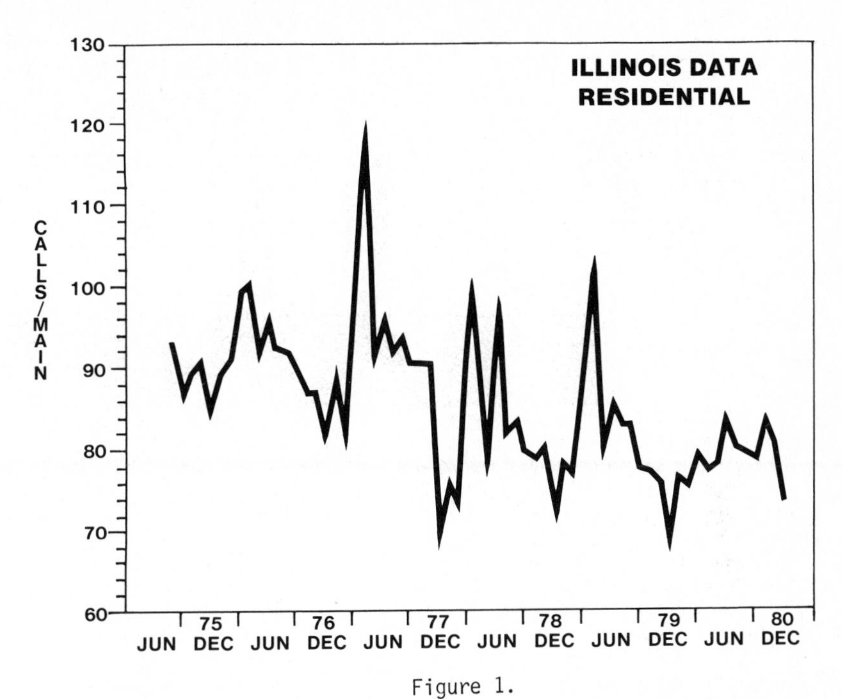

Figure 1.

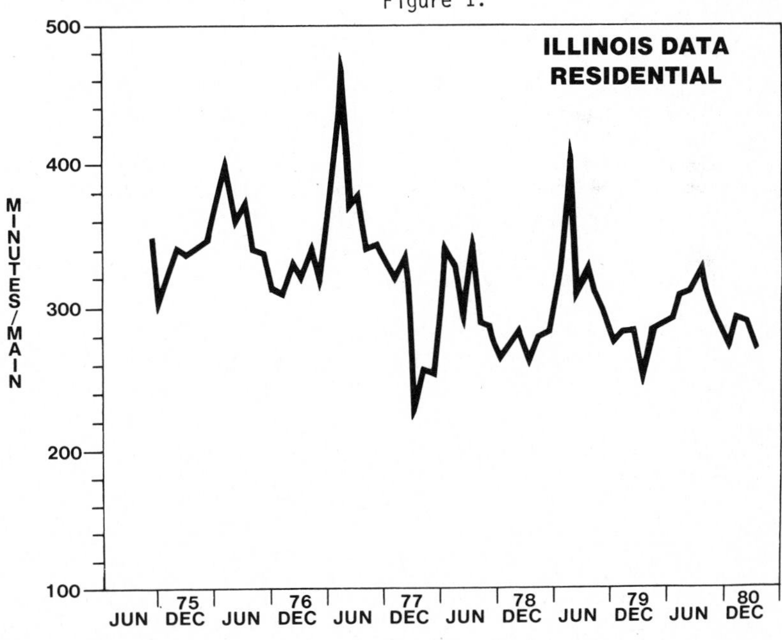

Figure 2.

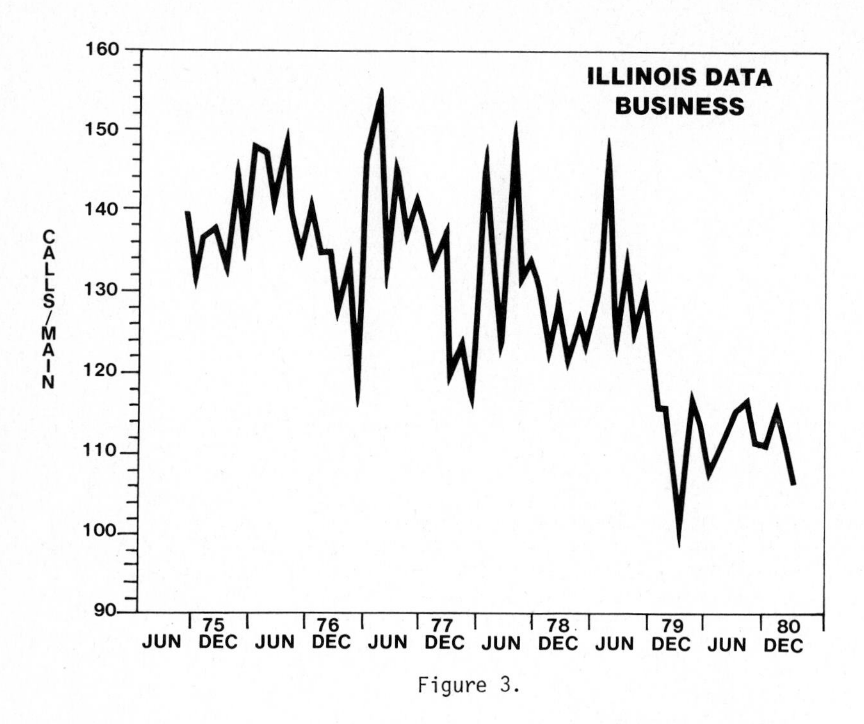

Figure 3.

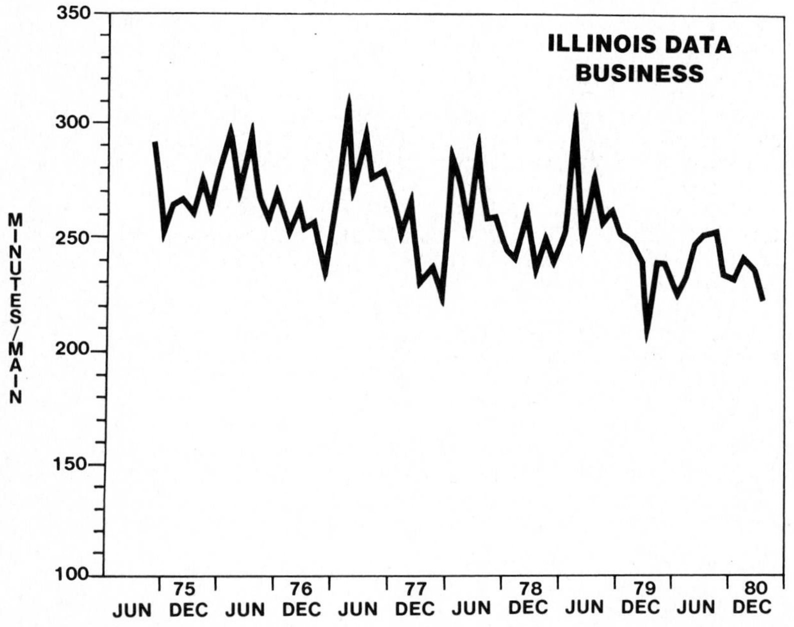

Figure 4.

Economic Analysis of Telecommunications:
Theory and Applications
L. Courville, A. de Fontenay and R. Dobell (eds.)

IDENTIFYING TARIFF INDUCED SHIFTS IN THE SUBSCRIBER DISTRIBUTION OF LOCAL TELEPHONE USAGE

T.F. Wong

Bell Laboratories

I. Introduction

Most residential and many business subscribers to local telephone service are charged at a flat rate (FR). Within a prescribed local calling area, the customers can make as many calls and talk as long as they like for a fixed monthly fee. Recently, actions by regulatory commissions, pressure from consumer organizations and changes in the economic environment of the telephone business (Garfinkel 1979) have caused consideration of a move to measured telephone service (MS). In the MS environment customers will be billed for their actual local usage over some allowance. Usage charges may depend on the frequency, duration, distance and time of day/day of week of local calls, much like long distance calls are charged for today. When such FR to MS conversion is undertaken, customer usage characteristics are expected to change. These changes will be a crucial input in evaluating the impact of such rate conversion, for the amount of local usage is a major determinant of the resulting telephone company revenues, expenses, and required investments.

Unlike many demand studies, analyses of usage changes at the aggregate or market level may not be sufficient in this case. When the tariff contains a usage allowance, or if the bill rendered to the customer as a function of usage is otherwise nonlinear, aggregate demand is insufficient knowledge to calculate resulting revenues. In addition, there is often interest in such questions as what percent of the customers have higher/lower telephone bills as a result of the conversion. Consequently, past studies (Pavarini 1978) of customer usage change in response to FR to MS tariff changes have modeled the usage under MS individual customers as a function of their previous usage under FR, price per unit usage in the new MS tariff and other economic variables. This disaggregate modeling requires a large customer usage data base and is complicated to execute. Analysis at this level of detail, however, may not be necessary for many useful studies. Since often all that is required is knowledge of the subscriber distribution of usage (and changes in it induced by the conversion), analyses aimed specifically at quantifying usage distributions might be considered. Such an approach is described below.

The new (post-conversion) usage density function, $g(x_M)$, is related to the old density,
$f(x_F)$, by

(1) $$g(x_M) = \int G(X_M/x_F) f(x_F)\, dx_F$$

where the variates x_M and x_F are usages, say in minutes/month, under MS and FR respectively. The conditional probability density G allows for a stochastic transformation, i.e. one where customers previously at the same usage level are allowed to react differently. The integration in x_F offers the prospect for relatively simple relations among (x_F) and $g(x_M)$. The crucial point is that these simpler relations are sufficient for many of the study purposes discussed before.

In this note, I shall discuss a procedure for direct identification of the shift of the usage distribution, in a situation where the usage distributions before and after the tariff conversion are both assumed to have a Weibull form.

I first explain why I use a Weibull distribution. It was observed (Pavarini 1978) that the local telephone usage distribution under FR can be represented by a truncated powernormal distribution. In section 2, I shall use the result that with the proper selection of parameters, a Weibull distribution can approximate a powernormal distribution. This result makes Weibull a plausible representation of the FR usage distribution. It has the added advantage of being simpler to manipulate analytically than the powernormal.

In section 3, I shall present an exploratory analysis of the subscriber distribution of local telephone usage data from a Denver, Colorado flat-to-measured conversion in 1971 called METROPAC (Metropolitan Preferred Area Calling). I found that a Weibull distribution is adequate to describe the data both under FR and under MS. Details of the statistical analyses and goodness-of-fit test are given in Appendix A.

Assuming that the usage distributions before and after the tariff conversion are represented by a Weibull function, two transformations were found that will conserve the Weibull form. These transformations also imply simple relations among Weibull parameters that can be tested (section 4). The METROPAC data was found to be consistent with this idea (distribution in total connect time seems to follow a power transformation whereas distribution in frequency follows a linear transformation) and have simple identification of the shift of telephone usage distribution in response to tariff change.

Further discussion of this procedure is presented in section 5.

Although I have limited my discussion so far to the specific case of the change of local telephone usage characteristics in response to an FR to MS conversion, these procedure and results could potentially have application to other fields, e.g. utility services like electricity and water or quality assurance testing. Whenever the before and after distribution of some measure are characterized by a Weibull (or equivalently by a powernormal distribution, see section 2), simple relations could be found which determine the response of the system to the change (tariff rates or the testing procedure in the above examples).

2. Equivalence of Weibull and Powernormal Distribution Functions

A Weibull distribution has the cumulative probability distribution function of the variate x (e.g. a customer's usage in minutes/month)

(2) $$Pw\ (x;\alpha,\beta) = 1 - e^{-(\frac{x}{\alpha})^{\beta}}$$

whereas the powernormal has[1]

(3) $$P_{PN}\ (x;\lambda,\mu,\sigma) = \phi(x) = \frac{1}{\sqrt{2\pi}} \int_{-\infty}^{t} e^{-\frac{1}{2}z^2} dz, t = \frac{x^{\lambda} - \mu}{\sigma}$$

where α and β in eq. (2), λ , μ and σ in eq. (3) are constant parameters. [When λ=1, eq. (3) is just the normal distribution P_N].

It was observed that the Weibull distribution in (x/α) is similar in shape to a powernormal distribution with $\lambda = \frac{1}{3.60}\beta = 0.28\beta$ (Johnson 1970).

Suppose the usage data can be represented by a Weibull distribution, and we have estimated the parameters α and β. The above result shows that the the data in the form of $(\frac{x}{\alpha})^{0.28\beta}$ will be equally well represented by a normal distribution, with μ=0.901 and σ=0.278. Comparing with eq. (3)

(4) $$t = \frac{(\frac{x}{\alpha})^{0.28\beta} - 0.901}{0.278}$$

$$= \frac{x^{0.28\beta} - 0.901(\alpha)^{0.28\beta}}{0.278(\alpha)^{0.28\beta}}$$

We conclude that the usage data can be represented by a powernormal distribution with λ= 0.28β, and a fixed mean to standard error ratio of 3.24.

Conversely, if the telephone usage distribution can be represented by a powernormal function with a mean to standard error ratio of 3.24, a Weibull distribution will be an equally good representation (Dubey 1967). In the following section we shall study the hypothesis that the Weibull function is the underlying distribution for usage distribution under both FR and MS tariffs.

3. Exploratory Analyses of METROPAC Usage Data

Metropolitan Preferred Area calling (METROPAC) is an optional offering in five exchanges in the state of Colorado. These exchanges are situated just outside the border of the Denver metropolitan local calling area. Under the standard FR tariff, residence customers in the five subject exchanges reached only those terminals in their own exchange for fixed monthly charges in the range of $4.05 to $5.05.

In response to demands by subscribers residing outside the metro area to be included in a larger flat rate area, the optional METROPAC offering was created in July 1969. A subscriber opting for the service paid an extra fixed charge ($6.25 to $7.30, depending upon his exchange) and in return was able to reach all exchanges within a thirty mile radius at flat rate. There was no limitation on the number or duration of outgoing calls.

The high market penetration coupled with the high usage of the METROPAC subscribers placed considerable burden on the network. In June of 1970, Mountain Bell of Colorado filed to convert METROPAC to a measured offering. Under the new tariff, subscribers would pay a fixed charge for METROPAC ($5) for which they would receive an allowance measured in total connect time of 60 minutes. Each minute over the allowance was charged at $0.08. Measured METROPAC became effective in January of 1971.

Individual subscribers' monthly usages in minutes per month were available over two four-month periods, April through July in 1970 under FR and the same months in 1971 under MS. To smooth out the month-to-month variation, each subscribers' usage had been averaged over these four months. In other words, I have two samples (FR and MS) of subscriber distribution of usage. A total of 383 permanent residence customers, who subscribed to both FR and MS METROPAC, were tracked.

Assuming that the underlying distribution is a Weibull function, I used the Maximum Likelihood Method (MLE) (Leone 1960) to estimate the scale parameter α and the shape parameter β(eq. 2), separately for the before and after usage distribution. The results and the estimated standard errors are summarized in table I of Appendix A. I used S* statistics (Dubey 1966) to test the goodness of fit and the hypothesis that α and β equal the MLE estimated values. I used Chi-square statistics (Hahn 1967) to test the goodness-of-fit and the hypothesis that Weibull is the underlying distribution. Details of these statistical analyses are contained in Appendix A.

The probability density (p.d.) of the subscriber distribution data, in the form of a histogram, and the fitted Weibull distribution are shown in Fig. 1 (Appendix A). The Weibull function provides an adequate description, except for two bins near the peak of the density, of the subscriber distribution of usage both before and after MS conversion.

Individual subscribers' usages can be measured in frequency (calls per month), although the total connect time is a more natural choice for the METROPAC tariff. I have another two independent samples (FR and MS) of the distribution of subscriber usage measured in calls/month. I repeated the statistical analyses in Appendix A. Corresponding results are summarized in table II and Figure 2.

4. Shift of Usage Distribution

Suppose the usage distribution before and after a tariff change (e.g. FR to MS) can be represented by a Weibull function, with parameters $\alpha_F, \beta_F; \alpha_M, \beta_M$ respectively. That is (Burlington 1970),

$$P_W(x_F; \alpha_F, \beta_F) = 1 - e^{-\left[\frac{x_F}{\alpha_F}\right]^{\beta_F}} \tag{5}$$

$$P_W(x_M; \alpha_M, \beta_M) = 1 - e^{-\left[\frac{x_M}{\alpha_M}\right]^{\beta_M}} \tag{6}$$

where P_W is the cumulative probability and x is customer usage, for example in minutes/month.

Given this hypothetical situation, we search for possible relations among the flat and measured Weibull parameters. The first step is to look for classes of variable transformation that will conserve the Weibull form.

4.1 Power Transformation

Let us rewrite eq. (5) by raising x_F and α_F to a constant power δ. To maintain the equality[2] we divide β_F by the same constant.

$$(7) \qquad P_W(x_F;\alpha_F,\beta_F) = 1 - e^{-\left[\frac{(x_F)^\delta}{(\alpha_F)^\delta}\right]^{\frac{\beta_F}{\delta}}}$$

$$= P_W\left[(x_F)^\delta;\ (\alpha_F)^\delta,\ (\frac{\beta_F}{\delta})\right]$$

where δ is a constant (independent of usage) not yet specified. We get another Weibull distribution with different variate and parameters.

From eq. (7) if we assume a "power" transformation (which is non-trivial) and identify

$$(8) \qquad x_M = (x_F)^\delta;\ \delta \neq 1$$

and if the usage distribution under FR can be fitted by a Weibull function (section 3), then the usage distribution under MS will also follow a Weibull function, with new parameters

$$(9) \qquad \alpha_M = (\alpha_F)^\delta$$

$$(10) \qquad \beta_M = \frac{1}{\delta}\beta_F$$

These equations can be rewritten

$$(11) \qquad \delta = \frac{\ln\alpha_M}{\ln\alpha_F}$$

$$(12) \qquad \delta = \frac{\beta_F}{\beta_M}$$

and should be tested: The RHS of eqs. (11) and (12) can be calculated from the estimated parameters and their values should be compared for equality. If the data is consistent with eqs. (11) and (12), we have a simple way to identify and forecast the shift of usage distribution in terms of the change in Weibull parameters [eqs. (9) and (10)].

The constant δ would be expected to depend on the tariff parameters, income, the level of other prices and perhaps other factors at the time of the tariff change and in a particular location. In principle if we have several tariff change experiments, the structure of δ can be mapped out in detail.

For a Weibull distribution (Burlington 1970)

$$\mu = \alpha \ \Gamma(1 + \frac{1}{\beta}) \tag{13}$$

For usage distributions similar to those reported by Pavarini (1978), reasonable values of β are in the range 0.9 to 2.0. For these β, the gamma function attains values around unity (1.05 to 0.89),

$$\mu_M \simeq \alpha_M, \ \mu_F \simeq \alpha_F \tag{14}$$

and

$$\frac{\ln \mu_M}{\ln \mu_F} \simeq \frac{\ln \alpha_M}{\ln \alpha_F} \tag{15}$$

From the results of the exploratory analyses of the METROPAC data (Table I of Appendix A), we find that

$$\frac{\ln \alpha_M}{\ln \alpha_F} = \begin{matrix} 0.77 \\ (0.01) \end{matrix} \tag{16}$$

$$\frac{\beta_F}{\beta_M} = \begin{matrix} 0.89 \\ (0.05) \end{matrix} \tag{17}$$

$$\frac{\ln \mu_M}{\ln \mu_F} = 0.76 \tag{18}$$

where α and μ are measured in minutes/month and the values in brackets are estimated standard errors. These values are in rough agreement with eqs. (11), (12) and (15). Comparing with eqs. (11) and (12), the weighed (by inverse variance) estimate of δ is

$$\delta = \begin{matrix} 0.77 \\ (0.01) \end{matrix} \tag{19}$$

4.2 Linear Transformation

There is another simple transformation that will preserve the Weibull function. If we multiply x_F and α_F by the same constant ε, eq. (5)

becomes

$$P_W(x_F; \alpha_F, \beta_F) = 1 - e^{(\frac{\varepsilon x_F}{\varepsilon \alpha_F})^{\beta_F}}$$

(20) $$= P_W[(\varepsilon x_F);(\varepsilon \alpha_F),\beta_F]$$

The linear transformation

(21) $$x_M = \varepsilon x_F;\ \varepsilon \neq 1$$

implies

(22) $$\alpha_M = \alpha \varepsilon_F$$

(23) $$\beta_M = \beta_F$$

Eqs. (13) and (23) give

(24) $$\frac{\mu_M}{\alpha_M} = \frac{\mu_F}{\alpha_F}$$

Finally, eqs. (23) and (24) can be rearranged

(25) $$\frac{\beta_F}{\beta_M} = 1$$

(26) $$\frac{\alpha_M}{\alpha_F} = \frac{\mu_M}{\mu_F}$$

and should be tested. The LHS can be calculated from the estimated parameters and the RHS of eq. (26) can be calculated from the mean usage data.

From the results of the analyses of the METROPAC data in table II of Appendix A, we find that

(27) $$\frac{\alpha_M}{\alpha_F} = \begin{matrix} 0.37 \\ (0.02) \end{matrix}$$

(28) $$\frac{\beta_F}{\beta_M} = \begin{matrix} 1.01 \\ (0.06) \end{matrix}$$

(29) $$\frac{\mu_M}{\mu_F} = 0.36$$

where α and μ here are measured in calls/month. These results are in good agreement with eqs. (25) and (26). The estimated ε [eq. (22)] is

(30) $$\hat{\varepsilon} = \underset{(0.02)}{0.37}$$

Distributions in total connect time are related to distributions in frequency in a non-trivial way [the individual subscriber's total connect time is the product of his call frequency and his average holding time in that month]. I can not offer any explanation of why these distributions follow different transformations. The observation that their pre- and post-conversion parameters are so simply related is just amusing.

4.3 Comparison with Pavarini's Repression Model

Pavarini (1978) studied individual customer usage in rsponse to FR to MS tariff change. He attempted to construct a demand function of a household, and to identify the relevant explanatory variables. In his preliminary study, he concluded that

(31) $$\hat{P} = \frac{P}{(P_e)^{0.5}Y^{0.5}}$$

is the appropriate variable, where p is the unit charge in the MS tariff, p_e is a price index for other goods, and y is the household disposable income. The form of (31) was motivated by the budget constraint and it captures the effect of both inflation and regional difference in income. [Effects of other demographic factors are presumably reflected in the FR usage]. We could use consumer price index (CPI) to measure p_e and medium household income (MHI) to measure income in a relatively homogeneous neighbourhood.

The repression indexes, δ and ε defined above, are subject to the boundary condition

(32) $$\left\{ \begin{array}{l} \delta = 1 \\ \varepsilon = 1 \end{array} \right. \quad \text{when } p = 0$$

as they, by definition, measure the change of usage in response to FR (p= 0) to MS conversion.
In the absence of other data, I assume a linear dependency of δ and ε in p. Results in eqs. (19) and (30) then give

(33) $$\delta = 1 - \frac{(3166 \pm 138)}{(CPI)^{0.5}(MHI)^{0.5}} P$$

(34) $$\varepsilon = 1 - \frac{8671 \pm 275}{(CPI)^{0.5}(MHI)^{0.5}} P$$

where I have used CPI = 122 [Labor 1978] and MHI = $9938 [Ziprofile 1980] for the Denver suburb (Zipcode 80501) in 1971.

These empirical results, eqs. (33) and (34), provide a complete quantification of the shift of subscriber distribution. They should be tested against data from other conversion experiments when available.

5. **Discussion**

I have proposed a procedure to directly study the shift of the subscriber usage distribution in response to a tariff change. In the case of a Weibull-to-Weibull shift, simple relations among the parameters could be found. The METROPAC data was found to be consistent roughly with these ideas. Eqs. (11) and (12) or (25) and (26) should be tested in detail with data from local telephone tariff change experiments.

The transformations in eqs. (8) and (21) may not be the only ones[3] that conserve the Weibull distribution. It will be useful to conduct a systematic search for a complete set of possible tranformations that preserves the distribution form. This however is beyond the scope of this short note.

The approach presented here is partially supported by empirical observation. It would be interesting to construct a model, with economic theory and consumer behavior built in, from which quantities such as distribution functions can be generated directly.

Alternatively, results obtained here could be used to guide an ambitious model development. One could construct an individual demand function under a general tariff. The model, when partially aggregated, should reproduce the shape of subscriber distribution that I observed. In this approach distribution parameters would be explicit functions of the tariff and other economic variables. Then the model should be estimated on data and its performance in predicting usage change verified.

I have presented a specific situation and found simple relations to hold between post and pre conversion distributions. This result indicates that a distribution-to-distribution study is a potentially rewarding approach. There are many systems which can be described by a Weibull, or powernormal distribution function, and our simple relations, eqs. (9) and (10) or (22) and (23) may be relevant in identifying such systems' response to external stimuli.

APPENDIX A

I shall summarize the statistical analyses of the subscriber distributions of usage, before and after MS conversion, of the 383 permanent subscribers in METROPAC. The background of this MS conversion was discussed in section 3.

a) Assuming that the underlying distribution is a Weibull function, with parameters and (eq. 2), I used the Maximum Likelihood Method (Leone 1960) to estimate the unknown parameters. They are the solutions to

$$\text{(A1)}\quad \frac{n}{\beta} - n\,\frac{\Sigma (x_i)^{\beta} \ln x_i}{\Sigma (x_i)^{\beta}} + \Sigma \ln x_i = 0$$

(A2) $$\alpha^{\beta} = \frac{\Sigma(x_i)^{\beta}}{n}$$

where n, the sample size, equals 383.

The values of $(\alpha_F,\beta_F),(\alpha_M,\beta_M)$ and their estimated standard errors (in brackets), for the FR and MS distribution respectively, are listed in table I. The associated variance was estimated by (Engelhardt 1977)

(A3) $$\mathrm{Var}(\alpha) = 1.1624\ (\frac{\alpha}{\beta})^2$$

$$\mathrm{Var}(\beta) = 0.6482\beta^2$$

b) To test the goodness of fit and the hypothesis that α and β equal the MLE estimated values, Dubey (1966) suggested a test function which involves all n observations

(A4) $$S* = 2 \sum_{i=1}^{n} (\frac{x_i}{\alpha})^{\beta},\ n = 383$$

Under the null hypothesis this function obeys the Chi-square distribution with 2 n (n_D = 766) degrees of freedom. For such a large n_D, the χ^2 distribution is well approximated by a normal distribution with variate (Abramovitz 1964).

(A5) $$\frac{\chi^2 - n_D}{\sqrt{2n_D}}$$

I followed this procedure and calculated S^* in table I. The estimated values cannot be rejected at 57% confidence level.

c) To test the goodness-of-fit and the hypothesis that Weibull is the underlying distribution, I calculated the Chi-square statistics after dividing the data into 41 cells. The boundary of these cells were determined such that every one of them have the expected probability of 1/41 or 9.34 observations in our sample. (Hahn 1967).

The χ^2 values, together with the corresponding confidence level (Harter 1964) are listed in table I. The low confidence level is perhaps not suprising. It is well known (Hahn 1967) that the more data there are (here n=383) the better are the chance of rejecting any model.

I plot the probability density of the usage data (histogram) and the fitted Weibull distribution in Fig. 1. The Weibull functions fail to reproduce the magnitude of the bins near the peak. They do provide an adequate description of the data before and after MS conversion.

d) The above analyses were repeated for the FR and MS subscriber distribution in frequency (calls per month). Results are summarized in table II and Fig. 2.

Table I

Estimated Parameters and Test Functions of the Assumed Weibull Distribution of Usages in Total Connect Time

	μ (MINS/MONTH)	α (MINS/MONTH)	β	$\chi^2(n_D=38)$	C.L.	$S^*(n_D=766)$	C.L.
FR	484.13	541.97 (20.04)	1.49 (0.06)	51.09	7.50%	758.86	57%
MS	112.75	127.79 (4.22)	1.67 (0.07)	78.28	0.02%	759.51	57%

FR = Individual customer usage averaged over 4 months (April to July 1970) under Flat Rate.

MS = Individual customer usage averaged over 4 months (April to July 1971) under Measured Service.

C.L. = Confidence level

Table II

Estimated Parameters and Test Functions of the Assumed Weibull Distribution of Usages in Frequency

	μ (CALLS/ MONTH)	α (CALLS/ MONTH)	β	$\chi^2(n_D=38)$	C.L.	$S^*(n_D=766)$	C.L.
FR	70.95	79.73 (2.48)	1.77 (0.07)	33.74	66.0%	772.00	44%
MS	25.88	29.33 (0.92)	1.75 (0.07)	53.66	4.5%	761.60	54%

FR = Individual customer usage averaged over 4 months (April to July 1970) under Flat Rate

MS = Individual customer usage averaged over 4 months (April to July 1971) under Measured Service

C.L. = Confidence level

Although the use of a Weibull function as the representation of usage distribution was motivated by Pavarini's result and the "equivalence" of Weibull and powernormal. The significance of such assumption was directly tested by the goodness-of-fit tests as discussed in the paper. The reader may notice that the confidence level (according to chi-square statistic)

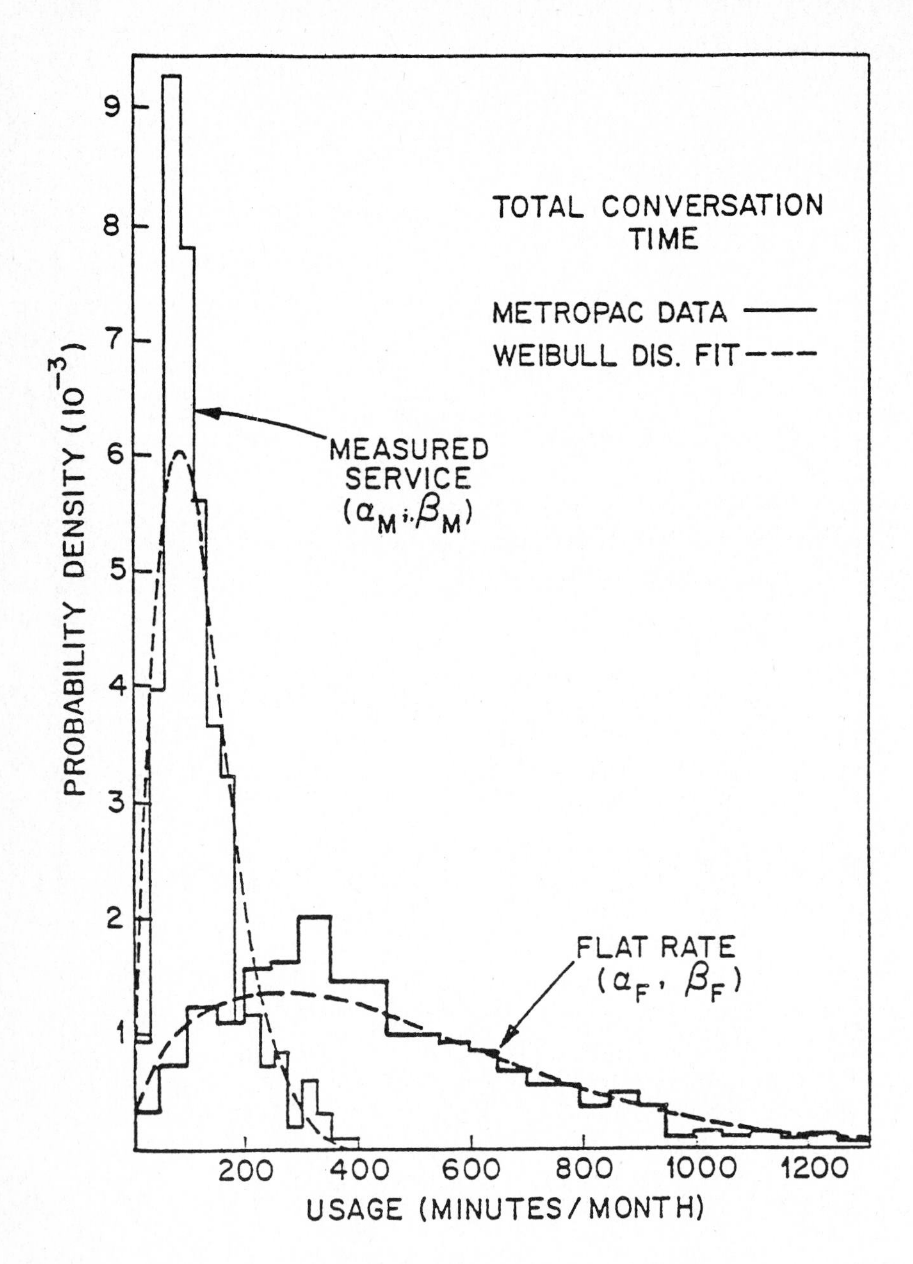

Fig. 1. Comparison of the Metropac usage distribution data and the maximum likelihood fit by a Weibull function.

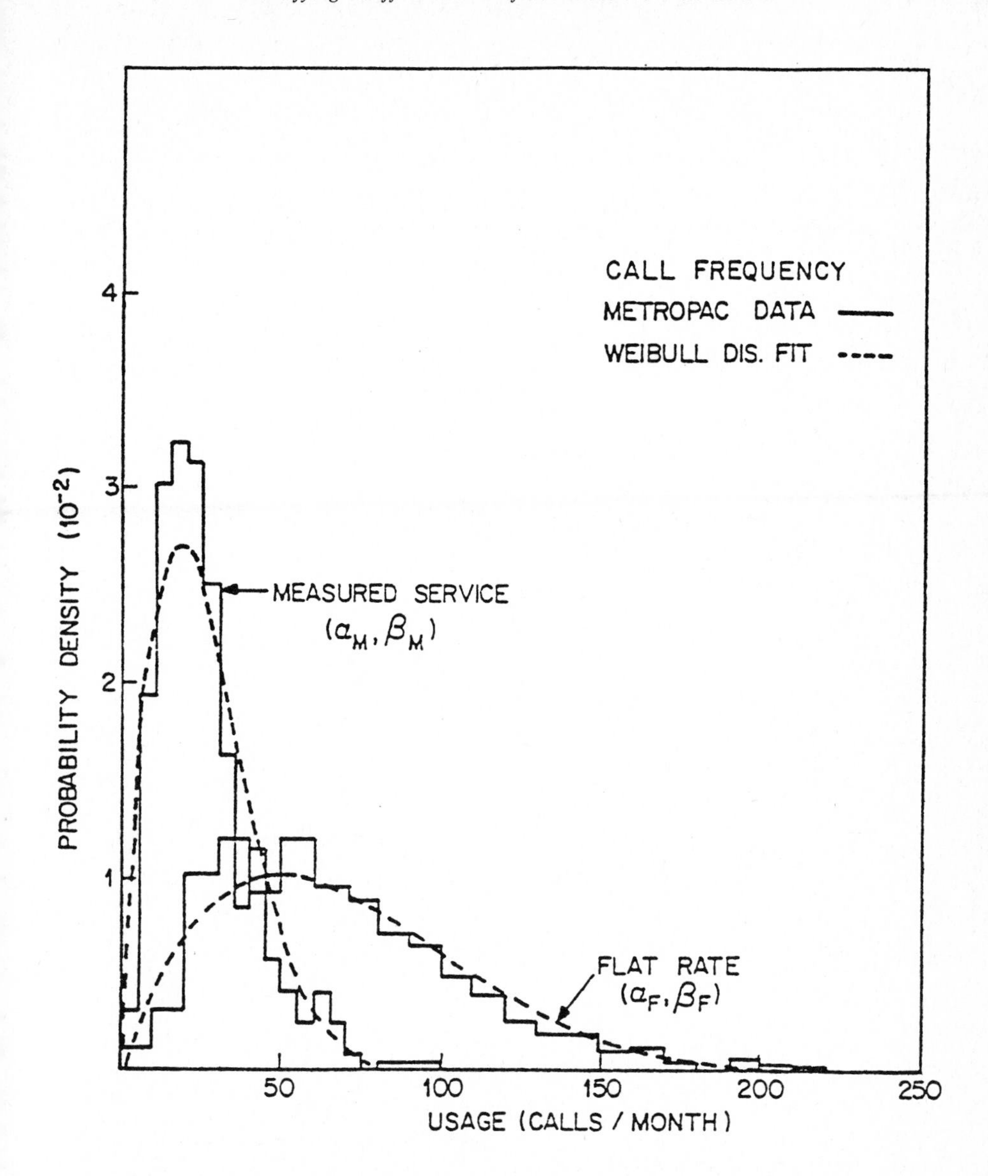

Fig. 2. Comparison of the Metropac usage distribution data and the maximum likelihood fit by a Weibull function.

ranges from a low of 0.02% to a high of 66% (tables I and II).

I have reanalyzed the same METROPAC data with a lognormal function. The new confidence levels are very high and the results and MLE fits are summarized in figure 3.

It can be shown that transformations in eqs. (8) and (21) also conserve the lognormal functional form, with the (transformed) mean and standard deviation (μ and α) related in simple ways.

Power transformation (total connect time data)

$$\mu_M = \delta\mu_F \qquad \sigma_M = \delta\sigma_F$$

and a consistency condition

$$\underset{(0.005)}{\overset{\text{(Data)}}{0.134}} = \frac{\sigma_M}{\mu_M} = \frac{\sigma_F}{\mu_F} = \underset{(0.005)}{\overset{\text{(Data)}}{0.127}}$$

Linear transformation (call frequency data)

$$\mu_M = \mu_F + \ln\varepsilon$$

and a consistency condition

$$\underset{(0.022)}{\overset{\text{(Data)}}{0.614}} = \sigma_M = \sigma_F = \underset{(0.023)}{\overset{\text{(Data)}}{0.631}}$$

In both cases, the consistency conditions are well satisfied. Furthermore the estimated repression indices are "robust" in the sense that the new values (based on lognormal fit)

$$\delta = \underset{(0.007)}{0.767}$$

$$\hat{\varepsilon} = \underset{(0.096)}{0.366}$$

are almost identical to the old results (egs. 19 and 30).

Finally, the procedures discussed in this paper have been applied successfully to a mandatory MS conversion implemented in Ohio in 1978. This result and other studies will be presented elsewhere.

FOOTNOTES:

[1] i.e. Y - Power-Normal (λ, μ, σ) iff $Y^{1/\lambda}$ - Normal (μ,σ). Customers do not have negative usage. A truncated form $\frac{\Phi(\chi)-\Phi(0)}{1-\Phi(0)}$ should be used instead of eq. (3). This approximation is justified a posteriori by the observation that the origin is 3.24 times σ away

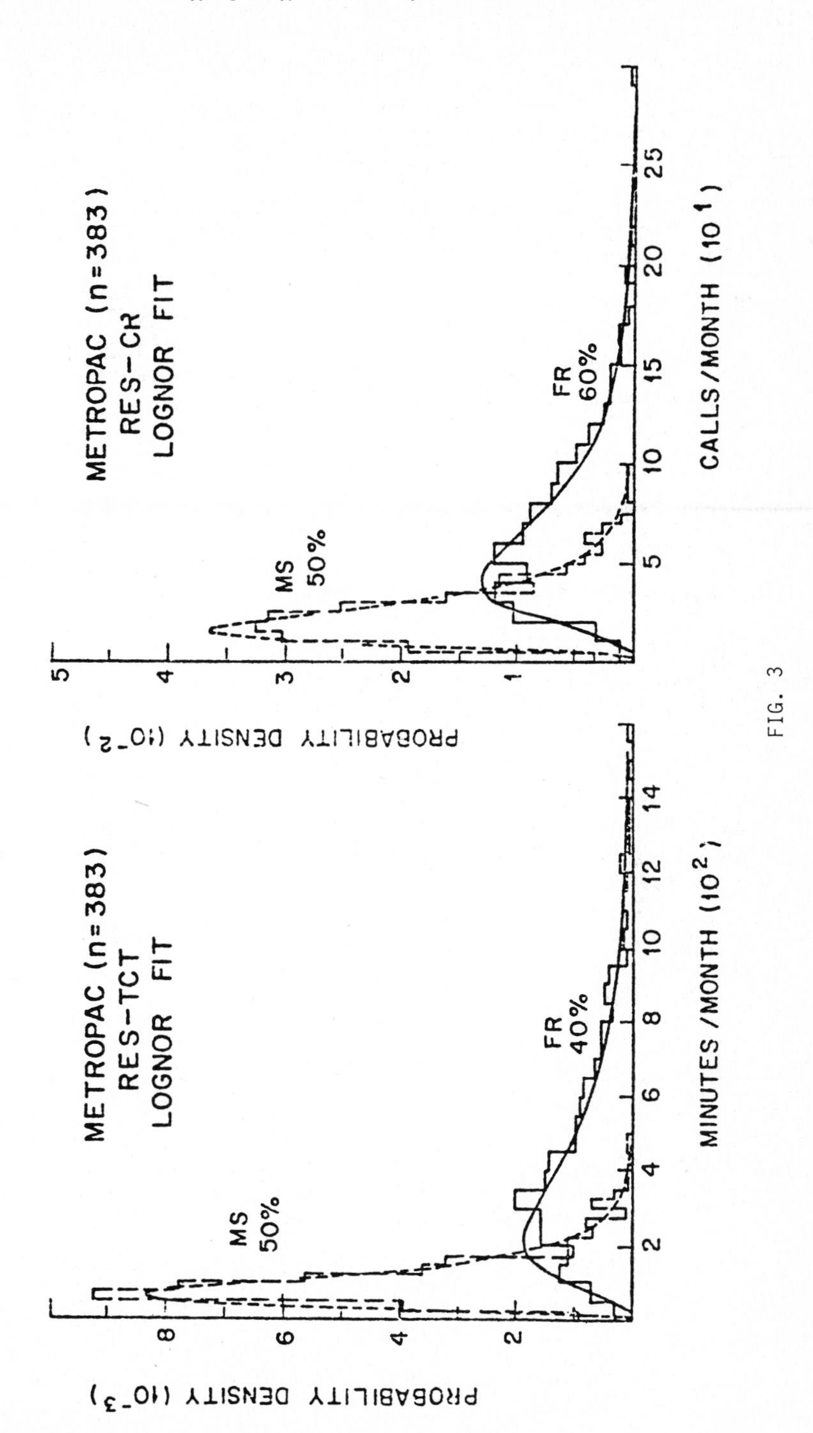
METROPAC (n=383)
RES-TCT
LOGNOR FIT
MS
50%
FR
40%
PROBABILITY DENSITY (10^-3)
MINUTES/MONTH (10^2)
METROPAC (n=383)
RES-CR
LOGNOR FIT
MS
50%
FR
60%
PROBABILITY DENSITY (10^-2)
CALLS/MONTH (10^1)

FIG. 3

from the mean or $\Phi(0)$ = 0.000.6 which is negligible.

[2] Eq. (7) requires the equality of the cumulative probabilities at variates X_F and X_M respectively. If this requirement is relaxed, then eqs. (9) and (10) need not be true.

[3] The discrete transformation, minimum value of a Weibull, gives rise to a Weibull distribution. It is not helpful in the problems that we are concerned.

REFERENCES:

[1] Abramovitz, M. and Stegun, I.A., Handbook of Mathematical Functions, Nat. Bureau of Standards, Applied Math., Series, 55 (1964).

[2] Burlington, R.S. and May, D.C., Handbook of Probability and Statistics with Tables, (McGraw Hill 1976).

[3] Dubey, S.D., Some Test Functions for the Parameters of the Weibull Distribution, Naval Research Logistics Quarterly 13 113 (1966).

[4] Dubey, S.D., Revised Tables for Asymtotic Efficiencies of the Moment Estimators for the Parameters of the Weibull Laws, Naval Research Logistics Quarterly, 14 261 (1967).

[5] Engelhart, M. and Bain, L.J., Simplified Statistical Parameters for the Weibull or Extreme-Value Distribution, Technometrics 19 323 (1977).

[6] Garfinkel, L. and Linhart, P.B., The Transition to Local Measured Telephone Service, Public Utilities Fortnightly (August 16, 1979).

[7] Hahn, G.I.and Shapiro, S.S., Statistical Models in Engineering, (Johy Wiley 1967).

[8] Harter, H.L., New Tables of the Incomplete Gamma-function Ratio and of Percentage Points of the Chi-square and Beta Distributions, Aerospace Research Lab., U.S. Air Force, Dayton, Ohio (1964).

[9] Johnson, N.I. and Kotz, S., Continuous Univariate Distributions I, Houghton Miffin, Boston (1970).

[10] Labor, Dept. of Monthly Labor Review, Bureau of Labor Statistics, Washington D.C. (December 1978) 105.

[11] Leone, F.C., Rutenberg, Y.H. and Topp, C.W., Order Statistics and Estimators for the Weibull Distributions Report #1026, Stat. Lab., Case Institute of Technology, Cleveland, Ohio (1960).

[12] Pavarini, C., The Effect of Flat-to-Measured Rate Conversions on Local Telephone Usage in J.T. Wenders (ed.) Pricing in Regulated Industries: Theory and Application II Mountain States Telephone (1978).

[13] Ziprofile, Donnellay Marketing, Stamford, Connecticut (1980).

Economic Analysis of Telecommunications:
Theory and Applications
L. Courville, A. de Fontenay and R. Dobell (eds.)

ECONOMIC ANALYSIS OF A MEASURED SERVICE OPTION

R. E. Dansby

Bell Laboratories
Murray Hill, New Jersey 07974

This paper examines the impact of an ex post billing option on a public utility's profit and its customers' net surplus, given that an optional two-part tariff is initially used to price the utility's service. The ex post billing option allows a customer to pay a premium so that their bill may be computed each billing period according to the optional two-part tariff which yields the smallest expenditure. It is shown that a customer's benefit from the ex post option depends on the variation in their usage and on the ex post premium. For some demand distributions, the ex post tariff option is pareto superior to the simple flat-measured option, i.e., the ex post tariff option will increase each consumer's surplus without reducing profits of the firm. However, for many demand distributions, use of the ex post option will reduce profits and aggregate welfare.

1. Introduction

Traditionally, telephone utilities have primarily used flat rate tariffs to price local exchange service.[2] However, in recent years, the utilities have seen the need to change the method used to price these services. Several considerations have led to this perceived need for new pricing techniques.[3] Among the primary considerations are: (1) flat rate tariffs do not permit the utilities to base customers' bills on their usage; (2) measurement of customers' usage is now economically feasible, though this has not always been the case; and (3) competition has dictated that all utility services be priced in relation to their cost. These considerations have led utilities to seek alternatives to the traditional flat rate tariffs for local exchange service.

For example, GTE (General Telephone and Electronics) has chosen to implement mandatory measured tariffs for local exchange service;[4] which requires that all customers in a given exchange be charged on a measured rather than flat rate basis. The measured service tariff requires payment of a fixed monthly charge plus a usage charge per unit of local calling. In some cases, a measured service tariff includes a specified number of call allowances; no usage charge is imposed on local usage which does not exceed the call allowance. An alternative scheme is typically used by Bell System companies; customers are offered the option of having their usage billed under either a measured service tariff or flat rate. Each customer must choose the tariff under which their usage is subsequently to be billed.

From the customer's point of view, the merits of the alternative tariffs may be judged in terms of their consumer surplus and income distributional consequences. Under mandatory flat-to-measured tariff conversions, high volume customers will tend to be losers while low volume customers will gain from mandatory measured service implementation. A desirable aspect of optional flat rate-measured is that customers are given the option to choose the tariff which best suits their needs. Customers whose usage is low and would achieve a larger net surplus by being billed under the measured tariff can choose that option. Because of customers' ability, under optional flat rate-measured, 'to select the tariff that best suits them, all customers can in theory be made better off than is possible with either flat rate or mandatory measured.[5]

However, when given such tariff options, it has been observed that customers frequently choose a tariff option which does not yield the minimum possible bill based on actual usage.[6] That is, customers often choose a tariff which is not cost minimizing, ex post. Several explanations for this empirical observation have been advanced. For example, some researchers have suggested that customers may be reluctant to accept measured service because of their long experience and familiarity with flat rate, e.g. habit formation. On the other hand, customers may choose flat rate, when measured would be cheaper ex post, because of the prospect of a high bill under measured, i.e. customers are risk averse. There are many possible explanations for the divergence between a customer's ex ante optimal tariff and the ex post optimal tariff, even if customers are making economically rational tariff choices.

In this paper, we focus on one factor that can create a situation in which customers may be better off with ex post billing: intertemporal variations in customers' usage. It is shown that the ex post billing option discussed in Mitchell [1980] will improve a customer's net surplus to an extent which depends on the variation in the customer's usage and on the ex post premium. For *some* demand distributions the ex post tariff option is pareto superior to the simple flat-measured option. That is, for some demand distributions an ex post tariff option will increase consumers' surplus without reducing profits of the firm. However, it is important to note that for many demand distributions, use of the ex post option will reduce profits and aggregate welfare. Section II characterizes the customers who would benefit from ex post billing. Section III explores the cost and profit consequences of ex post billing. These results are derived under the assumption that customers use expense as the criterion for choice among optional tariffs. Section IV derives similar results when customers use a consumer surplus criterion for tariff selection. Concluding remarks are stated in Section V.

2. Consumers' Expenditures and Firm's Revenue

In this section, we examine the impact on consumers of optional flat rate-measured tariffs and ex post billing. The focus is on the potential consumer benefits of ex post billing as compared to the benefits of optional flat rate-measured service (FR-MS) tariffs.

Suppose that F is the monthly charge prescribed for flat rate, local exchange customers. The measured rate tariff has a monthly charge ρ and a price per unit of usage, P. In this intertemporal scenario, customers are given an option between flat rate and measured service; however, customers may choose between tariffs only once every n periods. Therefore, a customer who chooses flat rate will incur a total bill of

$$E^f = n\mathsf{F} \tag{1}$$

during the n billing periods. A customer whose usage is $q_i(P)$, $i=1,\ldots,n$, during the n periods will incur a total bill of

$$E^m = n\rho + P\sum_{i=1}^{n} q_i(P) \tag{2}$$

if the measured service option is chosen. Therefore, if customers choose the option which minimizes their expenses, they will incur a charge

$$E^0 = Min\left[n\mathsf{F}, n\rho + P\sum_{i=1}^{n} q_i(P)\right] = Min\,[E^f, E^m] \tag{3}$$

The ex post billing option allows a customer to pay a premium P so that their bill may be computed each billing period according to the ex ante tariff which yields the smallest expenditure. That is, under the ex post option, a customer whose usage is $q_i(P)$, $i=1,\ldots,n$, would incur the expense

$$E^e = \mathsf{P} + \sum_{i=1}^{n} Min[\mathsf{F}, \rho + Pq_i(P)] \tag{4}$$

The customer's choice among the optional FR-MS and ex post billing will clearly depend on the distribution of usage across billing periods. If a customer's usage dictates that the same tariff be chosen in each period, then the customer will have no incentive to choose ex post billing. If the ex post premium is positive then a customer will prefer ex post billing only if there is sufficient variation in usage. These ideas are made precise in what follows. Suppose that

$$M^+(P) = \{i: \ q_i(P) \geq (\mathsf{F}-\rho)/P\} ,$$

$$M^-(P) = \{i: \ q_i(P) < (\mathsf{F}-\rho)/P\} . \tag{5}$$

Hence, at usage price P, M^+ is the set of billing periods in which an individual's usage exceeds or equals $(\mathsf{F}-\rho)/P$. Thus, for the m^+ billing periods in M^+, the customer's expenditure will be smaller under flat rate than under the measured service tariff. Similarly, for the m^- billing periods in M^-, the customer's expenditure will be smaller if usage is billed under the measured tariff. The average usage in the M^+ and M^- periods are

$$\bar{q}^+ \equiv \sum_{M^+} q_i/m^+ ,$$

and

$$\bar{q}^- \equiv \sum_{M^-} q_i/m^- .$$

respectively. It follows that[7]

Proposition I: If $\mathsf{P} > 0$ then $E^e < E^0$ only if M^+ and M^- are non-empty.

That is, if the ex post premium is positive, then the expenditure under the ex post tariff is smaller than the expenditure under the ex ante FR-MS option only if there are some billing periods in which flat rate is the minimum cost option and others in which measured is the minimum cost option.

It also follows that if there is no ex post premium, then a customer's expenditure under the ex post option will always be at least as small as their expenditure under the minimum cost ex ante option.[8]

Proposition II: If $\mathsf{P} = 0$, then $E^e < E^0$ if M^+ and M^- are non-empty, while $E^e = E^0$ if M^+ or M^- is empty.

This result says that ex post billing is valuable to any customer whose usage varies across periods, if no premium is charged for this option. However, it should be clear that the value, to any customer, of the ex post billing option decreases as the ex post premium rises. But even if no ex post premium is charged, a customer whose demand has no intertemporal variation will be indifferent between ex post billing and optional FR-MS. The latter result presumes of course that each customer is able to compute the minimum cost ex ante option.

The relative expense to the customer, under the alternative tariff options, depends crucially on the distribution of demand. Thus, it will be useful to characterize the demand distributions for which a given tariff option is cost minimizing. For this purpose, suppose that

$$X = \left\{(m^+,\bar{q}^+,\bar{q}^-):\ m^-\left[\frac{\mathsf{F}-\rho}{P} - \bar{q}^-\right] \leq \mathsf{P}/P\right\}, \tag{6a}$$

$$Y = \left\{(m^+,\bar{q}^+,\bar{q}^-):\ m^+\left[\bar{q}^+ - \frac{\mathsf{F}-\rho}{P}\right] < \mathsf{P}/P\right\}, \tag{6b}$$

and

$$Z = \{(m^+,\bar{q}^+,\bar{q}^-):\ \bar{q} \geq (\mathsf{F}-\rho)/P\}\ . \tag{6c}$$

where $\bar{q}$ is the average usage in all periods, i.e. $\bar{q} \equiv \frac{1}{n}\{m^+\bar{q}^+ + m^-\bar{q}^-\}$. The demand distributions in Z are ones for which flat rate is the minimum cost tariff under the optional FR-MS. The demand distributions in Z', the complement of Z, are ones for which measured service is the minimum cost tariff under optional FR-MS.[9]

The distributions for which flat rate is the minimum cost tariff under the ex post option are those for which $E^f < E^m$ and $E^f < E^e$. The distributions that satisfy $E^f < E^m$ are those in Z, while the distributions satisfying $E^f < E^e$ are those in X. Therefore, the demand distributions for which flat rate is the minimum cost tariff under the ex post option are those in $X \cap Z$. Similarly, the demand distributions for which measured service is the minimum cost tariff under the ex post option are those in $Y \cap Z'$; since $E^m < E^f$ in Z' and $E^m < E^e$ for distributions in Y. Finally, the ex post billing option is the cost minimum for distributions in $X' \cap Y'$, where X' and Y' are respectively the complements of X and Y.[10] The foregoing relationships among demand distributions and cost minimizing tariffs are summarized in Table 1.

TABLE 1

Cost Minimizing Tariffs

Tariff Option	If Demand Distribution Is In	then Cost Minimizing Tariff Is
FR-MS	Z	E^f, Flat Rate
	Z'	E^m, Measured Service
FR-MS or Ex Post Billing	$X \cap Z$	E^f, Flat Rate
	$Y \cap Z'$	E^m, Measured Service
	$X' \cap Y'$	E^e, Ex Post Billing

With these definitions, Proposition II may be restated simply as[11]

Proposition IIA: If $\mathbf{P} = 0$ then X and Y are empty sets; hence all demand distributions are in $X' \cap Y' = Z \cup Z'$.

Customers who are billed according to the flat rate tariff under optional FR-MS will incur a smaller expense under the ex post option in periods when usage does not exceed $(\mathbf{F}-\rho)/P$. Customers who are billed according to the measured tariff under optional FR-MS will incur a smaller expense under the ex post option in periods when usage exceeds $(\mathbf{F}-\rho)/P$. Thus, implementation of the ex post option, with a zero premium, may be beneficial to both flat rate and measured customers. Moreover, no customer is made worse off compared to the ex ante option.

If the ex post premium is positive, then not all customers will benefit if the underlying tariff parameters remain at their status quo levels. The next results provide insight concerning the customer groups that would benefit from implementation of an ex post option with a positive premium. First, consider the effect on customers whose expense is smaller when billed on a flat rate basis under optional FR-MS; these are the customers whose demand distributions are in Z. Note that the set Z is independent of $\mathbf{P}$. On the other hand, the number of elements in $X' \cap Y'$ is a non-increasing function of the ex post premium. To be precise, if $h(m^+,\bar{q}^+,\bar{q}^-)$ is the number of customers with demand distributions having parameters $(m^+,\bar{q}^+,\bar{q}^-)$, then the number of customers whose expense would be lower under the ex post option, i.e. the number of customers in $X' \cap Y'$, is

$$N(X' \cap Y') = \int_{X' \cap Y'} \sum_{m^+=0}^{N} h(m^+,\bar{q}^+,\bar{q}^-)d\bar{q}^+d\bar{q}^- \tag{7}$$

and does not increase with $\mathbf{P}$.[12] This property of $X' \cap Y'$ leads us to conclude that there must be a finite $\mathbf{P}$ such that $X' \cap Y'$ is a proper subset of Z.[13] In particular,[14]

Proposition III: If $\mathbf{P} \geq \underset{Z \cup Z'}{Max} [m^-](\mathbf{F}-\rho)$, then $X' \cap Y' \subset Z$.

Thus suppose that the ex post premium is greater than or equal to $(\mathbf{F}-\rho)$ times the maximum, over all demand distributions, of the number of periods in which an individual's demand is less than $(\mathbf{F}-\rho)/P$. Then the only beneficiaries of ex post billing will be customers whose expenditure is smallest under the flat rate tariff when they face the FR-MS option.

The fact that $X' \cap Y'$ is a non-increasing function of $\mathbf{P}$, and that $X' \cap Y' = \phi$ for some $\mathbf{P}$ also implies that there must exist an ex post premium for which customers in Z would not benefit from the ex post option.

Proposition IV: Suppose $\mathbf{P} > \bar{\mathbf{P}}$, where

$$\bar{\mathbf{P}} = Min \left[\underset{Z \cup Z'}{Max} \left\{ \frac{\mathbf{F}-\rho}{P} - \bar{q}^- \right\} Pm^-, \underset{Z \cup Z'}{Max} \left\{ \bar{q}^+ - \frac{\mathbf{F}-\rho}{P} \right\} Pm^+ \right], \tag{8}$$

then $X' \cap Y' = \phi$.

If $\mathbf{P} > \bar{\mathbf{P}}$, then each customers' usage cost saving from ex post billing will be smaller than the ex post premium. Therefore, if the ex post premium is greater than $\bar{\mathbf{P}}$, there will be no customer whose total bill is smaller under the ex post option than it is under the FR-MS option.[15]

Compared to the expense incurred under the FR-MS option, propositions II-IV may be summarized as follows: If $\mathbf{P} = 0$ then all customers will have a lower bill under the ex post option. If $0 < \mathbf{P} < (\mathbf{F}-\rho)Max[m^-]$, then the customers who have a lower bill under the ex post option will include some (but not necessarily all) measured service customers, those in Z', and all flat rate customers, those in Z. If $\mathsf{P} \geq (\mathsf{F}-\rho) \underset{Z \cup Z'}{Max} [m^-]$ then the only customers who will have a lower bill under the ex post option are the flat rate customers, those in Z. Finally, if $\mathsf{P} > \bar{\mathsf{P}}$ then no customer will have a lower bill under the ex post option.

3. Profits of the Firm

These results also have important implications concerning properties of the revenue function under the ex post option. To facilitate comparison of the revenues earned under the alternative tariff options, let $I(g;A) = \int_A \sum_{m^+ \in A} gh \; d\bar{q}^+ d\bar{q}^-$ denote the integral of $gh(\cdot)$ over the region A. The revenue, R^0, generated under the FR-MS option is then

$$R^0 = I(n\mathsf{F};Z) + I(n[\rho + P\bar{q}];Z') \; . \tag{9}$$

R^0 is the total expense incurred by all flat rate customers, $I(n\mathsf{F},Z)$, plus the total expense incurred by all measured service customers, $I(n\{\rho+P\bar{q}\},Z')$. The revenue, R^e, earned under the ex post option is

$$R^e = I(n\mathsf{F};X \cap Z) + I(n[\rho+P\bar{q}];\; Y \cap Z')$$

$$+ I(\mathsf{P} + m^+\mathsf{F} + m^-[\rho+P\bar{q}^-];\; X' \cap Y') \tag{10}$$

Customers in $X \cap Z$ are those for whom flat rate is the minimum cost tariff and $I(n\mathsf{F};X \cap Z)$ is the total expense incurred by those customers. Similarly, measured service is the minimum cost tariff for customers whose demand distributions are in $Y \cap Z'$, the total expense incurred by this consumer group is $I(n[\rho+P\bar{q}];\; Y \cap Z')$. Note that $n[\rho+P\bar{q}]$ is the total expense incurred by a given customer over n billing periods. Finally, under the ex post option, a given customer incurs a total expense of $m^+\mathsf{F}$ for periods that are billed on a flat rate basis, a total expense of $m^-(\rho+P\bar{q}^-)$ for periods that are billed under the measured service tariff, and a premium P for the ex post billing privilege. Thus, the total expense incurred by customers under the ex post option is $I(\mathsf{P} + m^+\mathsf{F} + m^-(\rho+P\bar{q}^-);\; X' \cap Y')$.

It follows from the results in Propositions II-IV that if $\mathsf{P} = 0$ then $R^0 > R^e$, and if $\mathsf{P} > \bar{\mathsf{P}}$ then $R^0 = R^e$.[16] These properties of the revenue functions are evident upon noting that[17]

$$R^0 - R^e = I(n\mathsf{F};Z \cap [X \cap Z]') + I(n[\rho+P\bar{q}];\; Z' \cap [Y \cap Z']') \tag{11}$$

$$- I(\mathsf{P} + m^+\mathsf{F} + m^-[\rho+P\bar{q}^-];\; X' \cap Y')$$

If $\mathsf{P} = 0$ then $X = \phi$ and $Y = \phi$, thus $X' \cap Y' = Z \cup Z'$, but $n \geq m^+$ and $n(\rho+P\bar{q}) \geq m^-(\rho+P\bar{q}^-)$ with the strict inequality holding in some periods. Hence $\mathsf{P} = 0$ implies that $R^0 - R^e > 0$. If $\mathsf{P} > \bar{\mathsf{P}}$ then $X' = \phi$ and $Y' = \phi$, hence $X \cap Z = Z$ and $Y \cap Z' = Z'$, thus $R^e = R^0$. More generally, the last term in R^e is the total revenue earned from customers who are billed under the ex post tariff, see Eq. (4). Since no customer will incur a higher bill under the ex post option than is incurred under the FR-MS option, it follows that $R^e \leq R^0$ for all demand distributions and tariff parameters.

The total cost incurred by the firm offering local service under the FR-MS option is

$$C^0 = c + I(c^u n\bar{q}(0);Z) + I(n[c^u + c^m]\bar{q}(P);Z') \tag{12}$$

where c is a fixed cost, c^u is the cost per unit of usage and c^m is the cost of measurement per unit of usage. The second term in C^0 is the total cost of usage generated by flat rate customers and the third term in C^0 is the total cost of usage and measurement generated by measured service customers.

The total cost incurred by a firm offering local service under the ex post option is

$$
\begin{aligned}
C^e &= \bar{c} + I(c^u n\bar{q}(0); X \cap Z) + I(n[c^u + c^m]\bar{q}; Y \cap Z') \\
&\quad + I(c^u m^+ \bar{q}^+(0); X' \cap Y') + I(m^-[c^u + c^m]\bar{q}^-; x' \cap Y') \qquad (13) \\
&\quad + I(c^m m^+[\mathrm{F}-\rho]/P; X' \cap Y') \,.
\end{aligned}
$$

Here, $\bar{c}$ is the fixed cost incurred under the ex post option; in general, we would expect that $\bar{c} \geq c$. The second term in C^e is the total cost generated by customers who are served on a flat rate basis under the ex post option. The third term in C^e is the total cost of usage and measurement generated by customers who are served under the measured service tariff. The fourth term is the total cost of usage generated by customers on the ex post tariff in periods when they are billed ex post on a flat rate basis. The fifth term in C^e is the total cost of usage and measurement generated by customers on the ex post tariff in periods when they are billed ex post on a measured service basis. Finally, the sixth term is the measurement cost generated by customers under the ex post tariff in periods when they are billed on a flat rate basis though their usage must be measured until it equals or exceeds $(\mathrm{F}-\rho)/P$.

The total profit of the firm when the optional FR-MS tariff schedule is used is then

$$\pi^0 = R^0 - C^0 \,,$$

(see Eqs. (9) and (12)), which reduces to

$$\pi^0 = I(n[\mathrm{F} - c^u \bar{q}(0)]; Z) + I(n[\rho + \{P - c^u - c^m\}\bar{q}]; Z') - c \,. \qquad (14)$$

The first term is the net profit earned from flat rate customers under the optional FR-MS. Alternatively, the second term is the net profit earned from measured customers under the optional FR-MS. On the other hand, the firm earns the profit $\pi^e = R^e - C^e$ if service is provided under the ex post tariff option. From Eqs. (10) and (13), it follows that

$$
\begin{aligned}
\pi^e &= I(n[\mathrm{F} - c^u \bar{q}(0)]; X \cap Z) + I(n[\rho + \{P - c^u - c^m\}\bar{q}]; Y \cap Z') \qquad (15) \\
&\quad + I(\mathrm{P} + m^+[\mathrm{F} - c^u \bar{q}^+(0) - c^m(\mathrm{F}-\rho)/P] + m^-[\rho + \{P - c^u - c^m\}\bar{q}^-]; X' \cap Y') \\
&\quad - \bar{c} \,.
\end{aligned}
$$

When the ex post option is used, there will be three classes of customers: (1) those who are served in all periods under the flat rate tariff, (the first term in Eq. (15) describes the net profit earned from them); (2) customers who are served in all periods under the measured service tariff, (the second term is the net profit generated by them); and (3) customers who are served under the ex post tariff, i.e. they pay the ex post premium, and are billed in some periods on a flat rate basis while in other periods they are billed on a measured service basis, (the third term in Eq. (15) characterizes the net profit earned from this customer class).

As we have seen, when the firm offers an ex post option, its revenue will always be less than or equal to the revenue earned from a FR-MS option. This results from the fact that customers in the third category pay less for service than they would under an FR-MS option. It is also clear from Eqs. (12) and (13) that the firm may incur a larger total cost when the ex post option is offered because of the need to measure calls that ultimately are billed on a flat rate basis. These considerations lead to the question of whether profits of the firm can ever be higher when the ex post option is used than when service is provided under the FR-MS option. The answer depends crucially on the characteristics of the distribution of demand across billing periods and across customers. The answer also depends on the size of the ex post premium **P** and the parameters $(\mathbf{F},\rho,P)$ of the basic tariff.

We examine these questions by computing from Eqs. (14) and (15), the profits differential $\Delta\pi = \pi^0 - \pi^e$ which is given by

$$\begin{aligned}\Delta\pi = (\bar{c}-c) &+ I(n[\mathbf{F}-c^u\bar{q}(0)];\ Z \cap (X \cap Z)') \\ &+ I(n[\rho + (P-c^u-c^m)\bar{q}];\ Z' \cap (Y \cap Z')') \qquad (16)\\ &- I(\mathbf{P}+m^+[\mathbf{F}-c^u\ \bar{q}^+(0) - c^m(\frac{\mathbf{F}-\rho}{P})] \\ &+ m^-[\rho+(P-c^u-c^m)\bar{q}^-];\ X' \cap Y')\ .\end{aligned}$$

If $\mathbf{P} = 0$, it follows that $X = \phi$ and $Y = \phi$ hence

$$\begin{aligned}\Delta\pi = I(m^-[\mathbf{F} - c^u\bar{q}(0)] &+ m^+c^m(\frac{\mathbf{F}-\rho}{P});\ Z) \\ &+ I(m^+[\rho+(P-c^u-c^m)\bar{q}^+];\ Z')\end{aligned}$$

which must be positive if $\pi^0 > 0$. Hence $\mathbf{P} = 0$ implies that $\Delta\pi > 0$ and $\pi^0 > \pi^e$. On the other hand, if $\mathbf{P} > \bar{\mathbf{P}}$ then $X' = \phi$ and $Y' = \phi$ and

$$\Delta\pi = \pi^0 - \pi^e = \bar{c} - c \qquad (17)$$

Since we would generally expect the fixed costs of administration and billing to be higher under the ex post option, it follows that if $\mathbf{P} > \bar{\mathbf{P}}$, then $\pi^0 > \pi^e$. These considerations indicate that if there exists an ex post premium for which profits are larger under ex post billing than under optional FR-MS, then it must be in the interval $(0,\bar{\mathbf{P}})$.

Obviously, there are many demand distributions for which $\Delta\pi > 0$, i.e. $\pi^0 > \pi^e$. It is also possible to specify conditions on the demand distribution, tariff and cost parameters which imply that $\pi^e > \pi^0$. For example, if the marginal measurement cost is small or zero, then $\pi^e > \pi^0$ if the following conditions are satisfied:[18]

$$[(\bar{c}-c)/I(1;\ X' \cap Y')] < \mathbf{P} < \bar{\mathbf{P}}\ , \qquad (18a)$$

$$P = c^u + c^m\ , \qquad (18b)$$

$$\mathbf{F} = c^u I(\bar{q}(0);Z)/I(1;Z)\ , \qquad (18c)$$

$$\frac{I(\bar{q}(0);Z)}{I(1;Z)} > \frac{I(\bar{q}(0);Z \cap \{X \cap Z\}')}{I(1;Z \cap \{X \cap Z\}')} \,, \tag{18d}$$

and

$$I(n;Z' \cap \{Y \cap Z'\}') < I(m^{-};X' \cap Y') \,, \tag{18e}$$

The condition in (18a) requires the ex post premium to exceed the additional cost of administration and billing per customer served under the ex post option; but the ex post premium must be less than the maximum price that any customer is willing to pay for the ex post option. The condition (18b) says that measured service usage is priced at its marginal cost; while (18c) presumes that the flat rate monthly charge has been chosen so that under optional FR-MS the total revenue from flat rate customers equals the total variable cost of usage generated by them. The (18d) restriction on the demand distribution requires that the average usage of flat rate customers, when facing the FR-MS option, be smaller than the average usage of customers who would choose flat rate when facing the FR-MS option but would not choose flat rate when facing the ex post option. Finally, condition (18e) requires that the number of customer-periods, which are billed under the measured service tariff, must be larger under the ex post option than when the FR-MS option is used.

If these conditions held then implementation of the ex post option would be a pareto improvement over the FR-MS option, since customers' surplus and profit would increase. But there are many demand distributions for which the ex post option would not be a pareto improvement, since the firm's profits would decline. Moreover, one may easily construct cases in which the ex post option will decrease aggregate welfare; for example, if $\bar{p} = 0$ and $\bar{c}$ is much larger than c. (See summary in Table 2.) Therefore, we conclude that for status quo tariff parameters the ex post option will increase customers' surplus but whether it simultaneously increases aggregate welfare or the firm's profit becomes an empirical question.

TABLE 2

Who Would Benefit From
Ex Post Billing?

	IF	THEN
Consumers:	$\mathsf{P} = 0$	All Customers Benefit
	$0 < \mathsf{P} < n(F-\rho)$	Some Measured Service Customers and All Flat Rate Customers Benefit
	$n(F-\rho) < \mathsf{P} < \bar{\mathsf{P}}$	Only Flat Rate Customers Will Gain
	$\bar{\mathsf{P}} < \mathsf{P}$	No Customer Will Benefit
The Firm:	$\mathsf{P} = 0$	Firm's Profits will Decline
	$0 < \mathsf{P} < n(F-\rho)$	$\pi^0 - \pi^e$ Depends Crucially on Distribution of Demand Across Customers and Periods
	$n(F-\rho) < \mathsf{P} < \bar{\mathsf{P}}$	
	$\bar{\mathsf{P}} < \mathsf{P}$	Firm's Profits Will Decline

4. Consumer Surplus Choice Criterion

The results in Section II and III (which are summarized in Table 2) were derived under the assumption that each consumer chooses the tariff option which gives them the lowest total bill. An alternative choice criterion that consumers might use is to select the tariff option which yields the largest consumer surplus. However, even if consumers used the consumer surplus criterion for tariff selection, the principal results would be qualitatively identical to those summarized in Table 2. That is to say, if consumers made their tariff selection decisions on the basis of consumer surplus, then the benefits of ex post billing (as compared to the FR-MS option) would be summarized as follows:

1) If the ex post premium, P, were equal to zero then ex post billing will increase the consumer surplus of all customer classes but will decrease the firm's profit.

2) There exists a finite $\bar{\mathsf{P}}$ such that if $\mathsf{P} > \bar{\mathsf{P}}$ then ex post billing will decrease the firm's profits.

3) Ex post billing will increase the firm's profit only if $0 < \mathbf{P} < \bar{\mathbf{P}}$; however, there are demand distributions for which it is not possible to choose a value of $\mathbf{P}$ which increases profits.

4) If $0 < \mathbf{P} < \bar{\mathbf{P}}$, then customers who would choose flat rate under the FR-MS option will receive a larger relative share of the consumer surplus benefits of ex post billing as the ex post premium increases.

Therefore, the ex post billing option will benefit some consumers but will tend to adversely affect profits of the firm. Thus, in general, the ex post option will not be pareto superior to the FR-MS option, even if customers use the consumer surplus criterion to choose among tariff options.

These insights are derived by the same methods used in Sections II and III; hence for brevity the details are merely sketched here. Let $S_i(P) = \int_P^{\infty} q_i(y)dy$, i.e. $S_i(P)$ is an individual's consumer surplus in period i when usage is priced at P dollars per unit. The total consumer surplus derived by an individual during the n billing periods is $S(P) = \sum_{i=1}^{n} S_i(P)$. The net consumer surplus of an individual billed on a flat rate basis is therefore

$$S^f = S(0) - n\mathbf{F}$$

while the individual's net consumer surplus if billed under the measured service tariff is

$$S^m = S(P) - n\rho \, .$$

Therefore, an individual who uses the consumer surplus criterion for tariff selection would choose flat rate if $S^f \geq S^m$, i.e. if $(S(0) - S(P))/nP \geq (\mathbf{F}-\rho)/P$. Alternatively the individual would choose measured service if $S^m > S^f$. Let[19]

$$\hat{Z} = \{(q_1, \ldots, q_n) \colon S^f \geq S^m\} \, ,$$

then individuals having demand distributions in the set $\hat{Z}$ will choose flat rate under the FR-MS option while individuals having demand distributions in the complement of $\hat{Z}$, i.e. in $\hat{Z}'$, will choose measured service.

Under ex post billing, an individual would achieve a net surplus of

$$S^e = \sum_{i=1}^{n} Max[S_i(P) + Pq_i(P) - Min\{\mathbf{F},\rho + Pq_i(P)\}; S_i(0) - Min\{\mathbf{F},\rho + pq_i(0)\}]$$

This expression can be simplified upon noting that no rational consumer would consume $q_i(P)$ even though $Min[\mathbf{F},\rho + Pq_i(P)] > Min[\mathbf{F},\rho + Pq_i(0)]$. Hence there are three possible combinations of consumption levels and ex post billing rate: (1) a customer consumes $q_i(P)$ and $Min[\mathbf{F},\rho + Pq_i(P)] < Min[\mathbf{F},\rho + Pq_i(0)]$; (2) a customer consumes $q_i(0)$ while $Min[\mathbf{F},\rho + Pq_i(P)] > Min[\mathbf{F},\rho + Pq_i(0)]$; and (3) a customer consumes $q_i(0)$ even though $Min[\mathbf{F},\rho + Pq_i(P)] < Min[\mathbf{F},\rho + Pq_i(0)]$.[20] With this information the billing periods can be separated into three sets, based on the classifications in (1)-(3) above, that accomplish the same purpose as the sets M^+ and M^- used in Sections II and III. Having simplified the expression for S^e by using these sets we define

$$\hat{X} = \{(q_1, \ldots, q_n) \colon S^f > S^e - \mathbf{P}\}$$

and

$$\hat{Y} = \{(q_1, \ldots, q_n):\ S^m > S^e - \mathsf{P}\}$$

Therefore, when customers use the consumer surplus criterion to choose among tariff options, the tariff option chosen for different demand distributions is summarized in what follows. Under the FR-MS option, a customer chooses: (1) flat rate if $(q_1, \ldots, q_n) \in \hat{Z}$; and (2) measured service if $(q_1, \ldots, q_n) \in \hat{Z}'$. Under the ex post option a customer chooses: (1) flat rate if $(q_1, \ldots, q_n) \in \hat{X} \cap \hat{Z}$; (2) measured service if $(q_1, \ldots, q_n) \in \hat{Y} \cap \hat{Z}'$ and ex post billing if $(q_1, \ldots, q_n) \in \hat{X}' \cap \hat{Y}'$.[21] (Note the similarity of these results to the tariff choices summarized in Table 1.) With these definitions the analysis proceeds exactly as developed in Sections II and III.

5. Conclusions

In this paper, it has been shown that the ex post billing option discussed in Mitchell [1980] will not, in general, be a pareto improvement over the simple flat rate-measured service tariff option. This result was shown to be true when consumers choose among alternative tariffs on the basis of expense or consumer surplus. In general, the ex post option will increase consumers' benefits but may decrease profits of the firm. It was demonstrated, however, that there are some demand distributions for which the ex post option would increase the firm's profit.

Footnotes

1. The author's interest in the subject of this paper was stimulated by discussion with John C. Panzar (Bell Labs) and Bridger M. Mitchell (Rand Corp.); both influenced the selection of issues addressed in this paper. I adhere to the usual error liability protocol.

2. For example, in 1950, 81.3% of residence-local exchange customers in the Bell System received local service under flat rate tariffs.

3. See Garfinkel and Linhart [1979].

4. See Park, et al. [1981].

5. Faulhaber and Panzar [1977] show that offering more tariff options usually is pareto optimal.

6. See Pavarini [1979].

7. Suppose that M^+ is empty, then $M^- = \{i{=}1,\ldots,n\}$ which implies that $\bar{q} = \sum_{i=1}^{n} q_i/n < (\mathsf{F}-\rho)/P$, thus $E^0 = n(\rho+P\bar{q})$ and $E^e = \mathsf{P} + n(\rho+P\bar{q})$ consequently, $E^0 < E^e$ if $\mathsf{P} > 0$. Alternatively, if M^- is empty, then $M^+ = \{i{=}1,\ldots,n\}$ which implies that $\bar{q} \geq (\mathsf{F}-\rho)/P$, thus $E^0 = n\mathsf{F}$ and $E^e = \mathsf{P} + n\mathsf{F}$, hence $E^0 < E^e$ if $\mathsf{P} > 0$. Therefore, if either M^+ or M^- is empty then E^e cannot be less than E^0 if $\mathsf{P} > 0$. It follows that $E^e < E^0$ for $\mathsf{P} > 0$ only if M^+ and M^- are non-empty.

8. Suppose $\mathsf{P} = 0$, then from Eq. (4) it follows that M^+ and M^- being non-empty implies that $E^e = m^+\mathsf{F} + m^-(\rho+P\bar{q}^-)$. If $E^0 = n\mathsf{F} \equiv (m^+ + m^-)\mathsf{F}$, then $E^e < E^0$ since $(\rho+P\bar{q}^-) < \mathsf{F}$ by definition of M^-. Similarly, if $E^0 = n\rho + P\sum_i q_i \equiv m^+(\rho+P\bar{q}^+) + m^-(\rho+P\bar{q}^-)$ then $E^e < E^0$ since $\mathsf{F} < (\rho+P\bar{q}^+)$ by definition of M^+. If M^+ is empty then $E^0 = E^m$ and $E^e = E^m$ similarly if M^- is

empty then $E^0 = E^f$ and $E^e = E^f$.

9. Under optional FR-MS, flat rate is the minimum cost option if and only if $E^f < E^m$, see Eqs. (1) and (2); which implies that $n\mathsf{F} \leq n\rho + Pn\bar{q}(P)$ and is equivalent to $\bar{q} \geq (\mathsf{F}-\rho)/P$, the boundary constraint of Z. The complement of Z is $Z' \equiv \{(m^+,\bar{q}^+,\bar{q}^-)$: $\bar{q} < (\mathsf{F}-\rho)/P\}$ and is the set of distributions for which $E^m < E^f$.

10. The boundaries of the sets X and Y are derived as follows.

$$E^f < E^e \; iff \; n\mathsf{F} < \mathsf{P} + \sum_i Min[\mathsf{F},\rho + Pq_i(P)] \equiv \mathsf{P} + m^+\mathsf{F} + m^-(\rho+P\bar{q}^-)$$

which is equivalent to $m^-\mathsf{F} < \mathsf{P} + m^-(\rho+P\bar{q}^-)$; the definition of X follows from this inequality. Y is the set for which $E^m < E^e$, thus it is defined by $n(\rho+P\bar{q}) < \mathsf{P} + m^+\mathsf{F} + m^-(\rho+P\bar{q}^-)$ which is equivalent to $m^+(\rho+P\bar{q}^+) < \mathsf{P} + m^+\mathsf{F}$. Thus, $E^e < E^f$ and $E^e < E^m$ iff $(m^+,\bar{q}^+,\bar{q}^-) \in X' \cap Y'$.

11. If $\mathsf{P} = 0$ then $X = \{(m^+,\bar{q}^+,\bar{q}^-)$: $m^-[(\mathsf{F}-\rho)/P - \bar{q}^-] \leq 0\}$ but $\bar{q}^- < (\mathsf{F}-\rho)/P$ by definition, thus $X = \phi$ the empty set. Similarly, $\mathsf{P} = 0$ implies that $Y = \{(m^+,\bar{q}^+,\bar{q}^-)$: $m^+[\bar{q}^+ - (\mathsf{F}-\rho)/P] < 0\}$, while by definition $\bar{q}^+ > (\mathsf{F}-\rho)/P$ thus $Y = \phi$. Since X and Y are both empty, their complements must both be equal to the universal set; therefore, $X' \cap Y'$ must be equal to the universal set of demand distributions.

12. From Eqs. (6a) and (6b) it follows that

$$X' = \{(m^+,\bar{q}^+,\bar{q}^-): \; m^-[(\mathsf{F}-\rho)/P - \bar{q}^-] > \mathsf{P}/P\}$$

and

$$Y' = \{(m^+,\bar{q}^+,\bar{q}^-): \; m^+[\bar{q}^+ - (\mathsf{F}-\rho)/P] \geq \mathsf{P}/P\} .$$

Therefore, the number of elements in X' and Y', respectively, must be a non-increasing function of P; consequently, the same must be true of $X' \cap Y'$.

13. Since $N(X' \cap Y')$ is a non-increasing function of P, and is strictly decreasing for all regions in which the distribution $h(\cdot)$ has compact support, it follows that there exists a finite P such that $X' \cap Y' = \phi$ and $N(X' \cap Y') = 0$, if usage is finite. Since $\phi \subset Z$, there is a P for which $X' \cap Y' \subset Z$.

14. Any demand distribution $(m^+,\bar{q}^+,\bar{q}^-)$ which is in $X' \cap Y'$ must satisfy

$$m^+\bar{q}^+ - m^-\bar{q}^- \geq 2\,\mathsf{P}/P + (m^+ - m^-)(\mathsf{F}-\rho)/P .$$

this follows from adding the bounds which specify the sets X' and Y' see Footnote 11. If $\mathsf{P} \geq m^-(\mathsf{F}-\rho)$, then $2\,\mathsf{P}/P + (m^+ - m^-)(\mathsf{F}-\rho)/P \geq [2\bar{m}^- + \bar{m}^+ - m^-](\mathsf{F}-\rho)/P = n(\mathsf{F}-\rho)/P$. Thus $\mathsf{P} \geq m^-(\mathsf{F}-\rho)$ implies that $m^+\bar{q}^+ - m^-\bar{q}^- \geq n(\mathsf{F}-\rho)/P$, and consequently that $m^+\bar{q}^+ + m^-\bar{q}^- \geq n(\mathsf{F}-\rho)/P$. Distributions which satisfy the last inequality are in Z. Hence, if $\mathsf{P} \geq (\mathsf{F}-\rho)Max\{m^-\}$ and if $(m^-,\bar{q}^+,\bar{q}^-) \in X' \cap Y'$ it follows that $(m^+,\bar{q}^+,\bar{q}^-) \in Z$, i.e. $X' \cap Y' \subset Z$. Since $X' \cap Y' \subset Z$ if $\mathsf{P} \geq (\mathsf{F}-\rho)Max[m^-]$ the same must be true if $\mathsf{P} \geq n(\mathsf{F}-\rho)$, since $n > Max[m^-]$.

15. From Footnote 11, it follows that $X' = \phi$ and $Y' = \phi$, respectively, if $\mathbf{P} > \max_{Z \cup Z'} \left\{ \frac{\mathsf{F} - \rho}{P} - \bar{q}^{-} \right\} m^{-} P$ and $\mathbf{P} > \max_{Z \cup Z'} \left\{ \bar{q}^{+} - \frac{\mathsf{F} - \rho}{P} \right\} m^{+} P$. Since $\mathbf{P}$ is defined to be the minimum of these two quantities, Proposition IV follows immediately, since $X' \cap Y' = \phi$ if $X' = \phi$ or $Y' = \phi$.

16. Recall that customers are presumed to always choose the minimum cost tariff, given their demand distribution and the tariff option they face.

17. This expression for $R^0 - R^e$ was derived by using the following properties of the function I: (1) $I(f;A) + I(g;A) = I(f+g;A)$; and (2) $I(f;A) + I(f;B) = I(f;A \cup B)$. Further, it is noted that by definition $X \cap Z \subseteq Z$, $Y \cap Z' \subseteq Z'$, and that if $B \subseteq A$ then $I(f;A) - I(f;B) = I(f;A \cap B')$.

18. From Eq. (17) it follows that if $P = c^u + c^m$ and $\mathsf{F} = c^u I(\bar{q}(0);Z)/I(1;Z)$ then $I(n[\mathsf{F} - c^u \bar{q}(0)];Z - \{X \cap Z\})$ is negative if condition (18d) holds; (18a) and (18e) then insure that the remaining terms in $\Delta\pi$ have a net value which is negative. Therefore, conditions (18a)-(18e) imply that $\pi^e > \pi^0$.

19. Here it is assumed that $S^f > 0$ for all relevant demand distributions.

20. These considerations imply that more customer-periods are billed under flat rate when customers use the consumer surplus criterion than when the minimum expense criterion is used for tariff selection.

21. With this notation, $\tilde{\mathbf{P}}$ is defined as

$$\tilde{\mathbf{P}} = Min \left[\underset{\hat{Z} \cup \hat{Z}'}{Max} \{S^e - S^f\}, \ \underset{\hat{Z} \cup \hat{Z}'}{Max} \{S^e - S^m\} \right].$$

Hence from considerations similar to those in Footnote 14, it follows that $\hat{X}' \cap \hat{Y}' = \phi$ if $\mathbf{P} > \tilde{\mathbf{P}}$.

References

Faulhaber, G. R. and J. C. Panzar, "Optimal Two-Part Tariffs with Self-Selection," Bell Labs Economics Discussion Paper No. 74, January 1977.

Garfinkel, L. and P. B. Linhart, "The Transition to Local Measured Service," *Public Utilities Fortnightly*, Vol. 104, No. 4, (August, 1979).

Mitchell, B., "Economic Issues in Usage-Sensitive Pricing," Report P-6530, The Rand Corp., September, 1980.

Park, R. E., B. M. Mitchell, B. M. Wetzel, and J. H. Alleman, "Charging for Local Telephone Calls: How Household Characteristics Affect the Number of Calls in the GTE Illinois Experiment," The Rand Corporation, R-2535-NSF, March 1981.

Pavarini, Carl, "The Effect of Flat-to-Measured Rate Conversions on Local Telephone Usage," in J. Wenders (eds.), *Pricing in Regulated Industries: Theory and Application*, Mountain States Telephone and Telegraph Company, Denver, 1979.

Economic Analysis of Telecommunications:
Theory and Applications
L. Courville, A. de Fontenay and R. Dobell (eds.)

LOCAL TELEPHONE COSTS AND THE DESIGN OF RATE STRUCTURES

B. MITCHELL

Rand Corporation

1. Introduction

A well-developed body of economic theory is available to guide the setting of prices for the multi-product regulated firm. Economic efficiency can be increased by designing rate structures that incorporate the basic principles developed from this theory. These principles call for provisionally pricing each of the firm's outputs at its marginal cost, testing to determine whether such rates satisfy a specified budget constraint (e.g., revenues = costs), and then suitably modifying the marginal-cost rates in order to satisfy the constraint. Most commonly, the trial rates produce insufficient revenue, and then rates must be raised according to the Ramsey rule--prices are increased above marginal costs in inverse proportion to the individual price elasticities of demand. This paper applies ratemaking theory to the design of rate structures for local telephone calls that efficiently reflect the costs of the local network.

The principal costs of supplying local telephone calls are embodied in the switching capacity of a local central office (exchange) and the trunking capacity that connects local offices together. (The dedicated local loop connecting the subscriber to the office is, of course, essential but its cost is independent of telephone usage). Several operating companies which are proposing to introduce local measured service are conducting special studies that will gather information about these costs, and how they vary with maximum loads. How should these costs, which relate to specific items of network equipment, be used to develop prices for telephone calls, which are commonly classified in terms of the hour at which they are placed and the distance between subscribers?

The markets for telephone services are characterized by high capacity-related equipment costs and very low variable (traffic) costs, the joint use of some equipment by several outputs, and the grouping together of several outputs that are charged a common price. The following sections develop a series of simple models that successively incorporate these basic elements. Throughout the paper I make several simplifying assumptions:

- the unit of output is a "call" of fixed duration, and there is a uniform rate of demand during any given period
- demand for a given output depends only on its own price, so that there are no temporal or spatial cross-elasticities
- all costs are due to providing capacity to meet the maximum rates of output and capacity can be constructed at constant returns to scale
- the costs of connecting subscribers to the telephone network via local loops are recovered in fixed monthly charges.

2. A Single Exchange

All telephone calls originate and terminate in one exchange and this output is produced using a single component of switching capacity. There are only two commodities, x^1 and x^2, the number of calls made in two equal-length periods ("day" and "night"). This situation is a version of the well-known Boiteux-Steiner peak-load pricing problem in which a homogeneous resource with a maximum capacity is available to produce output in each period.

The economic structure of the one-exchange telephone call market is given by the rate of demand functions for the two periods

$$(1) \quad x^1 = x^1(p^1), \qquad x^2 = x^2(p^2)$$

the required capacity

$$(2) \quad K = \max_t (x^t),$$

and total cost

$$(3) \quad C = \beta K$$

where β is the per-call unit capacity cost.

The marginal cost of a commodity is the change in total cost that results from a one-unit increase or decrease in the production of that commodity. In the long run, capacity can be adjusted to meet maximum demand. Therefore if $x^1 > x^2$ at the observed prices, the marginal costs are

$$mc^1 = \partial C/\partial x^1 = \beta$$

$$(4) \quad mc^2 = \partial C/\partial x^2 = 0.$$

This basic model can be used to illustrate several approaches to pricing telephone service. Each rate structure can be evaluated in terms of its effect on economic welfare, measured by the sum of consumers' and producer's surplus

$$(5) \quad W = CS + PS = \sum_t \int_{p_t}^{\infty} x^t(\xi^t)\,d\xi^t + \sum_t p^t x^t - \beta \max_t (x^t) .$$

A. Flat-rate pricing

Set $p^1 = p^2 = 0$. The total (usage) cost of local telephone service is recovered by increasing the fixed monthly chage per subscriber. Such flat-rate pricing seems inefficient. But because prices are zero, equipment to measure the number of calls is not needed, and the resulting saving in resources can outweigh the gains of per-call charges. Nevertheless, in order to focus on the design of usage-sensitive rate structures I will neglect measurement costs in this paper (For a comparison of benefits and costs under flat versus measured rates in a simple case, see Mitchell, 1980).

B. Average-cost pricing

Set $p^1 = p^2 = p^* =$ average cost $= C/(x^1+x^2)$. Charging a positive price p^* per call is seemingly more efficient than flat-rate pricing. In period 1 capacity is a scarce resource; the reduced demand due to the positive price will reduce calling and therefore capacity and total costs. But calling will also be reduced in period 2, even though excess capacity is available. Compared to flat-rate pricing the net result can be either a gain or loss in welfare.

In Fig. 1 the reduction in calling from flat-rate levels, $x^1(0)$ and $x^2(0)$, to average-price levels, $x^1(p^*)$ and $x^2(p^*)$ reduces capacity costs by $\beta[x^1(0)-x^1(p^*)]$, shown by areas $S^1 + T + U$. At the same time consumer surplus is reduced by S^1 in period 1 and S^2 in period 2. Thus, as illustrated in Fig. 1a, welfare is increased if $S^1 + T + U > S^1 + S^2$. If the demand curve is linear, $S^1 = T$, and a welfare gain occurs when $S^1 + U > S^2$. In contrast, Fig. 1b shows a relatively more elastic off-peak demand. In this case $S^1 + U < S^2$; average-cost pricing imposes greater welfare losses in the off-peak market than it achieves in net savings in the peak period.

C. Peak-load pricing with a firm peak

Set $p^1 = \beta$, $p^2 = 0$. In period 1, the marginal cost of an additional call is the marginal cost of increasing capacity, β. So long as demand in period 2, at a zero price, is less than period 1 output, the marginal cost of an additional call in period 2 is zero. These rates are optimal. Moreover, because capacity is produced at constant returns to scale, average cost and marginal cost are equal per unit of peak output. Therefore, these marginal-cost prices exactly recover total costs.

Figure 1

Welfare Effects of Average-cost Pricing

(a) Increased welfare: $S^1 + U > S^2$

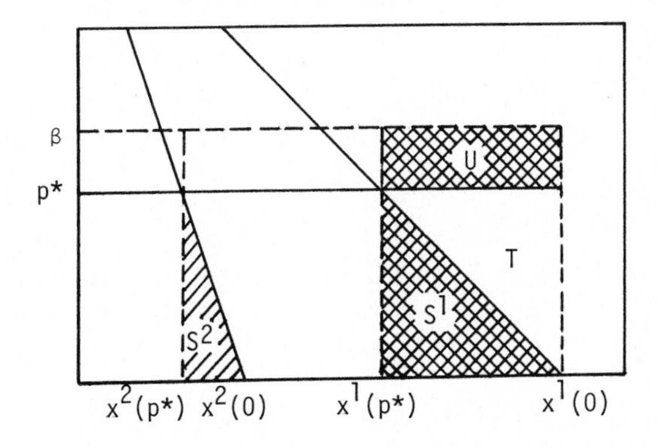

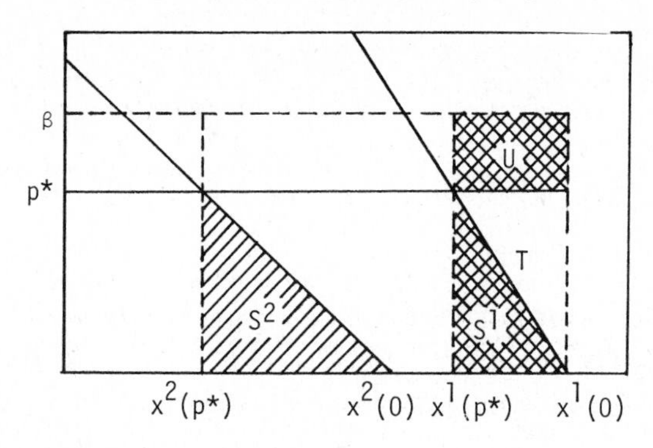

D. Peak-load pricing with a shifting peak

Set $p^1 > 0$, $p^2 < 0$, $p^1 + p^2 = \beta$. If the previous rate structure, with $p^2 = 0$, would cause the period 2 demand to exceed period 1 demand the result is a "shifting peak." In this case a positive period 2 price is necessary to equalize demands ($x^1 = x^2$) in both periods. The optimal rates are those that simultaneously (a) bring about this joint peak, and (b) sum to the marginal costs of capacity.

In the joint peak case the maginal cost of a commodity depends on whether its output is increased or decreased. An increase of 1 unit of either x^1 or x^2 requires adding a full unit of capacity and therefore has a marginal cost of β; but a decrease in either output permits no saving in capacity and has a zero marginal cost. However, the optimal prices of the joint peak case may be interpreted as the marginal opportunity cost of output in each period when capacity is fixed. (See Hirshleifer, 1958). The opportunity cost of supplying a marginal call in period 1 is the value of the most valuable alternative that must be foregone--the withdrawal of one period-1 call worth p^1 from some other subscriber. Similarly, the opportunity cost in period 2 is p^2. And the sum of subscribers' marginal valuations of capacity, $p^1 + p^2$, must equal the marginal cost of expanding capacity, β. Except where explicitly noted below, I assume hereafter that a "firm peak" exists at the rate structures under consideration.

This one-exchange model of the peak-load pricing problem yields clear-cut guidelines for ratesetting:

- price should be highest in the period with the maximum demand;
- price should exclude capacity costs in a period that has excess capacity;
- optimal prices are equal to marginal costs.

This conventional economic wisdom is an extensive abstraction from the complexities of actual regulated industries. When expanded to include fuel costs, it is perhaps most nearly applicable to the pricing of electricity, an industry in which the bulk of the fixed resources take the form of central generating and transmission capacity which is needed by all consumers.

However, the technology of the telephone industry corresponds less accurately to this paradigm. Instead, capacity is distributed throughout the network in a large number of separate facilities, each of which is available to serve only certain types of calls. To better characterize these aspects of telephone technology, I examine successively more detailed models.

3. Several Isolated Exchanges

In each exchange, subscribers place and receive calls only within the exchange. If each exchange has its own rate structure, the pricing problem is that of the previous model. But in practice, a single rate structure must be designed for an entire group of exchanges--for example, all exchanges within one state. (In discussing this paper William Vickrey points out that such restrictions on the rate structure would be avoided if the telephone company could signal the price to the subscriber at the time he placed his call. Indeed, such dynamic pricing, when combined with equipment to automatically forward one-way messages, promises substantial improvements over static time-of-day rate structures).

To illustrate this case, it is sufficient to consider just two exchanges, A and B, with demands

(6) $x^t_A(p^t_A), \quad x_B(p^t_B) \quad t=1,2$

and capacities

(7) $K_A = \max\ (x^t_A), \quad K_B = \max\ (x^t_B)$.

The total cost of local telephone usage is

(8) $C = \beta_A K_A + \beta_B K_B$.

Restricted Rate Structures

Because the exchanges are grouped the rates must satisfy the restrictions

(9) $p^1_A = p^1_B\ ,\ p^2_A = p^2_B$.

Of course, a common rate structure for both exchanges that is based either on flat rates or on an average price that applies in both periods will satisfy these restrictions. These cases are much like those considered for a single exchange.

Optimal restricted peak-load rate structures can be determined by maximizing the welfare function (5) subject to the pricing restrictions (9). In general, the optimal prices are weighted averages of the marginal costs of the individual commodities

(10) $p^t = [b^t_A/b^t_A+b^t_B)]\ mc^t_A + [b^t_B/(b^t_A+b^t_B)]\ mc^t_B$
where

(11) $b^t_j = \partial x^t_j/\partial p^t \quad j = A, B$

It is important to note that the weights for the marginal costs are composed of the slopes of the demand curves, not the number of calls. When a common price must be charged for two commodities with differing marginal costs, some loss of efficiency must result. For example, suppose the common price were set equal to the marginal cost in market A. Then the gap between this price and marginal cost in market B would cause a distortion given by the familiar welfare triangle with area proportional to the slope of the demand curve in that market. Bringing the price closer to mc_B will reduce that loss but create one in market A. The best balance of gain and loss depends on the demand changes in each market, as shown by equation (10).

A key result of restricting the admissible set of rate structures is that the optimal pricing rules can no longer be stated solely in terms of cost data, i.e., set price equal to marginal cost. Instead, as shown in equation (10), demand data, in the form of slopes or elasticities, are commingled into the pricing rule. Two types of peak-load cases need to be considered.

A. Same Peak Period in Each Exchange

With maximum demand in period 1 in both exchanges, marginal costs are

$$mc^1_A = \beta_A, \quad mc^1_B = \beta_B$$

$$mc^2_A = mc^2_B = 0. \tag{12}$$

Thus, the optimal rates are

$$p^1 = [b^1_A/(b^1_A+b^1_B)] \quad {}_A + [b^1_B/(b^1_A+b^1_B)]\,\beta_B$$

$$p^2 = 0\ . \tag{13}$$

The requirement that the exchanges be grouped for ratemaking imposes a particular type of data aggregation. Although there are four separate commodities, the admissible rate structure distinguishes only two types of output--total period-1 demand and total period-2 demand (the number of daytime calls and the number of nighttime calls throughout the state). For this case the optimal price for period-1 calls is a weighted average of the per-unit capacity costs in each exchange; in period 2 each exchange has idle capacity and the price is therefore zero.

Because the weights for the capacity costs are the slopes of the demand curves in each period, not the number of calls, this rate structure will not (except by chance) exactly recover total costs when capacity costs vary by exchange. To satisfy the revenue constraint (without resorting to a fixed charge), one or both prices must be adjusted. The best feasible rate structure would modify these prices, taking the demand elasticities in each market at each period into account. As a result, a positive off-peak price could be efficient if demand is relatively inelastic in that period.

2. Peak Periods Vary by Exchange

Suppose that in exchange A the maximum demand occurs in the first period, whereas in exchange B demand is maximal in period 2. In this case marginal costs are

$$mc^1_A = \beta_A, \quad mc^1_B = 0$$
$$(14)$$
$$mc^2_A = 0, \quad mc^2_B = \beta_B$$

and the optimal rates are

$$(15) \quad p^1 = [b^1_A/(b^1_A+b^1_B)] \, \beta_A, \quad p^2 = [b^2_B/(b^2_A+b^2_B)] \, \beta_B \, .$$

In period 1, the price is a fraction of the marginal capacity cost in exchange A. The relationship for period 2 price is similar. In each case the proportions depend on the demand slopes of the commodities in each exchange. Again, these optimal prices will not generally satisfy the budget constraint and the best feasible prices would modify these rates on the basis of demand elasticities.

Quantity-weighted marginal costs

A feasible method of meeting the budget constraint is to construct the prices using quantity weights in place of slope weights in the previous formulas. Let

$$(16) \quad \theta^t_A = x^t_A/(x^t_A+x^t_B), \quad \theta^t_B = x^t_B/(x^t_A+x^t_B) = 1 - \theta^t_A \qquad t = 1, 2$$

be the proportions of the grouped outputs that occur in each exchange in period t. For case 1 (same peak period) set

$$(17) \quad p^1 = \theta^1_A\beta_A + \theta^1_B\beta_B, \quad p^2 = 0 \, .$$

For case 2 (different peak periods) set

$$(18) \quad p^1 = \theta^1_A\beta_A, \quad p^2 = \theta^2_B\beta_B \, .$$

These rates, based only on quantity information, can be given an informative interpretation in terms of suitably defined marginal costs.

Marginal cost of a group of commodities

When commodities are grouped it is not immediately apparent just what the "marginal cost" of the aggregate is. To define its marginal cost we must specify how each of the components of the group changes when the group itself changes by one "unit."

One plausible definition is to specify that the quantities of each commodity in the group vary proportionately. Thus for a change dx^t in the group quantity let the components change by

$$(19) \quad dx^t_A = \theta^t_A dx^t, \quad dx^t_B = \theta^t_B dx^t \, .$$

The marginal cost of the grouped output in period t is then

(20) $$mc^t = \partial C/\partial x = \theta^t_A mc^t_A + \theta^t_B mc^t_B \quad t = 1, 2.$$

Thus, the group marginal cost is defined as the quantity-weighted average of the individual commodity marginal costs.

With constant returns to scale, the rates (equation (17) or (18)) based on this measure of marginal costs are feasible. And in one special case they will be optimal--when the individual commodities that make up a group have the same elasticities of demand. To see this, write the equation (15) for the optimal restricted rates in terms of elasticities

(21) $$p^t = \frac{\eta^t_A x^t_A}{\eta^t_A x^t_A + \eta^t_B x^t_B} mc^t_A + \frac{\eta^t_B x^t_B}{\eta^t_A x^t_A + \eta^t_B x^t_B} mc^t_B$$

When $\eta_A = \eta_B$, the weights for the terms mc^t_A and mc^t_B are just the quantity weights θ^t_A, θ^t_B. In this case, commodities are homogeneous in terms of demand, and the optimal pricing rule requires only cost date.

4. A Network Of Exchanges

Each exchange has intra-exchange calling as in the previous model. In addition, there are inter-exchange (AB) calls which make use of capacity in the originating and terminating exchanges and also require a third capacity component--trunking facilities that connect the exchanges. The key feature of this model is the introduction of joint production, which occurs when local exchange switching capacity is shared by two different commodities.

Demands are

(22) $$\begin{aligned} x^t_A &= x^t_A(p^t_A) \\ x^t_B &= x^t_B(p^t_B) \\ x^t_{AB} &= x^t_{AB}(p^t_{AB}) . \end{aligned}$$

The capacity constraints are

(23) $$\begin{aligned} K_A &= \max\ (x^t_A + x^t_{AB}) \\ K_B &= \max\ (x^t_B + x^t_{AB}) \\ K_{AB} &= \max\ (x^t_{AB}) \end{aligned}$$

where I assume each inter-exchange call requires the switching capacity of an intra-exchange call in each of the two exchanges as well as inter-exchange trunking. Total costs are then

(24) $$C = \beta_A K_A + \beta_B K_B + \beta_{AB} K_{AB} .$$

The optimal prices are obtained from the Kuhn-Tucker conditions of the mathematical program (See, e.g. Littlechild, 1970). The prices are

$$p^1_A = \mu^1_A, \quad p^2_A = \mu^2_A$$

$$p^1_B = \mu^1_B, \quad p^2_B = \mu^2_B \tag{25}$$

$$p^1_{AB} = \mu^1_A + \mu^1_{AB}, \quad p^2_{AB} = \mu^2_B + \mu^2_A$$

where μ^t_j is the dual variable in period t for capacity of type j. The central result is that even when there are as many prices as commodities, the technological interdependence of the separate markets destroys the simple correspondence between maximum demands and maximum prices. However, with firm peaks, the optimal prices are equal to the marginal costs of the individual commodities.

For example, suppose that exchange A and inter-exchange calls are day peaking ($x^1_A > x^2_A$ and $x^1_{AB} > x^2_{AB}$) and exchange B is night peaking ($x^1_B < x^2_B$). The optimal prices will be

$$p^1_A = \beta_A, \quad p^2_A = 0$$

$$p^1_B = 0, \quad p^2_B = \beta_B \tag{26}$$

$$p^1_{AB} = \beta_A + \beta_{AB}, \quad p^2_{AB} = \beta_B .$$

Thus the inter-exchange calls should pay positive prices in both periods, not only in their peak (t=1) period. Moreover, despite the fact that AB calling is highest in period 1, p^2_{AB} could mathematically exceed p^1_{AB}, although this is unlikely in the particular example of local and inter-exchange calls.

Restricted rate structures

Here we reach the "realistic" case for telephone ratemaking. In practice, rates might well be restricted to be the same for all intra-exchange calls at a given time of day throughout a region or state, with separate rates applying for inter-exchange calls. Frequently, however, the same "off-peak" percentage discount is applied to both types of calls. In this case the restrictions are

$$p^t_A = p^t_B, \quad t = 1, 2$$

$$p^2_{AB} = \lambda p^1_{AB} \quad \text{where } \lambda = p^2_A/p^1_A = p^2_B/p^1_B . \tag{27}$$

Effectively, there are three rate parameters--the mean levels of the intra-exchange and inter-exchange rates and the percentage discount in the off-peak period.

In principle, the mathematical program for the welfare-maximizing network prices can be solved for any specified constraints on the rate structure. For a small problem--such as this example--this is quite feasible. But for realistic situations, the dimensions of the problem are substantially greater. M, rather than two, exchanges must be considered. Because the rate of demand varies over both the daily and weekly cycle, N distinct periods must be analyzed. And there are several levels of inter-exchange calls, conventionally grouped according to distance bands.

5. Evaluating The Efficiency of Telephone Rate Structures

A practical approach is to use the structure of the demand and cost model to evaluate the welfare effects of alternative rate structures without attempting to achieve a global optinum. This approach should be undertaken at two levels.

1. A Given Rate Structure

A particular rate structure specifies a grouping of commodities into time periods, distance bands, and perhaps geographic areas. The quantity-weighted marginal costs of each grouped output can be calculated by proportionately incrementing demands of each commodity in the group. (For example, if the peak period is 8 a.m. - 5 p.m., weekdays, the traffic load curve at those hours can be increased by a constant percentage). By calculating the "Ramsey number" of each group k at current prices and output levels

(28) $R_k = \eta_k(p_k - mc_k)/p_k$

the group marginal costs can be compared with prices (See Willig and Bailey, 1977). If rates are optimal (given the rate structure), all of the Ramsey numbers will be equal. If not, welfare can be increased by raising rates for groups with low Ramsey numbers and reducing rates for high R^k values.

2. Alternative Rate Structures

Some redesigning of the rate structure may yield welfare gains at least as large as those achievable by adjusting rate levels. Two closely related questions must be investigated--the number of different prices to charge, and the particular commodities to be included in each group. For example, local telephone rates might be limited to two price levels throughout the week, with the particular hours that peak prices apply determining which telephone calls are grouped together.

Guidance for grouping commodities is provided by two results from the earlier analysis:

1. If, within each group, all commodities have the same marginal cost, then a group price equal to the common marginal cost will be (first-best) optimal. This will be true even if the commodities have different elasticities of demand.

2. If, within each group, all commodities have the same elasticity of demand, then price should be equal to the quantity-weighted average of the commodity marginal costs.

For the telephone network, the marginal costs of several commodities will be similar when they (a) have the same peak period, and (b) use equipment that has similar unit capacity costs. As for demand elasticities, they will perhaps be similar when exchanges are grouped by type of customer.

These general considerations suggest that efficient grouping will be promoted by combining commodities according to similarities in both marginal costs and demand elasticities. For example, alternatives to a proposed 8 a.m. - 5 p.m. peak period could be considered by comparing both the demand elasticities and marginal costs at, say, 6, 7, and 8 p.m. with those in earlier hours. Hours that clearly follow the earlier elasticity and cost pattern readily suggest an extended period for the time-of-day rate structure. A mixed pattern of elasticities and costs, however, would require evaluating different combinations of grouped hours.

To evaluate a change in the number of prices in the rate structure additional data are needed. Practical restrictions on the number of separate rates are presumably due to the "transactions costs" the subscriber must bear to cope with an increasingly detailed structure of rates, and the additional administrative complexity for the telephone company of calculating, defending and revising such rates. Measurements of transaction costs are not readily available. However, one can demonstrate the size of the efficiency gain that could be realized by adding an additional rate, or on the other hand, the efficiency cost of simplifying the rate structure in a specified manner. These values can then be compared to subjective assessments of the hassle of coping with rate structures of differing complexity.

6. Summary

The design of appropriate rate structures for local telephone calls should be determined by the technology and cost characteristics of the local network. Apart from the equipment dedicated exclusively to serve each subcriber, nearly all of the costs of local telephone service are due to providing capacity sufficient to meet maximum demands. Thus, some form of peak-load pricing is desirable. A uniform average-cost price at all hours may be less efficient than a flat-rate tariff which charges nothing per cal, even if metering were costless.

Switching and trunking capacity is distributed throughout the network and jointly used by different types of calls. As a result, optimal prices may be positive when demand is below the maximum level, and the highest rate need not occur at the hour of peak demand.

Realistic rate structures can have only a limited number of separate prices, requiring that individual commodities be aggregated into groups. An efficient rate structures will combine hours and routes that have similar marginal costs and demand elasticities.

REFERENCES:

[1] Hirshleifer, J., Peak Loads and Efficient Pricing: Comment, Quarterly Journal of Economics, 72 (1958).

[2] Littlechild, S.C., Peak-Load Pricing of Telephone Calls, Bell Journal of Economics, 1 (Autumn 1970).

Littlechild, S.C. and J. J. Rousseau, Pricing Policy of a U.S. Telephone Company, Journal of Public Economics, 4 (February 1975).

[3] Michell, B.M., Economic Issues in Usage-Sensitive Pricing, The Rand Corporation, P-6530-NSF (September 1980).

[4] Steiner, Peter O., Peak Loads and Efficient Pricing, Quarterly Journal of Economics, 71 (November 1957).

[5] Willig, R. D. and Elizabeth E. Bailey, Ramsey-Optimal Pricing of Long Distance Telephone Services, Pricing in Regulated Industries: Theory and Application, John T. Wenders (ed.). Paper presented at an Economic Seminar sponsored by The Mountain States Telephone and Telegraph Company (January 1977).

Economic Analysis of Telecommunications:
Theory and Applications
L. Courville, A. de Fontenay and R. Dobell (eds.)

LOCAL TELEPHONE PRICING: TWO-TARIFFS AND PRICE DISCRIMINATION

J.A. Brander and B.J. Spencer

Queen's University and Boston University

1. Introduction

Local telephone service in Canada is priced in a rather unusual way. Specifically, consumers pay a fixed monthly fee for service and then make as many local phone calls as they wish at no extra cost: expenditure is insensitive to usage. Whether or not such a policy ever was optimal, it seems unlikely to persist given recent technological advances: the cost of monitoring usage is very much lower than it was, and for large exchanges is now "small" compared with the production cost of a phone call, and the increasing use of telephone lines for computer transmissions implies that certain users will use the system so heavily that serious inefficiency may arise if usage sensitive prices are not set.[1]

From a regulator's point of view telephone companies are classic public utilities. The regulator's objective is to encourage efficient and equitable provision of telephone service while allowing a certain profit to the telephone company. This budget or profit constraint implies that, if production is being carried out under increasing returns to scale, departures from marginal cost pricing are required.

In considering departures from marginal cost pricing one principle that has emerged is the Ramsey principle: efficiency criteria imply that markups over marginal cost charged different groups or for different products should be related to elasticities of demand. Specifically, low elasticities should be associated with high markups. (Classic references on the Ramsey principle are Ramsey (1927) and Boiteux (1956). More recent work includes Baumol and Bradford (1970) on the multiproduct case, and Hartwick (1978) on price discrimination among groups consuming a single product).

In this paper we examine optimal two-part tariffs for a public utility that is capable of discriminating among different types of consumers. In particular we focus on the problem faced by a public enterprise like a telephone company, whose output is purchased both by consumers for final consumption and by firms as an input to production. Two-part tariffs are pricing structures that consist of an entry or license fee that must be paid before any consumption takes place, and a constant per unit price. Standard references on two part tariffs are Gabor (1955) and Oi (1971). Two-part tariffs can be thought of as special case of "nonuniform" or "nonlinear" or "quantity dependent" prices. (See Spence (1977, 1980), Willig (1978), and Mirman and Sibley (1980) for general treatments of nonuniform pricing).

There are several important features of the telephone industry that we ingore. The cyclical pattern of demand for telephone service over the day and week is particularly important. Also the multi-product aspect of telephone service and the existence of consumption externalities among consumers have generated considerable interest. These issues are set aside in this paper. As for general background to local telephone pricing, good references are Mitchell (1978) and Baude et al. (1979).

The optimal pricing rule for the case in which there are many groups of consumers is shown to be a two-stage Ramsey rule. The first term is the usual Ramsey term and in addition there is a correction term depending on elasticities with respect to the license fees. The same rule applies even when some users are firms who use the output of the utility as an input to further production, provided these firms are perfect competitors. If downstream firms are imperfectly competitive, efficiency considerations imply that they should normally be charged a lower price for inputs, other things equal.

The problem is to maximize the sum of producers' and consumers' surplus subject to the constraint that the public utility earn a certain minimum profit. Different groups of consumers can be identified but each group is internally hetereogeneous. The utility may use discriminating two-part tariffs; that is, each group can be charged a different two-part tariff. The two-part tariff for group i consists of a license fee L^i and a price p^i. Prices greater than marginal cost will mean that consumers who use the system will underconsume from a welfare point of view. On the other hand, positive license fees may exclude consumers whose consumption is socially desirable. The optimum will generally involve a mix of these two distortions. In addition, the different groups will be treated differently. If the utility is capable of breaking the consuming public into homogeneous groups prices can be set equal to marginal cost, license fees can be used to make up any deficit, and the first best outcome can be achieved. The problem arises because groups are internally heterogeneous.

There are n groups of consumers, indexed by the letter i, i = 1,..., n. Within each group consumers vary, but each consumer is identified as belonging to a particular group. For example, consumers may be indentified by the community in which they live. The two-part tariff facing group i is denoted (p^i, L^i). For any given tariff, group i can be divided into two subgroups: those who choose to consume, referred to as "members", and those who do not.

Let M^i = members in group i

S^i = surplus of members in group i

$$S = \sum_{i=1}^{n} S^i$$

We use surplus (areas under demand curves) as a measure of consumer welfare. The surplus measure is exact if consumer demands arise from utility functions of the U = u(x) + v where v is income spent on other goods. Alternatively, one can appeal to the approximation results of Willig (1976).

The good produced by the public utility is denoted X, and the consumption by group i is denoted X^i. Using μ to denote profit,

$$\Pi\ (P,L) = \Sigma P^iX^i + \Sigma L^iM^i - C(X(P,L))$$

where $P = (p^1,\ldots, P^n)$ and $L = (L^1,\ldots, L^n)$

Efficient pricing rules are found by maximizing the sum of producer and consumer surplus subject to the constraint that profits be non-negative. The Lagrangian function is

$$\ell(P,L,\lambda) = S(P,L) + \Pi(P,L) + \lambda\Pi(P,L)$$

The first order conditions are

$$S^i_p + (1 + \ )\Pi^i_p = 0$$

$$S^i_L + (1 + \ )\Pi^i_L = 0$$

S^i_p is the partial derivative of S with respect to P^i. By the usual "duality" result $S^i_p = -X^i$. Similarly $S^i_L = -M^i$. Intuitively, this last equality makes sense because an increase in the license fee reduces everyone's surplus by the license fee and forces marginal consumers out. However, marginal consumers have no surplus, so the only effect is that surplus to the group as a whole falls by M^i. Using C' to denote marginal cost, the first order conditions can be rewritten:

$$(1)\quad [(P^i - C')X^i_p + L^iM^i_p](1 + \lambda) + \lambda X^i = 0$$

$$(2)\quad [(P^i - C')X^i_L + L^iM^i_L](1 + \lambda) + \lambda M^i = 0$$

$$\lambda \geq 0,\ \Pi \geq 0,\ \lambda\Pi = 0.$$

The logic of the problem requires that $P^i \geq 0$, $L^i \geq 0$. We assume interior solutions. Corner solutions are possible but require a lot of explanation and contribute nothing substantial.

The first point to observe is that the groups of consumers are all connected through the Lagrange multiplier λ. If a non-distorting license fee can be imposed on one group, all groups should consume at the "first best" position with $P^i = C'$.

Proposition 1

If, at the optimum, $M^i_L = 0$ for some i (with $M^i > 0$) then $\lambda = 0$ and $p^i = C'$, for all j. If, in addition, $M^j_L \neq 0$ for some $j \neq i$, then $L^j = 0$.

Proff: See Brander and Spencer (1980).

Proposition (1) is concerned with the case in which, given the price and license fee, there are no marginal consumers for some group i. If $M^i_L = 0$ the license fee is (locally) a non-distorting lump sum transfer. The profit constraint is no longer binding (= 0) and all prices should be set equal to marginal cost. Also, license fees should be zero for any group for whom they would be distorting. With $P^j = C'$ and $L^j = 0$ all socially desirable consumption by group j takes place (provided second order conditions are satisfied). The distribution of license fees among groups with $M^j_L = 0$ is indeterminate. For the rest of the paper it is assumed that the profit constraint is binding: $\lambda > 0$.

2. Ramsey Rules for Final Consumption

If prices are to differ from marginal cost, the natural question to ask concerns how they should differ. The general insight, due to Ramsey (1927) is that large divergences should be associated with low elasticities. The reason is clear: when elasticities are low the deadweight welfare loss "triangle" is small compared to the transfer of surplus. Since the problem is to transfer surplus from consumers to the public utility at minimum deadweight loss it is not surprising that Ramsey rules arise. With two-part tariffs the Ramsey rules are more complicated, but the basic insights remain. To begin we consider the special case which corresponds to the pricing practice currently employed by Bell Canada: usage prices are set to zero. The problem is then to choose the optimal discriminating license fee.

Taking equation (2), setting $P = 0$, letting $\beta = \lambda/(1 + \lambda)$, and rearranging yields

$$-C'X^i_L + L^iM^i_L = \beta M^i$$

Letting the elasticity of membership with respect to the license fee, $M^i_L L^i/M^i$, be denote ε^i_L, and denoting the average consumption of marginal consumers, X^i_L/M^i_L, by x_i^* we get

$$L^i = C'x_i^*/(1 + \beta/\varepsilon^i_L) \quad (3)$$

This rule implies that low membership elasticities are associated with high license fees, as we might expect.

The more general case, in which discrimination over both a license fee and a price is allowed, is more complicated but a reasonable interpretation is possible. It is useful to define the following variables:

$e^i = p^i - C'$: the "excess price"

$\alpha^i = P^iX^i/L^iM^i$

$\eta^i_p = x^i_p P^i/X^i$: the price elasticity of demand by group i

$\varepsilon^i_p = M^i_p P^i/M^i$: the price elasticity of membership by group i

From first-order condition (1)

$$[(P^i - C')X^i_p + L^iM^i_p]/X^i = [(9p^j - C')X^j_p + L^jM^j_p]/X^j$$

Rewriting this in terms of elasticities,

$$[e^i\eta^i_p/P^i] + \varepsilon^i_p/\alpha^i = e^j\eta^j_p/P^j + \varepsilon^j_p/\alpha^j$$

and rearranging yields

$$(4) \qquad \frac{e^i/P^i}{e^j/P^j} = \eta_p^j/\eta_p^i + \frac{\alpha^i \varepsilon_p^j - \alpha^j \varepsilon_p^i}{\alpha^i \alpha^j \eta_p^i e^j/p^j}$$

The proportional markup follows a two-stage Ramsey rule. The first term is the standard Ramsey rule, as in Hartwick (1978) for the price discrimination case. The second term is a correction factor which depends on price elasticities of membership. As we might expect the Ramsey rule for the two-part tariff is more complicated than for the linear pricing case, but fortunately it can be expressed as the "original" term plus a correction which may be positive or negative, and which will be close to zero if the groups are similar.

3. Local Telephone Service as a Joint Output

(This section is based on comments of George Hariton). Some observers feel that local telephone service is best thought of as two joint outputs: access to the telephone network and usage within the local calling area. In this case the cost function should be thought of as depending on arguments X and M separately:

$$C = C[X(P,L),\ M(P,L)]$$

Using C_X and C_M to denote partial derivatives the first order conditions corresponding to (1) and (2) are

$$(1') \qquad [(P^i - C_X)X^i{}_p + (L^i - C_M)M^i{}_p](1 + \lambda) + \lambda X^i = 0$$

$$(2') \qquad [(P^i - C_X)X^i{}_L + (L^i - C_M)M^i{}_L](1 + \lambda) + \lambda M^i = 0$$

The interpretation implies that having non-zero license fees is consistent with a first best pricing configuration. Specifically, $P^i = C_X$ and $L^i = C_M$ is the marginal cost pricing solution with the license fee being just the marginal cost of access.

Even if there is a real marginal cost associated with access, however, there may be large fixed costs that do not depend on total access or total usage at the margin, so that the profit constraint may still be binding and Ramsey rules would still be of interest. The analog to expression (4) is

$$(4') \qquad \frac{e^i/p^i}{e^i/p^i} = \frac{\eta_p^i}{\eta_p^i} + \frac{A^i \varepsilon_p^j - A^j \varepsilon_p^i}{A^i A^j \eta_p^i e^i/p^j}$$

where $\quad A^i = X^i P^i / M^i (L^i - C_M)$

This reduces to (4) if $C_M = 0$.

This approach is probably more accurate for local telephone service. It does not alter the nature of the results in any substantial way. We prefer however, to continue considering the usual two-part tariff problem, in which $C_M = 0$.

4. Optimal Two-Part Pricing of Inputs

The previous sections establish some results concerning the structure of optimal price-discriminating two-part tariffs when the public utility can discriminate among different heterogeneous groups of consumers. For telephone companies the main form of discrimination is between business and residential customers. This type of discrimination is a little different from the model considered so far, however, because local telephone service used by businesses is an input to production rather than a consumption good. There is a fairly standard presumption in taxation theory that inputs should be priced at marginal cost and that only final outputs should be taxed[2] so we might wonder what kind of changes to the pricing formulas are required.

One difference between public utility pricing and optimal taxation is that the utility is not in a position to place a tax on the final output of firms who use telephone services as inputs. Consequently, it is efficient for the utility to charge firms a price that differs from marginal cost.

The main result of this section is that if the downstream industry is competitive, the input should be priced just as if it were being demanded for final consumption. If the downstream industry is not competitive, lower prices are called for.

Some extensions to the model are required. For ease of notation we shall assume that there is only one downstream industry and one group of final consumers. The generalization to the many-industry, many-group case follows directly upon reinterpretation of the relevant scaler variables as vectors. We shall also refer to the good as "telephone service" although the analysis is not specific to telephones and applies to any public utility.

As before, X stands for total output of telephone service. X^1 and X^2 refer to consumption by final consumers and downstream industry respectively. P and R are the associated prices and L and T are the associated license fees. M and N are interpreted as the total number of telephone connections for final consumers and for the downstream industry respectively, and there is a monthly license fee for each connection. When discussing consumers it was implicitly assumed that each consumer had at most one telephone connection, but in the case of firms we want to allow explicitly for the possibility that a single firm might use several telephone lines and pay a license fee for each.

Output of the downstream industry is denoted Y and the associated price is q, so the inverse demand function is q(Y). S^1 is consumers' surplus from telephone service and S^2 is consumers' surplus from good Y. Demands for telephone service and good Y are assumed independent in the sense that they enter utility functions in an additively separable way. This allows simple addition of S^1 and S^2 and lets us ingore cross-elasticities between the two goods. Finally, is the profit of the public utility and B is the profit of the downstream industry.

The Lagrangian function is

$$\ell = S^1(P,L) + S^2(q(Y)) + (1 + \lambda)\Pi(P,R,L,T) + B(R,T)$$

The first order conditions concerning P and L are as in equation (1) and (2). Using either subscripts or primes to denote derivatives, the first order conditions concerning R and T are

(5) $\ell_R = S^2_q q' Y_R + (1 + \lambda)\Pi_R + B_R = 0$

(6) $\ell T = S^2_q q' Y_T + (1 + \lambda)\Pi_T + B_T = 0$

These expressions can be usefully rearranged making use of the following relationships. Let k = marginal cost in the downstream industry.

(7) $S^2_q = - Y$

(8) $B_R = (q - k)Y_R + Yq'Y_R - X^2$

(9) $\Pi_R = (R - C')X^2_R + X^2 + TN_R$

(10) If we let $\sigma = (q - k)Y_R/X^2$

then inserting (7), (8), (9) and (10) in (5) yields

(11) $[(R - C')X^2_R + Tn_R](L + \lambda) + (\lambda + \sigma)x^2 = 0$

Similarly, expression (6) becomes

(12) $[(P - C')X^2_T + TN_T](1 + \lambda) + (\lambda + \delta)N = 0$

where $\delta = (q-k)Y_T/N$

Expressions (11) and (12) have exactly the same form as (1) and (2) except for the factors σ and δ. However, for competitive industries, price q is set equal to marginal cost k. Therefore, for competitive downstream industries $\sigma = \delta = 0$. Thus the following proposition can be stated:

Proposition 2

If the downstream industry is perfectly competitive, the optimal discriminating two-part tariffs for final demanders and downstreams firms has exactly the same structure as in the case in which all demand is forfinal consumption. (This follows because the first order conditions for R and T have the same form as the first order conditions for P and L).

By proposition 2 the two-stage Ramsey formula given by (4) applies to price discrimination among competitive downstream industries (or firms) as well as to groups of final consumers. There is, however, one important difference between the input case and the case in which telephone services are for final demand only. Specifically, Proposition 1 does not hold for downstream competitive industries. Even if total consumption of telephone lines does not change as the license fee, or charge per line, rises, the entire difference between marginal and average cost should not be loaded onto the competitive industry. Because higher license fees are reflected in higher prices for the output of the downstream industry, such license fees would be distorting. ($N_T = 0$ does not imply $X^2_T = 0$ for firms, so the proof of Proposition 1 does not go through).

In any case if the downstream industry is not competitive a different set of prices and license fees should be used. Normally, imperfectly competitive firms should face lower mark-ups.

Proposition 3

Provided that telephone service is not an inferior factor of production, an imperfectly competitive firm should be charged a lower mark-up of price over marginal cost than a corresponding competitive firm or group of final consumers.

Proof:

Consider a fairly general setting in which there are many imperfectly competitive downstream industries. We imagine setting R^i and T^i for each downstream industry i. Similarly we have the corresponding variables q_i, X^i, Y^i, N^i, and B^i. Form (11) the markup over marginal cost is

$$R^i - C^i = \frac{-(\lambda + \sigma)X^i}{(1 + \lambda)X^i_R} - TN_R/XR^i$$

The difference between the competitive and imperfectly case is that σ is zero for competition but not otherwise. The definition of σ: $\sigma = (q^i - k^i)Y^i_R/X^i$ indicates that σ must be negative if telephone service is a normal factor ($Y^i_R < 0$ if X^i is normal). Since $\lambda X^i/(1 + \lambda)X^i_R < 0$, an imperfectly competitive industry should be charged a lower mark-up over marginal cost than should a corresponding competitive industry. This completes the proof.

Thus, other things equal, imperfectly competitive industries should be charged lower markups (and lower actual prices). This comes about because it is desirable to cut back the imperfectly competitive sector less than the competitive, since the imperfectly competitive sector is already producing "too little". If, however, the input is inferior, increasing its price will increase output so a higher markup should be charged.

It is often claimed that business firms have lower demand elasticities for telephone service than do final consumers. If so this would tend to offset the relative subsidy to imperfectly competitive firms vis a vis final consumers.

5. Concluding Remarks

Setting price equal to marginal cost is generally the "first-best" solution to optimal pricing problems. However, if a public utility is producing with decreasing average cost and is constrained to achieve some (non-negative) profit target, marginal cost pricing is not possible: departures from marginal cost pricing are required. Pure efficiency criteria imply the Ramsey principle: markups over marginal cost should be larger as the relevant elasticities are smaller.

The rigorous welfare-theoretic foundation of using efficiency criteria is based on one of two principles. Either there is an explicit social welfare function behind the scenes which some agent is maximizing through optimal income redistribution, or the Pareto criterion is being used in that compensating lump sum transfers are made so that everyone winds up better off (or at least no worse off).

Neither of these approaches is very practical. Indeed, if non-distorting transfers could be made easily, the problem of maximizing welfare subject to a profit constraint would not arise: prices could be set equal to marginal cost and lump sum transfers could be used to subsidize decreasing cost industries. In practice, acceptance of Ramsey pricing rules (as with cost-benefit analysis) is tantamount to adopting the policy of maximizing economic benefits "to whomsoever they may accrue".

The implicit assumption being made when efficiency measures (such as consumer and producer surplus) are used is that the social value of a dollar is the same for every consumer. Thus solutions to such problems may involve large pure transfers among consumers to achieve small efficiency gains. Furthermore, these transfers are likely to be regressive. The results of this paper suggest that such problems are probably more acute when two part tariffs are available and when downstream imperfect competition is taken into account.

Specifically, Proposition 1 shows, in extreme form, that a "captive" group of consumers will subsidize other consumers under Ramsey optimal two-part pricing. Somewhat more interestingly, Proposition 3 show that imperfectly competitive downstream firms would be relatively subsidized under Ramsey optimal pricing.

It is not surprising, therefore, that subsidy-free pricing has become a major concern in the study of regulation. Form the regulator's point of view pricing structures that involve large cross subsidies are inequitable, and from the point of view of the regulated firms themselves, subsidized structures are dangerous because those cases in which cross-subsidization is large are precisely the cases in which private firms are likely to find competitive entry attractive, leading to so-called "sustainability" problems. (Willig (1978) considers different equity concepts for telephone pricing, and Faulhaber (1979) and Rheaume (1981) consider subsidy-free pricing for the multi-product case).

All this suggests that pure Ramsey optimal pricing may not be desirable. Also, as shown by equations (4) and (4'), the formidable array of elasticities that must be calculated makes one wonders if such formulae would ever be useful in actually setting prices in any case. Finally, although the issue has not been taken up in this paper, there is reason to believe that efficiency gains from adopting Ramsey optimal pricing structures are likely to be small. One suspects that a fairly simple usage sensitive pricing structure incorporating peak load prices close to marginal costs and subsidy-free license fees to make up deficits would be the best approach.

FOOTNOTES:

[1] In the United States, of course, there is an additional rather pressing concern. Local service has traditionally been subsidized by high priced and profitable long distance service. Long distance service, however, is in competition with a growing private telecommunications industry so local service may be forced to cover its own costs. If so, usage sensitive prices promise to be an

attractive method of raising revenue and cutting costs.

[2] For example, Diamond and Mirrlees (1971) argue that inputs should be priced at marginal cost and that only final outputs should be taxed.

[3] See Ferguson (1969), especially Ch. 6.

[4] A discussion of two different concepts of equity can be found in Willig (1979).

REFERENCES:

[1] Baude, J.A., Cohen, G., Garfinkel, L., Krehmeyer, M., and Ogg, J., Perspectives on Local Measured Service, Proceedings of the Telecommunications Industry Workshop (Kansas City, Missouri, 1979).

[2] Baumol, W.J. and Bradford, D.R., Optimal Departures from Marginal Cost Pricing, The American Economic Review, Vol. 60, 3 (June 1970) 265-283.

[3] Boiteux, M., Sur la Gestion des Monopoles Publics Astreints a l'Equilibre Budgetaire, Econometrica 24 (1956) 22-40.

[4] Brander, J. and Spencer, B., Ramsey Optimal Two-Part Tariffs: The Case of Many Heterogeneous Groups, mimeo (1980).

[5] Diamond, P.A., A Many Person Ramsey Tax Rule, Journal of Public Economics, Vol. 4, 4 (November 1975) 335-342.

[6] Diamond, P.A., and Mirrless, J., Optimal Taxation and Public Production, American Economic Review 61 (March and June 1971) 8-27 and 261-78.

[7] Faulhaber, Cross-Subsidization in Public Enterprise Pricing, AT & T Working Paper (1979).

[8] Faulhaber and Panzar, Optimal Two-Part Tariffs with Self-Selection, Bell Laboratories Economic Discussion Paper 74 (1977).

[9] Feldstein, M.S., The Pricing of Public Intermediate Goods, Journal of Public Economics 1 (1972) 45-72.

[10] Feldstein, M.S., Equity and Efficiency in Public Sector Pricing: The Optimal Two-Part Tariff, Quarterly Journal of Economics, Vol. 86, 2 (May 1972) 175-187.

[11] Ferguson, C.E., The Neoclassical Theory of Production and Distribution (Cambridge University Press, 1969).

[12] Gabor, A., A Note on Block Tariffs, Review of Economic Studies 23 (1955) 32-41.

[13] Goldman, M.B., Leland, H., and Sibley, D., Optimal Nonuniform Prices, Bell Laboratories Economic Discussion Paper 100 (1977).

[14] Hartwick, J.M., Optimal Price Discrimination, Journal of Public

Economics, Vol. 9, 1 (February 1978) 83-89.

[15] Leland, H. and Meyer, R., Monopoly Pricing Structures with Imperfect Discrimination, The Bell Journal of Economics, Vol. 7, 2 (Autumn 1976) 449-462.

[16] Littlechild, S.C., Two-Part Tariffs and Consumption Externalities, The Bell Journal of Economics, Vol. 6, 2 (Autumn 1975) 661-670.

[17] Mirnam, L.J. and Sibley, D., Optimal Nonlinear Prices for Multiproduct monopolies, The Bell Journal of Economics, Vol. 11, 2 (Autumn 1980) 659-670.

[18] Mitchell, B.M., Optimal Pricing of Local Telephone Service, American Economic Review, Vol. 68, 4 (September 1978) 517-537.

[19] Mohring, H., The Peak Load Problem with Increasing Returns and Pricing Constraints, American Economic Review, Vol. 60, 4 (September 1970) 693-705.

[20] Ng, Y. and Weisser, M., Optimal Pricing with a Budget Constraint - The Case of the Two-Part Tariff, Review of Economic Studies 41 (July 1974) 337-345.

[21] Oi, W.Y., A Disneyland Dilemma: Two-Part Tariffs for a Mickey Mouse Monopoly, Quarterly Journal of Economics 85 (February 1971) 77-96.

[22] Ordover, J. and Panzar, J., On The Nonexistence of Pareto Superior Outlay Schedules, Bell Journal of Economics, Vol. 11, 1 (Spring 1980) 351-354.

[23] Ramsey, F., A Contribution to the Theory of Taxation, Economic Journal 37 (March 1927) 47-61.

[24] Rheaume, G.C., Welfare Optimization and Efficient Prices Under a Regulated Monopoly (This volume).

[25] Spence, A.M., Nonlinear Prices and Welfare, Journal of Public Economics, Vol. 8, 1 (1977) 1-18.

[26] Spence, A.M., Multi-Product Quantity Dependent Prices and Profitability Constraints, Review of Economic Studies 47 (October 1980) 821-842.

[27] Spencer, B. and Brander, J.A., Second Best Pricing of Publicly Produced Good: The Case of Downstream Imperfect Competition, mimeo (1981).

[28] Williamson, O., Peak Load Pricing and Optimal Capacity with Indivisible Constraints, American Economic Review, Vol. 56, 4 (September 1966) 810-827.

[29] Willig, R.D., Consumers' Surplus without Apology, American Economic Review, Vol. 66, 4 (September 1976) 589-597.

[30] Willig, R.D., Pareto-Superior Nonlinear Outlay Schedules, Bell Journal of Economics, Vol. 9, 1 (Spring 1978) 56-69.

[31] Willig, R.D., Consumer Equity and Local Measured Service (Baude et al. 1979).

PART 3
WELFARE CONSIDERATIONS AND REGULATION

Pricing Implications

Economic Analysis of Telecommunications:
Theory and Applications
L. Courville, A. de Fontenay and R. Dobell (eds.)

WELFARE OPTIMAL SUBSIDY-FREE PRICES UNDER A REGULATED MONOPOLY

G.C. Rheaume

Bell Canada

1. Introduction

One particularly important issue in economic theory is optimal pricing. In the literature, two of the discussions on regulated monopolies have been Ramsey pricing and subsidy-free pricing. They are pricing rules that are related to costs and the problems of sustainability, predatory measures and inefficient competition. Ramsey pricing has focused on the efficient allocation of resources when a profit constraint is binding. Its sustainability has been demonstrated under strict cost conditions of a natural monopoly, by W.J. Baumol, E.E. Bailey, and R.D. Willig.[1] On the other hand, subsidy-free pricing has implied revenue conditions related to costs that have been associated with sustainability and the Pareto criterion. Such pricing was analysed extensively by G.R. Faulhaber.[2]

Ramsey prices are not necessarily compatible with subsidy-free prices. Such a comment has been mentioned in the literature[3] and examples of this possible incompatibility have been provided.[4] But the economic consequences from a Pareto criterion and other points of view have not been sufficiently explored.

The purpose of this theoretical paper is first, to discuss the analytical issue of incompatibility between Ramsey pricing and subsidy-free prices. Then a theoretical model is constructed to develop welfare optimal prices that are subsidy-free.

Ramsey pricing, cross-subsidization and subsidy-free prices are described in the first sections. Then, cross-subsidized Ramsey pricing leads to a discussion on the efficient allocation of resources and equity. Afterwards, the welfare model that satisfies allocative efficiency, anonymous equity[5] and the profit constraint is explored.

2. Major Assumptions

Because the following discussion is theoretical, some simplifying assumptions can be made. First, the industry is assumed to be a regulated monopoly under a profit constraint. Open entry is possible, at least for some outputs.

Second, the prices are linear. Linearity in the price structure is a constraint that could be impractical. But because of the theoretical simplicity of linear prices, they have been adopted for the purpose of this study. Services and outputs are used interchangeably although actual services should not be implied from them.

Third, there are no cross-elastic effects on demand and no externalities. Eliminating the cross-elasticities of demand and the externalities from the discussion simplifies the analysis and permits us to focus on the specific issues addressed in this paper.

The models presented in this paper use the concepts of consumer surplus, and of producer surplus or economic rent. Such concepts as measures of social welfare have been extensively debated in the literature of economic theory. They may not represent the true benefits of consumers and producers. Their assumptions have been defended and criticized. It is not the purpose of this paper to elaborate on the appropriateness of such concepts.

Services are defined according to consumer groups such that each service or subset of services is identified with a consumer group. Such a structure simplifies the discussion about subsidy-free prices and equity.

3. Ramsey Pricing

In this section, the Ramsey model is developed and discused. The model was originally elaborated in 1927 by F.P. Ramsey[6] as a solution to a taxation problem of obtaining given revenues. It then became popular in the literature on public financing. It is only later in the development of optimal price theory that the Ramsey rule was applied to a regulated monopoly under a profit constraint. Such an adaptation became popular in 1970 with the publication of an article entitled "Optimal Departures from Marginal Cost Pricing", written by W.J. Baumol and D.F. Bradford. But in the Ramsey pricing literature, cross-subsidization, competitive entry and predatory pricing were not directly discussed.

In the Ramsey model associated with public utilities, the regulated monopoly is assumed to have a binding profit constraint and the optimization goal is to set prices such that social welfare is maximized. The maximization process is to efficiently allocate resources without having the regulated monopoly incurring excess economic profits or economic losses.

Furthermore, as mentioned earlier, let us assume that there are no externalities that need to be incorporated in the maximization problem. There are also no cross-elastic effects of demand between services. Then, the consumer surplus is defined as follows:

$$(3.1) \qquad \sum_{i=1}^{n} \int_{0}^{q_i^*} p_i(q_i)\, dq_i - \sum_{i=1}^{n} p_i(q_i) q_i$$

Where p_i is the price of output i

q_i is the quantity of output i

$p_i(q_i)$ is the inverse demand function for output i, and

q_i^* is the optimal quantity of output i.

The producer surplus is defined as follows:

$$(3.2) \qquad \sum_{i=1}^{n} p_i(q_i)q_i - C(q)$$

Where q is a vector of outputs $(q_1, q_2, \ldots, q_n)$

$C(q)$ is the total opportunity cost to the regulated monopoly to produce the vector of output q

At prices $p_1, p_2, \ldots, p_n$, the quantity purchased of each output is equal to the quantity sold such that the welfare function is:

$$(3.3) \qquad \sum_{i=1}^{n} \int_0^{q_i^*} p_i(q_i)dq_i - C(q)$$

Therefore, the optimization problem is to maximize the welfare function (3.3) subject to a profit constraint that could be defined as:

$$(3.4) \qquad \sum_{i=1}^{n} p_i(q_i)q_i - C(q) \leq \Pi^*$$

where Π^* is the economic profit allowed.

The profit constraint is usually assumed to be between zero and the entry costs.[7]

The solutions to the problem are:

$$(3.5) \qquad p_i - \partial C/\partial q_i - \lambda((\partial p_i/\partial q_i)\, q_i + p_i - \partial C/\partial q_i) \leq 0 \qquad i = 1, 2, \ldots, n$$

$$(3.6) \qquad \sum_{i=1}^{n} p_i(q_i)q_i - C(q) - {}^* \leq 0$$

The inequalities become equalities if q_i's are strictly positive and the constraint is binding.

Transforming equation (3.5)

$$(3.7) \qquad p_i^* = \varepsilon_i\, MC_i/(\varepsilon_i + a)$$

Where p_i^* is the Ramsey price

ε_i is the price elasticity of demand for output i, $\varepsilon_i < 0$

MC_i is the marginal cost of output i

a is the Ramsey number, $a = -\lambda/(1 - \lambda)$

(For more infromation, see Appendix A).

Therefore, according to the Ramsey rule, the Pareto optimal solution under a profit constraint requires that the price of each output deviates from marginal cost according to the inverse elasticity of demand. Services with inelastic demands will have prices that deviate more from marginal cost than those with more elastic demands. Such a pricing policy assures that the quantities of each service are the least different from those under marginal cost pricing.

In the literature on Ramsey pricing, the weak inequalities of the Kuhn-Tucker conditions are assumed to be strict equalities. In other words, it is assumed that all services are provided. But, if instead of an equality, an inequality was assumed for a particular service (see Appendix B), it could still be an optimal solution where that particular service would not be offered. Provision versus non-provision of service at a particular price is usually not questioned in the Ramsey pricing literature.

4. Burden Test

In discussing Pareto optimal solutions, J. Rohlfs mentions the possibility of optimal pricing where some services are not provided.[8] To verify whether provision is the optimal solution, the "burden test" is suggested. It is defined at a particular set of prices, in this case, Ramsey prices, as follows: Suppose the service or subset of services considered was not offered by the multi-output monopoly while prices of the other services remain constant. If profits of the firm decrease from this action, then the service or subset of services passes the burden test.

If a service or subset of services passes the burden test at given prices, then its provision at optimal prices is Pareto superior to its non-provision. Such a test is similar to the subsidy-free concept.

The Pareto criterion applied in this paper and the related literature, refers to an industry not the overall economy. Therefore, it is necessary to assume that the industry's purchase of resources does not affect their allocation in the rest of the economy. Furthermore, the prices of services supplied in the industry are assumed to not affect significantly the output levels of products outside the industry.

5. The Theory of Subsidy-Free Rate Structures

The literature on cross-subsidization has defined Pareto optimal criteria for the issue of provision versus non-provision of a particular service or subset of services. It presents conditions of a subsidy-free rate structure based on the Pareto superiority principle of welfare economics.

A price structure is subsidy-free only if each customer prefers the state in which each service is provided by the multi-output firm to the state in which only some of the services are offered. Such a situation occurs because no consumer pays more for a service (or subset of services) with the provision of the other services than without, and the consumers of the other services are better off with the provision than non-provision by the multi-output enterprise. Therefore, at least some consumers are made better off and no consumer is made worse off with the provision than with the non-provision of each service by the multi-output firm. In other words, provision of service under a subsidy-free rate structure is Pareto superior to non-provision.

On the other hand, if there is cross-subsidization, then the consumers of the subsidizing service or subset of services are made worse off by the provision of the subsidized services than they would be without their provision. They are paying more for the consumption of the subsidizing services than they would if the subsidized services were not provided. Therefore, consumers of the subsidized services are made better off with the provision of these services than without by making the consumers of the subsidizing services worse off. In other words, provision of the subsidized services is not Pareto superior to their non-provision.

But non-provision of subsidized services is not Pareto superior to their provision. Under cross-subsidization, consumers of the subsidizing services are made better off by the non-provision of the subsidized services than with their provision, by making the consumers of the subsidized services worse off.

Since the consumers of the subsidizing services prefer non-provision of the subsidized services to provision while the consumers of the subsidized services prefer the opposite alternative, under cross-subsidization, provision is Pareto noncomparable to non-provision.[9] Such a solution can also be found using the Koopmans' efficiency ranking.[10]

In order to identify a subsidy-free rate structure from a welfare economics' point of view, the following two alternative criteria are developed in the theoretical literature:

a) Any service or proper subset of services offered by a multi-output firm is not providing a subsidy if its revenues do not exceed its "stand alone" costs.

Let N be the set of n services considered. For any proper subset of N, $S \subset N$, q^S is a n-vector such that:

$$q_i^S > 0 \quad , \quad i \in S$$

and $$q_i^S = 0 \quad , \quad i \notin S \quad \text{but } i \in N.$$

Then, any q^S is not providing a subsidy if:

$$\sum_{i=1}^{n} p_i q_i^S \leq C(q^S) \qquad (5.1)$$

Where $C(q^S)$ is the stand alone cost of q^S

p_i is the given price of service i

b) Equivalently, if the revenues of any service or proper subset of services offered by the enterprise are greater or equal to its incremental costs, then it is not being subsidized.

Any q^S is not receiving a subsidy if:

$$(5.2) \qquad \sum_{i=1}^{n} p_i q_i^s \geq C(q) - C(q^t)$$

where q^t is a n-vector such that

$q_i^t = 0 \quad i \in S$

$q_i^t > 0 \quad i \notin S$ but $i \in T \subset N$.

T is a proper subset of N where $S \cup T = N$ and $S \cap T = \phi$

Subsidy-free prices are not necessarily unique. On the contrary, there can be more than one subsidy-free price structure. Subsidy-free prices are such that, for any subset of outputs, its revenues must be greater than or equal to its incremental costs and smaller than or equal to its stand alone costs. If the incremental costs are significantly smaller than the stand alone costs for any subset of services, then a number of different subsidy-free price structures could be found.

Economies of scope are an essential requirement for the existence of subsidy-free prices when the regulated monopoly has a binding profit constraints. The following proof demonstrates the necessity of such cost characteristics.

For any S, $T \subset N$, $S \cap T = \phi$, $S \cup T = N$ there is a n price vector p such that

a) $\quad pq^s + pq^t - C(q^s + q^t) = 0$

where

q^s is a n output vector such that $q_i^s > 0$ if $i \in S$

and $q_i^s = 0$ if $i \in T$

q^t is a n output vector such that $q_i^t > 0$ if $i \in T$

and $q_i^t = 0$ if $i \in S$

Furthermore, let us assume that p is a subsidy-free price structure.

That is,

b) $\quad pq^s - C(q^s + q^t) + C(q^t) \geq 0$

and

c) $\quad pq^t - C(q^s + q^t) + C(q^s) \geq 0$

Subtracting (b) and (c) from (a),

$$C(q^s + q^t) - C(q^t) - C(q^s) \leq 0$$

Therefore, if there is a binding zero profit constraint, economies of scope are a necessary condition for the existence of subsidy-free prices from a welfare economics point of view.

It is possible for a regulated monopoly with a binding profit constraint to have the paradoxical situation where each service or subset of services is, at the same time, subsidizing and being subsidized, given the criteria of subsidy-free prices mentioned above.

For any S, $T \subset N$, $S \cap T = \phi$, $S \cup T = N$, there is a n price vector p such that

d) $pq^s + pq^t - C(q^s + q^t) = 0.$

Furthermore, let each subset have revenues greater than their stand alone costs. That is,

e) $pq^s - C(q^s) > 0$ and f) $pq^t - C(q^t) > 0.$

Subtracting e) from d), for example, g) $pq^t - C(q^s + q^t) + C(q^s) < 0$

On the other hand, if a) is subtracted from d), for example, then c) results.

Therefore, each subset of services is subsidizing and is being subsidized. In this paradoxical case, provision of service by the multi-output enterprise is not Pareto superior to non-provision. On the contrary, all consumers are worse off with the provision of service by the multi-output enterprise than they would be on a stand alone basis. They are paying more not because they are subsidizing other consumers but becuase they are subsidizing the multi-output enterprise such that it meets its revenue requirement. Hence, provision of service by the multi-output firm is Pareto inferior to provision of a stand alone basis.

The paradoxical case occurs only when there are diseconomies of scope.

Subtracting e) and f) from d), $C(q^s) + C(q^t) - C(q^s + q^t) < 0.$

If the price structure is not subsidy-free, that is, its criteria do not hold, then it implies cross-subsidization only if there are economies of scope. With the existence of diseconomies of scope, no subsidy-free price structure can be found and cross-subsidization may or may not occur.

6. Cross-Subsidization and Ramsey Prices

In order to have a subsidy-free price structure, each service and proper subset of services should not be subsidizing or be subsidized by the other services. Therefore, subsidy-free Ramsey prices must have revenues of each service and proper subset of services that are greater than or equal to their incremental costs and smaller than or equal to their stand alone costs. That is, for each $q^s \in S$ $S \cap T = \phi$, $S \cup T = N$,

$$(6.1) \quad C(q) - C(q^t) \leq \sum_{S} p_i^* q_1 \leq C(q^s)$$

where p_i^* is the Ramsey price for service i

$q \in N$

$q^t \in T$

For each output, it is possible to derive the boundaries for the Ramsey number in order to have subsidy-free Ramsey prices. For any output, q_i, $p_i^* = \varepsilon_i MC_i/(\varepsilon_i + a)$. Furthermore, let q^s be the vector of n outputs such that only one output i is greater than 0.

Then, from (6.1),

(6.2)

$$C(q) - C(7^t) \leq \frac{\varepsilon_i MC_i}{\varepsilon_i + a} q_i \leq C(q^s)$$

where $q^t \in T$, $S \cap T = c$, $S \cup T = N$.

Dividing inequalities (6.2) by q_i,

(6.3) $(C(q) - C(q^t))/q_i \leq \varepsilon_i MC_i/(\varepsilon_i + a) \leq C(q^s)/q_i$.

The left hand side of the inequalities can be defined as the average incremental cost of q_i (AIC_i). It is the additional cost per unit of output i incurred only for the supply of that output. The right hand side of the inequalities can be defined as the average stand alone cost of q_i (ASC_i). It is the cost per unit of output i, to produce it on a stand alone basis.

By a number of mathematical operations on inequalities (6.3), it is possible to derive the following:

(6.4) $\varepsilon_i (1 - MC_i/ASC_i) \leq -a \leq \varepsilon_i(1 - MC_i/AIC_i)$

The Ramsey number, a, must satisfy the boundaries for each output as a necessary condition for a subsidy-free solution. But, if the outputs have significantly different elasticities of demand, marginal to average cost ratios, or both, then Ramsey prices are likely to lead to cross-subsidization.

Let us assume that Ramsey pricing does induce cross-subsidization. As was demonstrated in the previous section, a cross-subsidized pricing structure implies that the provision of subsidized services is Pareto noncomparable to their non-provision if the strict Pareto criterion is applied.

Furthermore, cross-subsidized Ramsey prices have the potential of increasing inefficiency when a regulated monopoly is susceptible to competitive entry. When a public utility has natural monopoly characteristics but the prices do not reflect such cost advantages, entrants may compete, increasing the overall costs to supply the markets. Such a situation is likely to occur under cross-subsidization. The profits from the subsidizing services offer an incentive for firms to enter the respective markets by profitably offering those services at prices that are lower than or equal to those of the monopoly. The regulated natural monopoly which cannot realize its cost advantages because of cross-subsidization would likely lose at least a portion of the profitable market making cross-subsidized Ramsey pricing unsustainable and increasing the industry's overall production costs.

Cross-subsidized Ramsey prices could also produce another type of inefficiency. Cost characteristics could possibly indicate that competition for the subsidized services would be more efficient than monopoly. But the cross-subsidized Ramsey prices for such outputs would likely discourage competition. In such a case, the monopoly supply of the subsidized outputs would have production costs above the optimal market structure solution.

7. Allocative Efficiency and Equity

In order to compare economic states from an allocative efficiency point of view, the Pareto criterion and Koopmans' efficiency rule are widely used. The Pareto criterion enunciates that one economic state has a more efficient allocation of resources than another, if it can make at least one individual better off without making anyone else worse off. Two economic states are said to be Pareto noncomparable if neither of them has resource allocations more or less efficient than the other. Noncomparable resource allocations imply that either of them is neither Pareto superior or Pareto inferior to the other.

Koopmans' efficiency ranking states that a feasible vector of outputs is more efficient than another if it has at least a greater quantity of one output and no smaller quantity of the other outputs.[11] If, between two feasible vectors of outputs, none satisfies Koopmans' efficiency ranking, then they are noncomparable from an allocative efficiency point of view.

As was discussed earlier, provision of service is Pareto noncomparable to non-provision under cross-subsidization. For a regulated monopoly under similar conditions, Koopmans' efficiency rule also cannot establish the ranking between provision and non-provision under cross-subsidization.

Under a profit constraint, $\Pi^* = 0$, provision of service would give $p^s q^s + p^t q^t - C(q^s, q^t) = 0$ where $S \cap T = \phi$, $S \cup T = N$. Furthermore, let subset S be the set of subsidized services and subset T be the set of subsidizing services.

Then, $p^s q^s - C(q^s, q^t) + C(0, q^t) < 0$

and $p^t q^t - C(0, q^t) > 0$

Let the profit function, $\Pi(q^t)$, defined as $p^t q^t - C(0, q^t)$, be concabe to the prices p_i, for all $i \in T$, which is a usual property of the profit functions with no cross-elasticities. Furthermore, given the profit constraint, $\Pi^* = 0$, non-provision of the subsidized services would imply $\Pi(q^t) = 0$.

Since, under provision $\Pi(q^t) > 0$, there exists at least one price, p_i, $i \in T$, which is lower under non-provision that would satisfy $\Pi(q^t) = 0$.[12] Given the inverse relationship between quantity demanded and prices, and assuming that price elasticity of demand is not zero, then, for all $i \in T$, $q_i \geq q_i'$ and there exists at least one $q_i \geq q_i'$ where q_i is the level of output under non-provision of the subsidized services and q_i' is the level of output of the provision scenario. Therefore, under provision, the subsidized services have positive quantities but smaller quantities of the subsidizing services. On the other hand, under non-provision, the quantities of the subsidizing services are greater but the subsidized services have zero output levels. Therefore, provision is Koopmans' efficiently noncomparable to non-provision.

For subsidy-free prices, it has been demonstrated earlier that provision is Pareto superior to non-provision. It is more efficient from Koopmans' ranking criterion since no service or subset of services is provided less because of provision of the other services. Provision of the latter services may even permit or require (given the profit constraint) a reduction in the prices of the former service or subset of services, increasing their quantities (if they do not have zero price elasticities of demand).

The discussion above was on comparisons between provision and non-provision of service at particular prices. But it is also important to compare different price structures from an allocative efficiency point of view. Each price structure implies an economic state in which given quantities of outputs are being supplied. Furthermore, to each subset of outputs corresponds a consumer group. For the purpose of this paper, it is necessary to compare cross-subsidized Ramsey prices to subsidy-free prices.

Cross-subsidized Ramsey prices are Pareto noncomparable to a subsidy-free price structure. Consumers of the Ramsey subsidized services would lose if a subsidy-free rate structure was adopted while the subscribers to the Ramsey subsidizing services would gain by such a change. The gains and losses would be the opposite for a change from the subsidy-free rate structure to the cross-subsidized Ramsey prices.

In going from cross-subsidized Ramsey prices to subsidy-free prices, smaller quantities of the subsidized services are supplied at higher prices while greater quantities of the subsidized services are provided at lower prices, under a profit constrained regulated monopoly.[13] The inverse situation would occur from subsidy-free prices to cross-subsidization. Therefore, applying Koopmans' efficiency ranking, cross-subsidized Ramsey prices are noncomparable to subsidy-free prices.

Using either the Pareto criterion on the Koopmans' efficiency ranking, it is impossible to state that cross-subsidized Ramsey prices are more or less allocatively efficient than subsidy-free prices. They are noncomparable from the point of view of allocative efficiency.[14]

In the literature on welfare theory, compensation rules have often been discussed to provide an answer to such Pareto noncomparable states. There are a number of such rules that exist. The most stringent of them is the Scitovsky criterion. It can be defined as follows: If, from a comparison between economic states A and B, those that benefit from A can both potentially compensate the losers and still be in a better position than at B while those that benefit from B cannot do the same, then A is socially preferable to B.

Given the structure of the paper, the Scitovsky criterion can be applied in the following way. If, from subsidy-free prices to cross-subsidized Ramsey prices, the consumer surplus gain to the subsidized customers is greater than the consumer surplus loss to the subsidizing ones, then cross-subsidized Ramsey prices is socially preferable to the subsidy-free prices. The subsidized customers can compensate the total loss to the subsidizing consumers and still be better off, while the subsidizing cannot do the same. On the other hand, if the gain is smaller than the loss, subsidy-free prices is socially preferred to cross-subsidized Ramsey prices.

A number of unsolved theoretical issues arise from the compensation rule. First, it compares consumer benefits by using monetary value as a cardinal index of consumer welfare gains and losses. Such interpersonal comparisons of consumer welfare involve the same unsolved debate as the one for consumer surplus.[15]

Second, the potential compensation rule has greater problems than consumer surplus. The latter is a measurement that leads to interpersonal comparisons. The former has the additional criticisms of embodying personal value judgements. The compensation is potential and hence, one economic state compared to another has a redistribution of income where some consumers benefit, others lose. The rule makes the value judgement that one economic state and thus, one distribution of income is better than another because the winnings from the gainers are greater in monetary value than the losses of the losers. Since some consumers are better off while others are worse off, the compensation rule transcends the problem of allocation of resources and addresses the policy issues of distribution. Therefore, the compensation rule makes value judgements that transcend the scope of economics.[16]

One may argue that actual compensations would achieve the socially preferred state if they are adequate to make the losers at least as well off and the gainers better off. Then, such a case would be based on the Pareto criterion. But actual compensations can be made only through some form of taxation. Such an instrument may not be available or socially desirable.

Although cross-subsidized Ramsey prices and subsidy-free prices are noncomparable from an allocative efficiency point of view, they each imply a distributional principle. Since it is assumed that the producer has the same level of profit from each of the two alternatives,[17] distributional considerations are between consumer groups, each group identified by the subset of outputs they are willing to purchase.

In the context of this paper, equity is the set of rules that provides a distribution of outputs between consumers. There are at least two such distributional priciples that may be applied to a regulated monopoly: willingness-to-pay and anonymous equity.

The principle of willingness-to-pay describes a distribution of outputs as follows. Consumer groups that are willing to pay more for their subset of services should pay more. This implies that services associated with consumer groups that have a high willingness-to-pay should have relatively higher prices than those with consumers that have a low willingness-to-pay. Ramsey prices implicitly imply such a principle of equity. Since they are based on the inverse price elasticity of demand, prices depend on the willingness-to-pay.

Another principle of fairness is anonymous equity. It states that customers of each service or subset of services should pay at least the incremantal costs of its provision. No consumer groups should be subsidized or be subsidizing. Such a principle is equivalent to subsidy-free prices.

There may be conflicts between anonymous equity and the willingness-to-pay principle, for example, in the case of cross-subsidized Ramsey prices. As mentioned above, such prices are likely to lead to inefficient competition or could discourage efficient competition. Such circumstances would then result in an economic state that is less efficient than a subsidy-free rate structure.

Nevertheless, if the principle of willingness-to-pay is adopted, then Ramsey prices are welfare optimal prices that will satisfy allocative efficiency and meet such a form of equity. It would be socially preferable to any other price structure even if it implies cross-subsidization.

On the other hand, if anonymous equity is the principle of fairness, a welfare optimal subsidy-free price structure would then be socially preferable to cross-subsidized Ramsey prices. In order to establish such optimal prices, a welfare model needs to be developed. Such a model is found in the next section.

8. Welfare Optimal Subsidy-Free Prices

Let us attempt to define optimal prices that satisfy efficiency and anonymous equity according to technological and market considerations. To build the associated model, it is not possible to impose anonymous equity implicitly. Furthermore, it cannot be incorporated in the objective welfare function. Therefore, it can only be added by means of additional constraints.

The multi-output regulated monopoly is assumed to have a profit constraint. Furthermore, subsidy-free prices are necessary and sufficient to satisfy the principle of anonymous equity. Therefore, anonymous equity will be expressed in the model as subsidy-free prices constraints.

For the purpose of the analysis, let us assume there are only two services that have independent demands.

The welfare function can be defined as:

$$(8.1) \qquad \int_0^{q_1^*} p_1 dq_1 + \int_0^{q_2^*} p_2 dq_2 - C(q_1, q_2).$$

The profit constraint is described as follows:

$$(8.2) \qquad p_1(q_1)q_1 + p_2(q_2)q_2 - C(q_1, q_2) \leq \pi^*$$

where $\Pi^* = 0$[18] which implies that the constraint is necessarily binding.[19] Furthermore, in order that provision of service be Pareto superior to non-provision, subsidy-free prices are required. That is, for the two-output case,

$$(8.3) \quad -p_1(q_1)q_1 + C(q_1, q_2) - C(0, q_2) \leq 0$$

$$-p_2(q_2)q_2 + C(q_1, q_2) - C(q_1, 0) \leq 0$$

Inequalities (8.3) state that each service covers at least its incremental costs such that no service will be subsidized.

The welfare optimization model is the maximization of the welfare function (8.1) subject to the profit constraint (8.2) and the subsidy-free constraints (8.3). The Kuhn-Tucker conditions are:

$$(8.4) \quad \partial L/\partial q_i = p_i\,(1+(\lambda_i - \lambda)(1 + 1/\varepsilon_i)) - MC_i(1 + \lambda_i - \lambda) - \lambda_j\,(MC_i - MC_i')$$

$$\leq 0 \quad i,j = 1,2,\ i \neq j,\ \varepsilon_i < 0$$

where λ_i is the Lagrange multiplier of the subsidy-free constraint of output i,

λ is the Lagrange multiplier of the profit constraint,

MC_i is the marginal cost of output i when produced jointly with output j,

and MC_i' is the marginal cost of output i on a stand alone basis.

$$q_i\,\partial L/\partial q_i = 0, \qquad q_i \geq 0$$

$$(8.5) \quad \partial L/\partial \lambda_1 = p_1q_1 - C(q_1, q_2) + C(0, q_2) \geq 0$$

$$\lambda_1\,\partial L/\partial \lambda_1 = 0 \qquad \lambda_1 \geq 0$$

$$(8.6) \quad \partial L/\lambda\partial_2 = p_2q_2 - C(q_1, q_2) + C(q_1, 0) \geq 0$$

$$\lambda_2\,\partial L/\partial \lambda_2 = 0 \qquad \lambda_2 \geq 0$$

$$(8.7) \quad \partial L/\partial \lambda = \Pi^* - p_1q_1 - p_2q_2 + C(q_1, q_2) \geq 0$$

$$\lambda\,\partial L/\partial \lambda = 0 \qquad \lambda \geq 0$$

Initially, these conditions do not indicate a solution since we do not know which constraints are binding. But from this two-output model, it is possible to derive some conclusions.

Since the anonymous equity priciple is imposed in the model, provision of service is necessarily Pareto superior to non-provision.[20] Therefore, the outputs are strictly positive and $\partial L/\partial q_i = 0$

Reworking equation (8.4), the optimal prices are

$$(8.8) \quad p_i^* = \varepsilon_i\ MC_i/(\varepsilon_i + a_i) - a_{ij}\ \varepsilon_i(MC_i - MC_i')/(\varepsilon_i + a_i)$$

where $\varepsilon_i < 0$

$$a_i = (\lambda_i - \lambda)/(1 + \lambda_i - \lambda)$$

$$a_{ij} = \lambda j/(1 + \lambda_i - \lambda)$$

Furthermore, it is not possible that $\partial L/\partial\lambda_1 = \partial L/\partial\lambda_2 = \partial L/\partial\lambda = 0$ unless there are no strict economies of scope and the difference between the costs of providing the services together and the costs of providing them separately (i.e., each on a stand alone basis) is equal to the allowed profits, Π^*.[21] Therefore, if strict economies of scope are assumed, at least one constraint must not be binding.[22]

According to our assumptions, the profit constraint is satisfied as an equality. Therefore, either both subsidy-free constraints are strict inequalities or only one is an equality.

Case 1: $\lambda_i = 0, \quad i = 1,2$

If both subsidy-free constraints are strict inequalities, the problem becomes the standard Ramsey pricing rule where,

$$p_i^* = \frac{\varepsilon_i\ MC_i}{\varepsilon_i + a} \qquad \text{where } a = -\lambda/(1 - \lambda)$$

and $\varepsilon_i < 0$.

Case 2: $\lambda_i = 0,\ \lambda_j > 0,\ i,j = 1, 2$ and $i \neq j$

If constraint for service j is binding, then service j covers only its incremental costs from the fact that $\partial L/\partial\lambda_j = 0$. Therefore, in order to satisfy the profit constraint equality, service i must have revenues to cover its stand alone costs. Therefore, in such a situation, the optimal solution is to price the otherwise Ramsey subsidized service according to its average incremental costs and the otherwise Ramsey subsidizing service according to its average stand alone costs.

That is, for service j,

$$(8.9) \quad p_j^* = \varepsilon_j\ MC_j/(\varepsilon_j + a_j)$$

and

$$(8.10) \quad p_j^* q_j - C(q_i, q_j) + C(q_i, 0) = 0$$

or

$$(8.11) \quad p_j^* = (C(q_i, q_j) - C(q_i, 0))/q_j = AIC_j.$$

For service i,

(8.12) $p_i^* = \varepsilon_i \, MC_i/(\varepsilon_i + a) - a' \, \varepsilon_i \, (MC_i - MC_i')/(\varepsilon_i + a)$

where $a = -\lambda/(1 - \lambda)$

$a' = \lambda_j/(1 - \lambda)$

and

(8.13) $p_i^* q_i - C(q_i, 0) = 0$

or

(8.14) $p_i^* = C(q_i, 0)/q_i$

$= ASC_i.$

Therefore, a few interesting insights are found from this exercise. First, if the welfare optimum is an interior solution of the opportunity set, then it is the usual Ramsey pricing solution. Second, if the welfare optimum is on the boundary of the opportunity set, then the solution is at the intersection or tangency of the binding profit constraint and one of the subsidy-free constraints. Third, if Ramsey pricing leads to cross-subsidization, then the welfare model specified above suggests as a welfare optimal solution that the otherwise Ramsey subsidized service should cover only its incremental costs and the otherwise Ramsey subsidizing service should cover only its stand alone costs.

To generalize the welfare optimal subsidy-free model to n services, (n being greater than 2), a number of cumbersome problems occur. First, the number of constraints increase drastically. The formula to calculate such a number is the following:

$$\sum_{r 1}^{n-1} \frac{n!}{r! \, (n - r)!} + 1$$

where n is the number of services,

r is the number of services selected in a combination.

The first part of the formula is the number of subsidy-free constraints to take into account all possible proper subsets of services that could be produced separately. It is a summation of combinations of r services. The addition of one in the formula corresponds to the profit constraint.

Second, the number of different marginal costs required to calculate the optimal price follows the number of constraints. The total number of marginal costs for each service is:

$$0.5 \sum_{r=1}^{n-1} n!/(r! \, (n - r)!) + 1$$

where n and r are defined above.

For example, if there are three services, then the welfare model has seven constraints and four different marginal costs for each service in its pricing solution.

The general model is defined as follows:

(8.15) Maximize $\sum_{i=1}^{n} \int_{0}^{q_i^*} p_i dq_i - C(q)$

Subject to: $- \sum_{s} p_i(q_i)q_i + C(q) - C(q^t) \leq 0$

$i \in S,\ S \cap T = \phi,\ S \cup T = N$

$S,T = 1, 2, \ldots, \sum_{r=1}^{n-1} n!/r!\,(n - r)!$

and

$\sum_{i=1}^{n} p_i(q_i)q_i - C(q) \leq \Pi^*$

where q is a n-output vector.

Solving for the n services and the $\sum_{r=1}^{n-1} n!/r!\,(n - r)! + 1$ Lagrange multipliers, the following Kuhn-Tucker conditions are found:

(8.16) $\partial L/\partial q_i = p_i\,(1 + \sum_{s}\lambda_s - \lambda) + p_i/\varepsilon_i\,(\sum_{s}\lambda_s - \lambda)$

$- MC_i\,(1 + \sum_{s} \lambda_s - \lambda) - \lambda_t\,(MC_i - MC_i^s) \geq 0$

$i \in S,\ S \cap T = \phi,\ S \cup T = N,\ \varepsilon_i < 0$

$S,T = 1, 2, \ldots, \sum_{n=1}^{n-1} n!/r!\,(n - r)!$

where λ_t is a $0.5 \sum_{r=1}^{n-1} n!/r!\,(n - r)!$ row vector of Lagrange multipliers,

$(MC_i - MC_i^s)$ is a $0.5 \sum_{r=1}^{n-1} n!/r!\,(n - r)!$ column vector

λ_s is the Lagrange multiplier of the sth subsidy-free constraint

$$q_i \; \partial L/\partial q_i = 0, \qquad q_i \geq 0$$

MC_i is the marginal cost of $C(q)$

MC_i^S is the marginal cost of $C(q^S)$

$i = 1, 2, \ldots, N$

$$(8.17) \quad \partial L/\partial \lambda_s = \sum_S p_i(q_i)q_i - C(q) + C(q^t) \geq 0$$

$$s = 1, 2, \ldots, \sum_{r=1}^{n-1} n!/r!\,(n - r)!$$

$$\lambda_s \partial L/\partial \lambda_s = 0, \quad \lambda_s \geq 0$$

$$(8.18) \quad \partial L/\partial \lambda = \Pi^* - \sum_{i=1}^{n} p_i(q_i)q_i + C(q) \geq 0$$

$$\lambda \partial L/\partial \lambda = 0$$

Since under the constraints of the model, provision of service is Pareto superior to non-provision, all outputs are strictly positive at the optimal solution.

That is,

$$\partial L/\partial q_i = 0.$$

Therefore,

$$(8.19) \quad p_i^* = \varepsilon_i \; MC_i/(\varepsilon_i + a_i) + a_{it} \; (MC_i - MC_i^S) \; \varepsilon_i/(\varepsilon_i + a_i)$$

where $\varepsilon_i < 0$

$$a_i = (\sum_S \lambda_s - \lambda)/(1 + \sum_S \lambda_s - \lambda)$$

a_{it} is a $0.5 \sum_{r=1}^{n-1} n!/r!(n - r)!$ row vector of λ_t multiplied by the scalar $1/(1 + \sum_S \lambda_s - \lambda)$.

If none of the subsidy-free constraints are binding, that is, $\sum_S \lambda_s = \sum_t \lambda_t = 0$ for every $S, T \subset N$, then the usual Ramsey pricing rule is the optimal solution. In such a situation,

$$a = a_i = a_j = - \lambda/(1 - \lambda) \quad \text{for any } i, j = 1, 2, \ldots, n$$

and $a_{it} = 0$ such that

$$p_i^* = \varepsilon_i \; MC_i/(\varepsilon_i + a)$$

Furthermore, if there are strict economies of scope for any combination of proper subsets such that their union equal the total set of n outputs and the profit constraint is binding, then, at most, half the subsidy-free constraints are binding. Therefore, under strict economies of joint production, at most half the subsets of services will cover only their respective incremental costs.

Equivalently, if there are strict economies of scope and the profit constraint is binding, then there are at least half the subsidy-free constraints that are not binding. Therefore, at least half of the Lagrange multipliers are equal to zero.

In the multiple output case (n > 2), there are, however, a number of unsettled problems where more research is required. First, for those subsets of services which have binding subsidy-free constraints, the price of each output must be found such that each subset covers its incremental cost. Second, for those subsets of services which do not have binding subsidy-free constraints, the price of each output must be found such that each subset covers both its incremental cost and a proportion of the common costs. Third, further exploration of the existence of a unique solution is required when Ramsey pricing leads to cross-subsidization. Fourth, it is possible that no single welfare optimal subsidy-free pricing structure can be found. Such a situation occurs only if there are diseconomies of scope.

9. Conclusions

The discussion on welfare optimal pricing was based on the criterion of allocative efficiency and distributional principles of equity under regulated monopolies with a profit constraint. The Pareto criterion and Koopmans' efficiency ranking were used to test for allocative efficiency.

Cross-subsidized Ramsey prices were demonstrated to be Pareto noncomparable to subsidy-free prices because of the partial order of the Pareto criterion. They were also proven to be Koopmans' efficient noncomparable under the given assumptions.

Ramsey pricing has been proposed in the literature as a means of eliminating the producer loss under marginal cost pricing. It is a welfare optimal pricing rule that satisfies a profit constraint.

Ramsey pricing does not explicitly state the welfare gain or welfare loss between consumer groups. This paper demonstrated such welfare differences between cross-subsidized Ramsey pricing and a subsidy-free rate structure. A compensation rule to choose which alternative was socially preferable, was critically analysed. Then, equity principles were discussed.

Ramsey pricing implicitly has a willingness-to-pay equity principle. It states that consumers that are willing to pay more for their service or subset of services should pay more.

Anonymous equity enunciates that no consumer group should be subsidized by, or be subsidizing, the prices of services. It is synonymous to subsidy-free pricing.

The choice between a willingness-to-pay distributional principle and anonymous equity remains a policy issue. If the former is chosen, Ramsey prices are suggested as the welfare optimal solution. On the other hand, if anonymous equity is preferred, then welfare optimal subsidy-free prices would be the suggested solution.

The paper elaborated a welfare model to achieve anonymous equity efficiently. The producer meets its allowed profit as in Ramsey pricing. The model then focuses on the distribution of the services between consumer groups.

The two-output welfare optimal subsidy-free model gave interesting insights. First, it is equivalent to Ramsey prices if the latter are subsidy-free. Second, if Ramsey prices lead to cross-subsidization, then the otherwise Ramsey subsidized service would cover in this model its incremental costs while the otherwise Ramsey subsidizing service would cover its stand alone costs. Unfortunately, further research is required when there are more than two outputs. The solution becomes very complex to analyse in a general way.

If a profit constraint is binding, economies of scope are a necessary condition for the existence of subsidy-free prices. If there are diseconomies of scope, then provision of service by a multi-output firm is Pareto inferior to provision on a stant alone basis.

If legal barriers to entry are eliminated or become less stringent, cross-subsidized Ramsey prices, as defined in this paper, can encourage inefficient competition for the subsidizing services and can discourage competition where it could be socially desirable. On the other hand, subsidy-free prices may not completely eliminate the possibility of a non-optimal market structure, but it is a necessary condition to the sustainability of natural monopoly or to the existence of socially desirable competition.

APPENDIX A:

From inequality (3.5) on page 4,

$$\text{(A.1)} \quad p_i^* - \partial C/\partial q_i - \lambda \left((\partial p_i/\partial q_i)\, q_i + p_i^* - \partial C/\partial q_i \right) = 0 \qquad \text{if } x_i > 0$$

$$\text{(A.2)} \quad p_i^* - MC_i - \lambda\, p_i^* \left(1 + \frac{1}{\varepsilon_i}\right) - \lambda MC_i = 0$$

$$\text{(A.3)} \quad (1 - \lambda)\,(p_i^* - MC_i) = \lambda p_i^*/\varepsilon_i$$

$$\text{(A.4)} \quad p_i^* - MC_i = -a p_i/\varepsilon_i \qquad \text{where } a = -\lambda/(1 - \lambda)$$

$$\text{(A.5)} \quad (\varepsilon_i + a)\, p_i^*/\varepsilon_i = MC_i$$

$$\text{(A.6)} \quad p_i^* = \varepsilon_i MC_i/(\varepsilon_i + a)$$

APPENDIX B:

The Lagrange multiplier, λ, of Ramsey's welfare optimization problem can be interpreted as the change in social welfare for a given change in allowed profit. Let us specify $\lambda < 0$, that is, social welfare decreases as the allowed profit increases.

From inequality (3.5) on page 4,

$$(1 - \lambda)(p_i - MC_i) - \lambda p_i/\varepsilon_i \leq 0$$

where $\varepsilon_i < 0$.

Then,

$$p_i - MC_i - \lambda p_i/(1 - \lambda)\varepsilon_i \leq 0$$

since $1 - \lambda > 0$.

Let us specify, $a = -\lambda/(1 - \lambda) > 0$.

Thus,

$$p_i (\varepsilon_i + a)/\varepsilon_i < MC_i.$$

Therefore, if $|\varepsilon_i| > a$, then $p_i \leq \varepsilon_i MC_i/(\varepsilon_i + a)$

if $|\varepsilon_i| < a$, then $p_i \geq \varepsilon_i MC_i/(\varepsilon_i + a)$

if $|\varepsilon_i| = a$, then it is undefined.

FOOTNOTES:

[1] Baumol, W.J., Bailey, E.E. and Willig, R.D., Weak Invisible Hand Theorems on the Sustainability of Prices in a Multiproduct Monopoly, American Economic Review (June 1977).

[2] See Faulhaber's "Cross-Subsidization in Public Enterprise Pricing" and "Cross-Subsidization: Pricing in Public Enterprises".

[3] See G.R. Faulhaber, AER, December 1975; W.J. Baumol, E.E. Bailey and R.D. Willig, AER, June 1977.

[4] See G.R. Faulhaver, AER, December 1975.

[5] Anonymous equity is the principle of fairness where no consumer subsidizes or is subsidized by another consumer. A definition is provided in the paper.

[6] Ramsey's paper is entitled: "A Contribution to the Theory of Taxation".

[7] The allowed economic profit is usually equal to zero. But there are models that have a wider scope for the allowed profit. They specify that it is between zero and the entry costs when the latter are strictly positive.

[8] See J. Rohlfs' "Economically Efficient Bell System Pricing" pages 7-8.

[9] Cross-subsidization thus implies that the Pareto criterion would not

provide an argument for the firm to change its status on the provision of subsidized services. Based on the Pareto principle, it would neither eliminate subsidized services if they were already being provided nor supply subsidized services if they are not already being provided. However, such a situation does not imply that the status quo is Pareto superior to non-provision of existing subsidized services or to provision of new subsidized services. It is neither Pareto superior nor Pareto inferior to such alternatives: It is Pareto noncomparable to them.

[10] Such a ranking is based on output levels as found in Koopmans' Three Essays on the State of Economic Science.

[11] Koopmans' ranking criterion is based on a number of assumptions. In order that it applies to an industry, the assumptions need to be specially interpreted. First, it is assumed that the consumers of the services considered are not saturated by their consumption of these outputs. Second, the technology and resources available to the industry are supposed to have limitations. Thirs, the industry's purchase and allocation of resources are assumed not to affect significantly the rest of the economy.

[12] There would also exist at least one price, p_i, $i \in T$, which is higher under non-provision that would satisfy $\Pi(q^t) = 0$ given the concavity of the profit function. Since the latter would be welfare inferior to the existing price under cross-subsidization and to the lower price where $\Pi(q^t) = 0$, it is socially preferable under non-provision of q^s, to choose the lower price. J.C. Panzar and R.D. Willig refers to such a choice as the undominated zero profit price vector in their paper, "Free Entry and the Sustainability of Natural Monopoly". In the AER, December 1975, G.R. Faulhaber mentions that the lower price would be chosen because of increased consumer welfare and the indifference of the firm between the higher and lower prices unless, as Faulhaber states, the firm is malicious or ignorant.

[13] Such an analysis is based on a number of underlying assumptions. First, the profit functions are concave to prices. Second, the price elasticity of demand is strictly greater than minus infinity and strictly smaller than zero. Third, there exists cross-subsidized Ramsey prices. Fourth, there exists a subsidy-free price vector.

[14] This conclusion is based on the arguments presented above and their assumptions.

[15] Interpersonal comparisons of the compensation rule assume that the marginal utility of income for each individual is defined equal whatever the level of income and whichever the individual. Furthermore, the consumer welfare indicators can be expressed as a cardinal measurement for each individual and given the proper assumption of the marginal utility of income, interpersonal comparisons can be made. Alternative assumptions to measure variations in consumer welfare have been discussed in the literature of welfare economics. The debate remains unresolved as in the case of consumer surplus. For further information, one can refer to W.J.

Baumol's Welfare Economics and the Theory of the State, or J. Rothenberg's, The Measurement of Social Welfare.

[16] See W.J. Baumol, Welfare Economics and the Theory of the State.

[17] Since the level of profit is the same, that is, $\Pi = 0$, it is assumed that the producer is indifferent between the two alternatives.

[18] The allowed economic profit is assumed to be equal to zero because a strictly positive allowed profit is difficult to incorporate in the subsidy-free constraints. But it is not unusual to assume allowed profit to be equal to zero in a regulated monopoly. Such an assumption seems acceptable especially for the purpose of this paper.

[19] The profit constraint is necessarily binding because if the profits are less than zero, not all factors are being paid their opportunity cost. Hence, the producer would incur economic losses.

[20] This issue is discussed in section 5 above.

[21] In the particular case being considered, since $\Pi^* = 0$,

$$C(q_1, q_2) = C(q_1, 0) + C(0, q_2).$$

[22] If there are weak economies of scope and they are defined as $C(q_1, q_2) = C(q_1, 0) + C(0, q_2)$, then the welfare optimal subsidy-free solution would be that each output covers its incremental costs which are equal to its stand alone costs.

Economic Analysis of Telecommunications:
Theory and Applications
L. Courville, A. de Fontenay and R. Dobell (eds.)

EMPIRICAL EVALUATION OF CROSS-SUBSIDY TESTS FOR CANADIAN INTERREGIONAL TELECOMMUNICATIONS NETWORK[1]

C. Autin and G. LeBlanc

Departement d'Economique Universite Laval

1. Introduction

The central objective of this paper is the empirical evaluation of some cross-subsidy tests for various telecommunications services (mainly private lines and public message services) in Canada, using the National Planning and Policy Simulation Model (NPPS model). Loosely speaking, by cross-subsidization one means that somebody has to pay in full or in part for somebody else's consumption of a particular service. At the policy level, this problem has always been a subject of discussion. In particular, in the telecommunications industry, it has becoming more crucial given the new competitive-monopolistic environment in which the companies, both in Canada and in the U.S., must now operate.

At a more operational level, every cross-subsidy test necessitates the computation of some revenues and costs. To be meaningful, those computations must be made at a certain level of disaggregation and, consequently one must use a big machinery to do so. The model we used in order to compute these components of tests is the National Planning and Policy Simulation Model. This model has been fully described in [2]. Suffice here to say that it is a very disaggregated model, closed to the long distance network observed in Canada. It was built to evaluate the financial and economic impact, on each carrier, of scenarios mixing variations on technical, accounting and/or economic variables at the level of perception of managers, so no aggregate production function is utilized. The modular design is made of operations research and accounting sub-models. The switching network comprises 96 nodes and 373 links; the transmission network includes 219 transmission nodes and 239 transmission links; the tariffs data base is derived from the Trans Canada Telecommunication System (1977).

Section 2 and 3 of the paper present the theoretical cross-subsidy tests which will be applied and report on the empirical results of the various tests performed using the aforementioned model respectively. The cross-subsidy tests are derived from the game theory approach as explained for example in Faulhaber [5]. (As far as the authors are aware, it is the first time that this set of tests is applied in a serious manner). Two series of simulation are performed. In the first series of simulation two tests were applied (stand alone cost test and incremental cost test) for the following pairs of services: public message versus private line, short distance traffic versus long distance tool one, peak versus off-peak demand, regional versus adjacent versus nonadjacent plus U.S. traffics. Note that all simulations are based on the present demands for these services and consequently assume the current usage of the network. The second series of simulation has been done for the public message versus private line services only. But instead of assuming the present demand for these services, like in the first series of simulations, we formulated different hypotheses about the rate of growth of the demands and also about the allocation of the common costs, especially using the Shapley value as a way of splitting those costs. For these services, and combining these hypotheses, four tests have been applied: incremental costs, stand alone cost, generalized incremental cost and finally generalized stand alone cost tests. Finally a last section evaluates the empirical findings, discussed the main weaknesses of our approach and suggests some extensions.

2. Cross-subsidy Tests to Be Used in the NPPS Model

2.1 Description of Suitable Tests for Experimental Purpose

For the purpose of empirical calculations, four tests drawn from the game approach [5] appear relevant for testing cross-subsidization. A system (economy, carriers, ...) producing and distributing a set $N = \{1,\ldots,i,\ldots,n\}$ of n services is assumed. $R(\cdot)$ and $C(\cdot)$ are respectively the revenue and the cost functions defined for a service or a group of services.

The incremental-cost test (CT) is:

(1) $R(i) \geq C(N) - C(N-i)$, for any i in N.

The stand-alone test (SAT) is :

(2) $R(i) \leq C(i)$, for any i in N.

The generalized incremental-cost test (GICT) is:

(3) $R(S) \geq C(N) - C(N-S)$, for all subsets S of N.

The generalized stand-alone test (GSAT) is:

(4) $R(S) \leq C(S)$, for all subsets S of N.

It is worth remembering that if the carrier (subset) has to meet a zero-profit constraint, and if cross-elasticities are zero (hypothesis necessarily assumed when no "demand block" exists), then GICT is equivalent to GSAT.

Although not a test but a useful "fair" cost-allocation formula [11], the following will also be needed:

(5) $$F(i) = \sum_{\substack{G \subset N \\ i \in G}} \frac{(n-g)!(g-1)!}{n!} \{C(G) - C(G-1)\}$$

where G is a subset of size g of N and F(i) is the common cost part of service i.

The proposed theoretical tests involve sets of economic agents and sets of costs. The N.P.P.S. model has been designed to show a fine level of disaggregation for traffic as well as for facility costing. Therefore, it is possible to regroup the demands of the economic agents in a meaningful way and to compute some of the several types of incremental costs used in the cross-subsidization tests. However interpretations and simplifications are necessary to implement the theoretical tests.

a) Defining Meaningful Demand Subsets (or Services)

The theories postulate that any individual has perfect knowledge of the alternative subsets he can join and that he has communication and cooperating capacity. Also, any "subset" knows the cost of supplying its own demand. For the problem at hand, it is more realistic to postulate that intermediates (enterprises) regroup individual demands through their offering of services. The meaningful demand subsets are thus characterized by communication-streams involving: origin-destination, types of service, time of day, time of week. For instance a subset could be: "all public message traffic between 100 and 500 miles, from 8 to 18 hours in the business day". The computing cost of tests which present a combinatorial nature will force us to limit the number of demand subsets. Moreover, the regulating agencies already in place impose the regrouping in a limited number of services.

b) Hypothesis on Demand Reactions to Prices and Quality

Given the demand subsets, note that the empirical demand functions are unknown. For the time being, only requirements (in C.C.S. and number of circuits message or in number of circuits only for all other services) for a base period of projected requirements for future periods are available. Tests involving demand elasticities (direct and crossed) are thus out of reach and a fortiori so since the cross-subsidy theory in game theory terms is not well developed to include demand elasticities.

c) Data Availability for Costing the Services

The cost associated with a given demand subset is theoretically the minimum total cost to supply that subset alone. In real situations, the "initial state" must take care of the physical network and institutional organizations already in place. The actual network design will impose its structure of nodes and links and most of the incremental cost configurations. For stand-alone tests of relatively small demand subsets, the cost functions available in N.P.P.S. and network configurations will not be satisfactory, since the available network has not been designed for such demands.

d) Dynamic Aspect, Hidden and Explicit

Up to now, the N.P.P.S model, except the Accounting Block, computes results for one current year. The cross-subsidy theories above do not have time explicitly as a variable. One can always think of a typical year or of a planning horizon during which decisions are made but the computing of costs is quite different in each case. The latter case requires facility expansion features linked to forecast demands. Some conceptual development has been done along that line (see [1]) but no software is yet available.

Even if the one period method is retained, the hidden dynamic characteristics are represented by the existence of excess capacities which can be justified by the indivisibility of installed facilities and other economies of scale combined with growing demands. Therefore, the cost of excess capacities should be either excluded or imputed to the cost of the tested services. Several solutions will be proposed in the second serie of simulations.

2.2 Empirical Test Proposals

Although embedded cost scenarios can be run with the use of the Aging, Indexing and Depreciation programs, the costing concept retained for present computations is the incurred cost based on the reproduction cost. Since our asset valuation functions are of the "fixed cost" type, among many others, two obvious possibilities are available for each existing network element. Average cost or marginal cost (link or node) from which the incremental cost of a "service" (a requirement subset for the entire switching and transmission networks) is computed. The tests will be executed with both concepts.

Four cross-subsidy tests will be presented:

a) Public Messages Versus Private Lines

This is a recurrent question. Private lines should at least pay for their incremental cost

b) Origin-destination pairs less than or equal to 1 000 miles apart versus pairs more than 1 000 miles apart.

It is possible that very long lines were favored. Time did not permit the regrouping of mileage bands used in tariff tables using a clustering device as: a new (larger) mileage band is created if the tariffs that form it do not deviate from the average tariff by more than a fixed amount. By such reasoning, long distance calls can be approximately clustered in equi-tariff bands: 0 to 180 miles, 181 to 540 miles, 541 to 1 200 miles, over 1 200 miles.

c) Regional-adjacent-non-adjacent (including U.S.) traffics

Negociations between carriers distinguish these three types of traffic.

d) Peak-hour Traffic Versus Non Peak Traffic

A thorny question in economics is whether peak users are subsidized or not by off-peak users. A possible formulation of such a question may be the following: we know that the traffic matrices are dimensioned with respect to the peak demand (rather a kind of average peak demand). If we are given the information that the average demand is about 70% of the peak demand, what is the incremental cost from that average to the full 100%. And does the incremental revenue cover this cost? Alternatively, any percentage down from peak demand could be costed.

e) Full Allocation Versus "Fair" Formula

This is not a cross-subsidy test, but a comparison between two cost allocations: a full allocation based on usage and a "fair" formula allocation (5) a formula which is a weighted average of all possible incremental costs that a service can add when it joins all possible combinations of other services.

2.3 Treatment of Excess Capacities

The existence of excess capacities can be explained in several ways: simple planning error, redundancy for survivability, decreasing demand along a cycle or trend, indivisibility of optimal facilities associated with relatively small demands, growth reserve accumulated to protect against any large positive demand variation, growth reserve built to take advantage of economies of scale when the enterprise faces a sustained growing demand. In telecommunication networks, "protection" facilities and indivisibilities leading to economies of scale are frequently mentioned explanations that we can associate with rapidly growing demand. In other words, in such a dynamic setting, growth reserves will benefit futures as well as present customers. It is therefore important to impute at least part of the excess capacities to actual services.

In devising several methods to take account of excess capacity when computing incremental costs, we will initially allocate all excess capacity between services, first according to the "fair" formula approach and second, proportionally to utilization. Secondly, keeping in mind that allocation may be made according to game theory or to usage, we will distinguish pure excess capacity and growth reserve by introducing growth rates for services. A last method of treating excess capacity will propose some trade-off between present and future.

The five methods depicted below all obey the same pattern. The incremental cost to be used in the incremental-cost test will be modified. It will be the sum of the previously calculated incremental cost and a term representing a certain part of the excess capacity. Thus, the incremental cost IC(i) for service i will be:

$$IC(i) = C(N) - C(N-i) + EC(i)$$

where EC(i) is the value of excess capacity imputed to service i. Of course, the expression obtained is not a "true" incremental cost, but an "exhaustive" incremental cost in the sense that excess capacity is taken into consideration in the procedure. The methods described may be applied to any service obviously.

In the first two methods, the principle is the same. We admit that the cost of excess capacity must be supported by present customers, whether excess capacity is a growth reserve for the future of an incorrect forecast of future demand. We thus run the model with a specified demand and obtain the magnitude of excess capacity. This excess capacity may be priced on a marginal basis or with average coefficients. This procedure permits the cost of excess capacity to be obtained by link or node.

METHOD A: With the first method, we want to allocate the cost of excess capacity proportionally to the usage of the element. We then multiply the cost of excess capacity on each link by the relative usage of this link. To obtain the term EC(i) for service i, we proceed in the same way for all links.

This method puts the weight of the cost of excess capacity only on the shoulders of the present generation. Moreover, it is based on the actual relative utilization of the elements and this may be completely out of line with the future usages.

METHOD B: This method adopts the same approach as that employed in method in method A but allocates excess capacity according to a fairness and game theoretic view. We remember that the cost-allocation formula (5) allows a fair separation of common costs. We thus can allocate the cost of excess capacity in proportion to these game theoretic coefficients. The term EC(i) would then equal:

$$EC(i) = \left[F(i) / \sum_{i \in N} F(i) \right] EC(N)$$

where EC(N) is the cost of the total excess capacity for all services over the whole network.

This method, as well, puts the burden for excess capacity, on the present generation only.

The next four methods try to make a distinction between growth reserve which tends to meet an expanding demand as accurately as possible, and what is called pure excess capacity which is the surplus of capacity over the growth reserve. It is probable that the notion of pure excess capacity will require a new interpretation when multiplexing costs are more thoroughly understood. For the moment, we shall accept this concept.

In this perspective, we shall choose a moving horizon of three years since it is admitted that facility installations are anticipated for a period of at least two to six years. Hence, we run the model successively for three years, increasing the demand for each service according to a growth rate particular to each service and determined exogenously. This rate might be of the multiplicative form with $d_i(t) = a_i(1 + r_i)^t$, where $d_i(t)$ represents the demand of service i after a lapse of t years, a_i is the demand of service i presently, and r_i is the growth rate.

The philosophy of these two methods lies in the hypothesis that only growth reserve must be imputed to customers and then allocated between the services. Pure excess capacity must be supported only by the carrier.

Two cases are possible after three years. First, all the links are saturated. In that case, all excess capacity is growth reserve, and all excess capacity has to be separated between services. This possibility reduces to the first two methods previously discussed. In the second case, there is excess capacity on some or all links after having run the model with demand $d_i(3)$. Pure excess capacity is therefore present in the network and must be borne by the carrier. We need only allocate growth reserve in order to execute the cross-subsidy test. This method, however, necessitates some expansion features in the model since after each year some links could be saturated and block future growth even if ample excess capacity still existed on most of the links.

METHOD C: This method allocates the growth reserve only and does it on the basis of present utilization. It represents an improvement on method B since the burden imposed on present consumers corresponds only to their probable growing demand.

METHOD C': This method is very similar to method C but allocates growth reserve on the basis of future utilization.

METHOD D: This method looks like method C since it attempts to allocate only growth reserve. However, here the priciple on which separation is grounded is the "fair" allocation formula. The new incremental cost would be:

$$IC(i) = C(N) - C(N-i) + \left[F(i) / \sum_{i \in N} F(i) \right] GR(N)$$

where the F(i)'s constitute a fair separation of costs incurred by the present level of utilization and GR(N) is the value of the total growth reserve to be allocated.

METHOD D': This method is similar to method D but allocates growth reserve on the basis of future utilization.

3. Tests Based on Current Use of Equipment

3.1 Public Messages and Private Lines

Table 1 shows the total costs incurred in the switching and the transmission networks required to accomodate first public messages alone and secondly both services. The difference is the incremental cost of private lines. Since this service is not a switched service, there is obviously no incremental cost in the switching network. Also appearing in Table 1 are the revenues derived from the services considered as estimated in N.P.P.S. All figures are shown separately for Bell Canada and for the whole network, the raltionship between incremental costs and revenues not being always the same at the carrier level.

These comparisons must however be handled very carefully since estimated revenues and costs are not strictly comparable. As a matter of fact, revenues correspond to the part of the service generated in the carrier's territory while costs are those associated with satisfying the whole service over the said territory. For instance, the incremental cost of non-adjacent traffic for the Bell is constituted by the cost of originating, terminating and going through non-adjacent traffic, while calculated revenues are those generated by originating traffic only.

3.1.1 O-D pairs < 1 000 miles apart / O-D pairs > 1 000 miles apart

Three simulations were performed:

- one with all traffic between cities more than 1 000 miles apart;
- one with all traffic between cities less than 1 000 miles apart;
- one with both types of traffic.

Since destination/origin points in the U.S. are not precisely known, U.S. traffic was deliberately omitted from all three simulations.

Table 2 is very similar to Table 1 and yields the incremental costs of both types of traffic. It can be seen that for pairs > 1 000 miles apart revenues exceed incremental costs by a factor of about 17. For pairs

< 1 000 miles apart, the ratio is somewhat lower at about 7.

3.1.2 Regional/Adjacent/Non-adjacent and U.S. Traffic

Total incurred costs for each subset of services are shown in Table 3. Resulting incremental costs for each service or combination of two services appear on Table 4 where they are compared to corresponding revenues. In all cases, revenues are larger than incremental costs; in order words, all tests are passed. The ratio of revenues over incremental costs varies however quite substantially between simulations and between carriers as shown in the last column of Table 4.

3.1.3 Peak Hour Traffic/Non-peak Traffic

Traffic profiles during an average business day have the general form shown in Figure 1.

The network is dimensioned for the peak-hour traffic T and costs C(T). Should the peak-hour traffic be smaller, say T', a smaller cost would result C(T'). The test hence consists in comparing the incremental cost of peak-hour traffic (i.e. C(T) - C(T')) to the revenues it generates. These revenues are calculated by multiplying the shaded area of Figure 1 by the appropriate tariff. For this experiment, T' was arbitrarily set at 70% of T.

Total incurred costs for peak and reduced peak simulations are presented in Table 5. The incremental cost of peak traffic is derived in Table 6 and compared to its revenues.

It can be observed that once more incremental revenues largely exceed incremental costs.

3.1.4 Preliminary Comments on First Series of Tests

The tests presented so far indicate that the incremental cost is always satisfied. In addition, the ratio of revenues over incremental cost is so large that it could hardly be reduced to values inferior to 1 simply by improving certain approximations of the model.

It appears in certain instances that revenues exceed the stand-alone of a service (e.g. public messages). Strictly speaking, the stand-alone cost of a service should be representative of all facilities required to support this service. Consequently, the stand-alone cost of any long distance service would include the cost of the local network. Given its relative importance in the total plant, it becomes clear that stand-alone cost tests are also satisfied.

The most important point to notice however is the large discrepancy which exists between the cost of the existing transmission network and the part which is allocated to the various services or groups of services tested in this section. The total cost incurred in the toll network (as estimated in N.P.P.S.) is compared to the cost allocable to public messages and private lines in Table 7.

It becomes clear from this table that this difference has to be explained before any further tests are performed and we give below a list of possible contributing factors:

- Circuit requirements as estimated by dimensioning the switching network are far below those contained in the data base (14 000 vs 23 600). It must be remembered that the dimensioning algorithm is applied to traffic which is estimated based on limited data (traffic between 17 cities during two weeks of July 1971) and does not include WATS, TWX and data tranmission.

- It was mentioned earlier that costing the transmission network with the average cost formula is a poor approximation when the link loading is low.

It will be seen, see Table 8, that costing transmission facilities with the (fixed + marginal) cost approach would result in a total cost of $86.1 millions to be compared to $33.3 millions obtained with the average cost formulation (Table 7).

- It is known that cost trade-offs between multiplexing and channeling result in a channel loading which generally does not exceed 75%.

- A certain amount of unused equipment is included in the plant as a growth reserve.

- Finally, it must be remembered that the N.P.P.S. allocation procedure does not take survivability constraints into account and therefore yields an allocation which is cheaper than it would be in reality.

3.2 Tests Based on Prospective Use of Equipment

3.2.1 General

In view of the results presented in the previous section, a new series of tests was performed. It was decided to concentrate on the appropriate calculation of costs rather than on various splits of the services taken into account. All tests were consequently based on a public message/private lines separation. In order to improve estimation of costs and in line with the observations of section 3.1.4, the following rules were applied:

a) Transmission facilities were costed using the "fixed cost + marginal cost" approach.

b) The multiplexing plan was approximated by the formulation suggested that circuit requirements constitute integer number of groups, the loading of which does not exceed 75%.

c) Since no precise definition of the growth reserve is available, various policies were tested by which growth reserve was defined as the incremental cost associated with the growth of a service over 1, 2 and 3-year periods.

3.2.2 Description of Simulation Runs and Incremental Cost Tests

Five simulation runs were performed. The first one is based on present demand. The next three consider prospective demand 1, 2 and 3 years from now using:

- a 12% annual growth rate for public messages[2]
- an 18% annual growth rate for private lines[2].

To test the sensitivity of the results to growth rates, a fifth simulation was performed considering propective demand in year 3 but with a 10% annual growth rate for private lines.

The results of the five simulations are presented in Table 8. It must be remembered that, private lines being a non-switched service, only transmission costs have been analyzed.

One will also notice that the total cost of the transmission network increases with the length of the planning horizon since capacity had to be increased on a certain number of links in order to render the allocation feasible. The corresponding incremental cost of private lines can easily be derived from these results and is shown below:

Growth for x year (s) x:	0	1	2	3	3 (with less growth)
Incremental cost of private lines including growth reserve (in $10^6$$)	10.1	12.9	15.2	16.8	13.4

Although consideration of a 3-year growth reserve increases the incremental cost of private lines by 66%, the revised incremental cost figure still remains much smaller than revenues estimates of $41.6 millions.

3.2.3 Tests Based on Full Allocation of Costs

All tests performed so far have shown that the incremental costs of private lines is always covered by generated revenues. If one examines closely the total transmission cost of supporting both private lines and public messages, it can be broken down as follows:

	$ millions
Incremental cost of private lines	10.1
Incremental cost of public messages	36.3
Cost of equipment used jointly	39.7
Total transmission cost (excluding growth reserve)	86.1

If one further considers a 3 year growth reserve and compares all these costs to total costs of the existing transmission network one obtains a graph of the form shown in Figure 2, where surfaces are proportional to costs.

It becomes clear then that a definition of cross-subsidy based on incremental costs alone is not sufficient given the importance of the common costs and other non directly allocable costs and given that total costs must eventually be recovered.

Two questions then arise:

a) How should common cost be allocated?

b) Which common costs should be allocated, namely, should the cost associated with the so-called "pure excess" capacity be paid by the consumer or by the carrier. This depends obviously on the origin of this excess which could result from:

- deficiencies of the model (i.e. not enough traffic, no survivability constraints...);

- a larger planning horizon than used in our calculations (i.e. more than three years);

- a very safe and/or suboptimal planning of the network by the carriers;

- a mixture of the three above-mentioned factors.

This leads us to the application of cost separation formula (5).

Table 9 presents all data necessary to calculate cost allocations using the methods described above. The first three columns (Stand-alone cost, incremental cost and "fair" allocation of used capacity) are directly derived from Table 8. The allocation based on usage was obtained by the N.P.P.S. model.

Table 10 presents cost separations based on methods A and B. It can be noticed that for both methods, full allocated costs of private lines exceed the estimated revenues of $41,6 millions. It can also be seen that the "exhaustive" incremental cost (defined the true incremental cost plus a "fair share" of excess) yielded by method B for private lines also exceeds revenues (i.e. $44.6 vs $41.6 millions).

Cost allocations based on methods C and C' for various planning horizons are exhibited in Table 11. Both these methods do not allow for the estimation of an "exhaustive" incremental cost and only full allocations are computed. It can be seen however that revenues of private lines always exceed their fully allocated cost independently of the planning horizon and the method chosen.

Cost allocations based on methods D and D' for various planning horizons appear in Table 12. It can be seen that the "exhaustive" incremental cost of private lines never exceeds $15.2 millions while the fully allocated cost varies between $31.7 and $35.1 millions according to the planning horizon and the method selected. One can also notice that fully allocated costs based on the game theoretic approach always disfavour private lines when compared to allocations based on usage.

4. Conclusion

Going from theoretical test statements to empirical implementation with the logic of the N.P.P.S. model and the data at hand, the computations have shown that the generalized incremental tests (GICT) are passed for all partitions of services chosen in the tested examples. Moreover, in each example, if the sub-additivity hypothesis as well as the hypothesis asserting that the revenue of the "grand coalition" equals its cost, are true, the generalized stand alone tests (GSAT) are also passed without having to be computed. This somewhat reduces the problems since the actual network configuration and its associated costs are often not appropriate for a small service to stand alone. Therefore, if one is willing to accept the notion of cross-subsidy as described earlier, it follows that no such subsidy has been detected in our examples.

Another clear and interesting finding is the fact that incremental costs are often relatively small with respect to common cost. This could explain the large difference observed between the revenue generated by a subset of services and its incremental cost. As a further result, it should be noted that throughout the test series it has been recognized that a relatively large installed excess capacity was present in the network model over a normal three years growth reserve. However, some transmission links had been found to be saturated in the prospective use base tests.

In the course of this paper, we have outlined a number of model and formulation qualifications which could affect the outcome of tests performed.[3] However some sensitivity analyses have been done with the model. Among those sensitivity studies, let us mention the introduction of the demand by WATS and TWX, the take into account of some indivisibilities in the transmission network by costing it in integer numbers of channels, by increasing the annual growth rate of demand, and by modifying the treatment of the multiplexing cost and finally the introduction of survivability constraints. It can be shown (see [7]) that any of the factors considered could not, independently, invalidates the results of our private lines incremental cost test.

FOOTNOTES:

[1] A preliminary version has been presented to "Telecommunications in Canada: Economic Analysis of the Industry", March 4,5,6, 1981, Montreal. This project has been financed by the Canadian Department of Communications. It is a tripartite effort of : Department of Communications, Sores Inc., of Montreal and the Laboratoire d'Econometrie, Laval University. Mr. G. Henter from DOC was the project manager. Prof. T. Matuszewski contributed to the conceptualization of the model. Mr. J.P. Schaak from Sores Inc was responsible for the software. However the present authors remain the sole responsables for the interpretations given in the present paper.

[2] These rates were applied uniformly to all existing demand and no new demands were considered.

[3] Role of the cross-elasticities, the treatment of excess capacities in the tariff determination, the inter-carriers cross-subsidies, prospective costs and accountability for cross forecasting.

REFERENCES:

[1] Autin, C., St-Cyr, G., Expansion sur plusieurs periodes, Note technique, projet N.P.P.S., troisieme phase (April 1975).

[2] Autin, C., LeBlanc, G., A National Telecommunications Planning and

Policy Simulation Model, in Models and Decisions in National Economies, J.M.L. Janssen, L.F. Pau, A. Straszak (eds) (North-Holland Publishing Company, 1979).

[3] Baumol, W.J., Bradford, D.F., Optimal Departures from Marginal Cost Pricing, Annual Economic Review, 60 (1970) 265-283.

[4] Faulhatber, G.R., On Subsidization: Some Observations and Tentative Conclusions, Conference on Communication Policy Ressources, Office of Telecommunication Policy, Washington (1972).

[5] Faulhaber, G.R., Cross-subsidization: Pricing in Public Enterprises, Bell Lab. Econ., Discussion Paper (May 1975).

[6] Hazelwood, A., Optimum Pricing as Applied to Telephone Service, Review of Economic Studies, Vol. 18, 2 (1950-51) 67-78.

[7] Laboratoire d'Econometrie; Sores Inc., Development and Empirical Evaluation of Cross-subsidy Tests, Unpublished DOC (March 31, 1977).

[8] Littlechild, S.C., Marginal-cost Pricing with Joint Costs, Economic Journal, 80 (1970) 323-335.

[9] Littlechild, S.C., A Game-theoretic Approach to Public Utility Pricing, Western Economic Journal, Vol. 8, 2 (1970) 162-166.

[10] Loehman, E., Whinston, A., A New Theory of Pricing and Decision-making for Public Investment, The Bell Journal of Economics, Vol. 1, 2 (1971) 606-625.

[11] Loehman, E., Whinston, A., An Axiomatic Approach to Cost Allocation for Public Investment, Public Finance Quarterly, Vol. 2, 2 (1974) 236-251.

[12] Panzar, J.C., Willig, R.D., Free Entry and the Sustainability of Natural Monopoly, The Bell Journal of Economics, Vol. 8, 1 (Spring 1977).

[13] Shapley, L.S., Cores of Convex Games, International Review of Game Theory (1971) 11-26.

[14] Zajac, E.E., Some Preliminary Thoughts on Subsidization, Conference on Communication Policy Ressources Office of Telecommunication Policy, Washington (1972).

TABLE 1

Incremental Cost of Private Lines
(incurred costs and revenues in $ millions)

Service	Carrier	Switching	Transmission Cost[1]	Total	Estimated Revenues[2]
Public Messages	Bell	64.4	16.0	80.4	316.9
	Network	95.8	23.6	119.4	395.2
Both Services	Bell	64.4	21.9	86.3	341.5
	Network	95.8	33.3	129.1	436.7
Incremental Costs & Revenues of Private Lines	Bell	0	5.9	5.9	24.6
	Network	0	9.7	9.7	41.6

(1) Using average cost.
(2) US revenues excluded.

TABLE 2

Incremental Cost of O-D Pairs More or
Less Than 1 000 Miles Apart
($ millions)

Simulation and Carrier	Switching	Transmission Cost[1]	Total Cost	Estimated Revenues
Pairs < 1 000				
Bell	60.7	10.2	70.9	298.4
Network	90.7	13.8	104.5	352.9
Pairs > 1 000				
Bell	53.6	2.0	55.6	18.5
Network	83.6	4.7	88.3	42.2
Both Services				
Bell	61.7	11.6	73.3	316.9
Network	94.7	17.2	111.9	395.1
Incremental Costs & Revenues for Pairs < 1 000				
Bell	8.1	9.6	17.7	298.4
Network	11.1	12.5	23.6	352.9
Incremental Costs & Revenues for Pairs > 1 000				
Bell	1.0	1.4	2.4	18.5
Network	4.0	3.5	7.5	42.2

(1) Using average cost.

TABLE 3

Three-service Experiment
Total Incurred Costs and Revenues
($ millions)

Simulation & Carriers	Switching Cost	Transmission Cost(1)	Total Cost	Estimated Revenues(2)
Reg + Adj + N-Adj + US				
Bell	64.4	16.1	80.5	316.9
All carriers	95.8	23.5	119.3	395.2
Adj + N-Adj + US				
Bell	56.1	7.7	63.8	33.6
All carriers	87.5	14.6	102.1	96.1
Reg + N-Adj + US				
Bell	63.7	15.3	79.0	305.2
All carriers	94.7	21.4	116.1	352.9
Reg + Adj				
Bell	60.7	9.9	70.6	292.2
All carriers	90.7	12.5	103.2	341.4
N-Adj + US				
Bell	56.1	6.9	63.0	21.8
All carriers	86.6	12.4	99.0	53.8
Adj				
Bell	53.6	1.3	54.9	11.8
All carriers	83.6	3.3	86.9	42.3
Reg				
Bell	59.8	9.0	68.8	283.4
All carriers	89.9	10.0	99.9	299.1

(1) Using average cost.
(2) Excluding US

TABLE 4

Three-service Experiment
Incremental Costs
($ millions)

Subset test	Carriers	Incremental cost in switching	Incremental cost in transmission	Total incremental cost (2)	Revenues N.P.P.S. Estimates (1)	Ratio (1)/(2)
Adj +	Bell	4.6	7.1	11.7	33.6	2.9
Non-Adj + US	All	5.9	13.6	19.5	96.1	4.9
Reg +	Bell	10.8	14.8	25.6	305.2	11.9
Non-Adj + US	All	12.2	20.3	32.1	352.9	11.0
Reg +	Bell	8.3	9.2	17.5	292.2	16.7
Adj	All	9.2	13.2	22.4	341.4	15.2
Non-Adj + US	Bell	3.7	6.2	9.9	21.8	2.2
		5.1	11.1	16.2	53.8	3.3
Adj	Bell	.7	.8	1.5	11.8	7.9
	All	1.1	2.2	3.3	42.3	12.8
Reg	Bell	8.3	8.4	16.7	283.4	17.0
	All	8.3	9.1	17.4	299.1	17.2

TABLE 5

Total Incurred Costs
Peak/Off-peak Traffic
($ millions)

Simulations & Carrier	Switching Costs	Transmission Costs[1]	Total Costs
Peak			
Bell	64.4	16.1	80.5
All carriers	95.8	23.5	119.3
Reduced Peak			
Bell	61.5	13.0	74.5
All carriers	92.8	19.3	112.1

(1) Using average cost.

TABLE 6

Incremental Costs of Peak Traffic
($ millions)

Carrier	Switching	Transmission	Total[2]	Revenues[1] N.P.P.S. Estimates (*)	(1)/(2)
Bell	2.9	3.1	6.0	41.5	6.9
All carriers	3.0	4.2	4.2	51.8	7.2

(*) Excluding US.

TABLE 7

Comparison of Total Cost of Plant to Cost Allocable to Public Message and Private Lines ($ millions)

Switching	Total incurred cost of plant as estimated in N.P.P.S. (2)	Cost allocated to PM and PL (3)	(3)/(2) %
Switching Network	106.7	94.7	89
Transmission Network(1)	184.8	33.3(4)	18
Total	291.5	128.0	44

(1) Excluding channels used for video.
(4) It will be seen in Table 8 that when using the (fixed cost + marginal cost) formula this value becomes 86.1.

TABLE 8

Simulation Results
Incurred costs in $ millions

Simulation	Services considered	Incurred fixed cost	Incurred variable cost[2]	Total incurred cost	Cost of[1] excess capacity	Total cost of transmission NW (excluding channels used for video)
#1 Present demand	P.M. only	49.4	26.6	76.0	108.8	
	P.L. only	39.1	10.7	49.8	135.0	184.8
	Both services	49.4	36.7	86.1	98.7	
#2 Demand after one year of growth	P.M. only	49.4	28.6	78.0	107.8	
	P.L. only	39.1	12.2	51.3	134.5	185.8
	Both services	49.4	41.5	90.9	94.9	
#3 Demand after two years of growth	P.M. only	49.4	31.1	80.5	109.3	
	P.L. only	39.1	13.7	52.8	137.0	189.8
	Both services	49.4	46.2	95.7	94.1	
#4 Demand after three years of growth	P.M. only	49.4	34.1	83.5	108.2	
	P.L. only	39.1	15.6	54.7	137.0	191.7
	Both services	49.9	50.4	100.3	91.4	
#5 Demand after three years of growth (lower rate for P.L.)	P.M. only	49.4	34.1	83.5	107.8	
	P.L. only	39.1	13.4	52.5	138.8	191.3
	Both services	49.4	47.5	96.9	94.4	

(1) Including $9.2 millions for links not used at all.

(2) It can be seen that the incurred variable cost associated with both services is generally lower than the sum of individual variable costs. This results from the rounding procedure which, when applied to both services, results in requirements smaller than the sum of individual rounded requirements.

TABLE 9

Cost Allocation of Used Capacity
($ millions)

Simulations	Services	Stand alone cost	Incremental cost	"Fair" allocation of used capacity	Allocation of used capacity based on usage (derived from the N.P.P.S. model)
#1	P.M.	76.0	36.3	56.1	61.9
Present	P.L.	49.8	10.1	30.0	24.2
#2	P.M.	78.0	39.6	58.8	65.3
One year	P.L.	51.3	12.9	32.1	25.6
#3	P.M.	80.5	42.9	61.7	66.8
Two years	P.L.	52.8	15.2	34.0	28.9
#4	P.M.	83.5	45.6	64.5	68.3
Three years	P.L.	54.7	16.8	35.8	32.0
#5	P.M.	83.5	44.4	64.0	70.9
Three years slower growth for P.L.	P.L.	52.5	13.4	32.9	26.0

TABLE 10

Cost Allocations Based on Methods A and B
($ millions)

		P.M.	P.L.	P.M.	P.L.
Used capacity	Common costs	61.9	24.2	19.8	19.8
	Incremental cost			36.3	10.1
Unused capacity		71.0	27.7	64.3	34.5
Total		132.9	51.9	120.4	64.4

TABLE 11

Cost Allocations Based on Methods C and C'
($ millions)

		Method C		Method C'	
		P.M.	P.L.	P.M.	P.L.
One year	Used capacity	61.9	24.2	61.9	24.2
	Growth reserve	3.5	1.3	3.4	1.4
	Total	65.4	25.5	65.3	25.6
Two years	Used capacity	61.9	24.2	61.9	24.2
	Growth reserve	6.9	2.7	6.7	2.9
	Total	68.8	26.9	68.6	27.1
Three years	Used capacity	61.9	24.2	61.9	24.2
	Growth reserve	10.3	4.0	9.7	4.5
	Total	72.2	28.2	71.6	28.7
Three years (lower growth on P.L.)	Used capacity	61.9	24.2	61.9	24.2
	Growth reserve	7.8	3.0	7.9	2.9
	Total	69.7	27.2	69.8	27.1

Estimated revenues of private lines: 41.6

TABLE 12

Cost Allocation Based on Methods D and D'
($ millions)

Cost component	Planning horizon							
	One year		Two years		Three years		Three years (lower growth P.L.)	
	P.M.	P.L.	P.M.	P.L.	P.M.	P.L.	P.M.	P.L.
Used capacity(1)								
Common costs	19.8	19.8	19.8	19.8	19.8	19.8	19.8	19.8
Incremental cost	36.3	10.1	36.3	10.1	36.3	10.1	36.3	10.1
Total	56.1	30.0	56.1	30.0	56.1	30.0	56.1	30.0
Growth reserve according to D	3.1	1.7	6.3	3.3	9.3	4.9	7.0	3.8
Exhaustive incremental cost according to D	39.4	11.8	42.6	14.4	45.6	15.0	46.3	13.9
Growth reserve according to D'	3.1	1.7	6.2	3.4	9.1	5.1	7.1	3.7
Exhaustive incremental cost according to D'	39.4	11.8	42.5	14.5	45.4	15.2	43.4	13.8
Total allocable cost according to D	59.2	31.7	62.4	33.3	65.4	34.9	63.1	33.8
Total allocable cost according to D'	59.2	31.7	62.3	33.4	65.2	35.1	63.2	33.7

(1) Identical for both methods and independent of planning horizon.

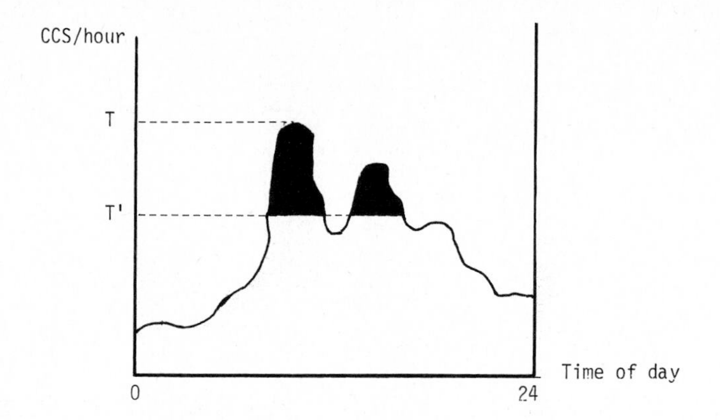

Fig. 1. Typical traffic profile.

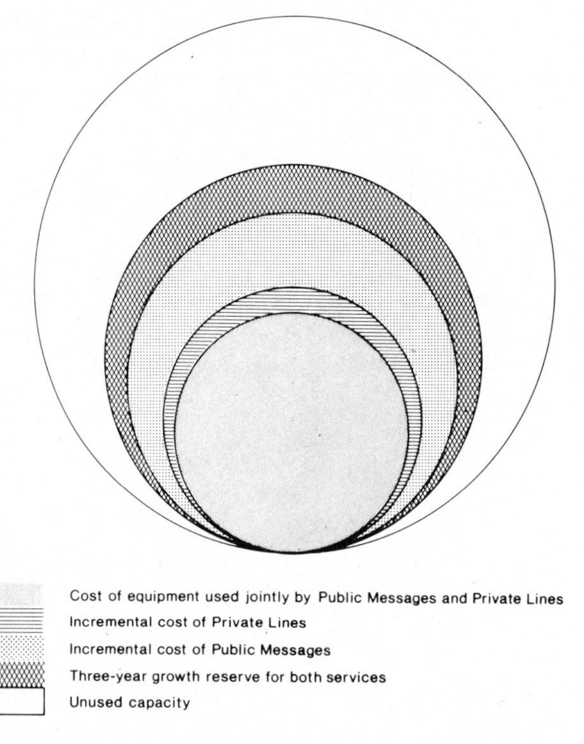

Fig. 2. Allocation of incurred costs in transmission network as estimated by N.P.P.S. (surfaces are proportional to estimated costs).

PART 3
WELFARE CONSIDERATIONS AND REGULATION

Investment Decisions Under Regulation

Economic Analysis of Telecommunications:
Theory and Applications
L. Courville, A. de Fontenay and R. Dobell (eds.)

TAXES, FINANCING AND INVESTMENT FOR A REGULATED FIRM

Jeffrey I. Bernstein

Department of Economics, Carleton University

1. Introduction

Firms operating under the purview of regulatory authority are subject to a variety of restrictions. Casual observation points out that approval of output price structures, expenses, the terms of bond and share issues, and capital accumulation plans, are all part of the domain of regulation. Although this is a complex process, in essence, regulation involves a constraint on the sources and uses of the firm's funds. The purpose of this paper is to formulate a regulatory constraint on the flow of funds, and to characterize how corporate behaviour is affected.

Regulation pertains, in part, to the investment decision undertaken by the firm, and so it would appear appropriate to formulate a dynamic model. However, it is noteworthy that most of the work on regulation is framed in a static manner. Since we can view the static equilibrium as the long-run solution to a dynamic process, then much is gained by the simplicity of the static model, when the intertemporal analysis leads to the same solution. Unfortunately, as Appelbaum and Harris [1] have demonstrated, the dynamic model of a rate of return regulated firm does not necessarily lead to the same equilibrium as the static one. Therefore, we begin by developing an intertemporal model of resource allocation for a flow of funds regulated firm, which builds on the theory of investment as developed by Lucas [2], and Treadway [13], among others.

It is now recognized that a firm's capital structure can affect the return to its shareholders, and in particular, the choice of investment projects (see Stiglitz [11], King [7] and Auerbach [2]). For a regulated firm, there is a further aspect to the interrelationship between real and financial decisions. If the authority's rule depends on the capital structure, the firm may be able to obtain additional benefits by judiciously selecting its financing instruments. The determinants of a regulated firm's capital structure have been analyzed by Elton and Gruber [3], [4], Jaffe and Mandelker [5], Sherman [10] and Taggart [12]. However, these papers do not integrate corporate investment with the capital structure decisions.

In this paper the firm finances investment through retentions, bond and share issues. The firm cannot instantaneously adjust to the long-run equilibrium capital stock, because installation is costly. These installation costs are internal, but separable from the production process. Regulation is not of the rate of return variety, but the rule appears as a limit of the corporation's flow of funds. In particular, there is a maximum amount by which the firm's revenues can exceed its committed funds, such as operating expenses, taxes, interest payments, dividends and the repayment of the principal on debt. This constraint effectively places an upper bound on the magnitude of internally generated funds that can be used for capital expansion programs.

The flow of funds constraint implies that the regulated firm may have to use external financing sources, even though, in the absence of regulation, internally generated funds are less expensive (at the margin). This means that regulated firms may exhibit a capital structure with a greater proportion of external financing compared to non-regulated firms. In addition, because the regulated firm is forced away from the cheapest source of funds, the marginal cost of capital for this firm may be higher than for its non-regulated counterpart. As a consequence, it is possible that a regulated firm may invest less and have a smaller capital stock.

Although in this paper we do not investigate the determination of the limit on the flow of funds, we do analyze the effects this limit exerts on the firm. In general, we find that the flow of funds constraint affects both the capital accumulation path and the capital structure. Section 2 of this paper contains a description of the model. In Section 3 we describe the alternative financing patterns for regulated and non-regulated firms. Sections 4 and 5 contain the analysis of the equilibrium and the dynamics, and lastly we conclude.

2. The Model

Consider a corporation whose flow of funds is delimited by

$$(1)\quad R(K_t) - A(I_t) - r_{bt}B_t - T_t + b_t - B_t + s_t - p_{It}I_t - D_t = 0$$

where R is the indirect variable profits function with $R' > 0$, $R'' < 0$, K_t is physical capital, A is the adjustment cost function $A(0) = A'(0)$, $A'(I) > 0$ for $I > 0$, r_{bt} is the interest rate on corporate bonds, B_t is the value of corporation bonds, T_t are corporate taxes, b_t is the value of new corporate bonds (we assume all bonds have a term to maturity of one period), s_t is the value of new shares, p_{It} is the price of investment and D_t are dividends.

Corporate taxes are

$$(2)\quad T_t = u_c[R(K_t) - A(I_t) - r_{bt}B_t - \delta p_{It}K_t]$$

where $0<u_c<1$ is the fixed corporate income tax rate, $0 < \delta < 1$ is the fixed rate of depreciation. Substituting for T_t from (2) into (1) yields,[1]

$$(3)\quad D = (1 - u_c)[R(K) - A(I) - r_bB] + u_c\delta p_IK + b - B + s - p_II$$

By their very nature dividends cannot be negative. Thus $D > 0$.

The firm is confronted with a financing constraint which restricts investment to be financed by issuing shares, bonds or by retentions, which we denote as E,

(4) $p_I I = b + s + E.$

We assume that physical investment is irreversible so $I \geqslant 0$. Moreover, because the firm is not a financial corporation, then it does not purchase bonds and consequently, $b \geqslant 0$. As will be explained below, we assume that share repurchasing is not possible, $s \geqslant 0$. Hence, from equation (4), retentions (E) are nonegative. We can define the net flow of funds available to the shareholders as F. These funds can either be retained by the firm or distributed as dividends. This means that $F=E+D \geqslant 0$.

The firm under consideration must satisfy a rule imposed upon it by the regulatory authority. The rule states that revenues can only exceed operating expenses, installation costs, interest payments, dividends and the repayment of the principal on debt by some fixed magnitude. This means that,

(5) $(1 - u_c)[R(K) - A(I) - r_b B] + u_c \delta p_I - B - D \leqslant E^r$

where E^r is the regulated value. There is an upper limit on the committed net flow of funds. Another way to see the implications from the constraint is by combining equations (3), (4) and (5). We find that

(6) $E \leqslant E^r.$

Regulation limits the amount of internally generated funds that can be used for capital accumulation. We assume that E^r is fixed for all $t \geqslant 0$.[2]

The regulatory constraint illustrates that there is, at any period, a trade-off between investment and dividend payments. Given the capital stock and the value of outstanding debt, and in the absence of new bond and share issues, from (3), (4) and (5).

$$dD = -[(1 - u_c)A' + p_I]dI.$$

An increase by $1 of dividends leads to a decrease in investment by the post tax marginal cost of investment. The only way the firm can break the trade-off between dividends and investment is by issuing bonds and shares.[3]

The rate of return on corporate shares is defined as

(7) $r = (D/p_s N_s) + (1 - u_g)\dot{p}_s/(1 - u_p)p_s,$

where p_s is the share price, N_s is the number of shares, u_g is the fixed capital gains income tax rate, and u_p is the fixed dividend income tax rate.[4] The present institutional setting is governed by $0 < u_g < u_p < 1$.[5] The rate of return equals the sum of dividends per share adjusted for any net of tax capital gains or losses. Defining

$S = p_s N_s$, the $\dot{S} = \dot{p}_s N_s + p_s \dot{N}_s$ and with $s = p_s \dot{N}_s$, equation (7) becomes

$$(8)\quad \dot{S} - arS = -a(D-s/a)$$

where $a = (1-u_p)/(1-u_g) < 1$. Solving for the initial share value from (8), we find

$$(9)\quad S(0) = \int_0^{\infty} e^{-\int_0^t \rho(z)dz} a(D-s/a)dt$$

where $\rho = ar$. The initial share value depends on the dividends derived from ownership minus the dilution from the issuance of new shares. An interesting result emerges from (9). Since $a<1$ then a \$1 increase in D increases S(0) by less than a \$1 decrease in s. Thus the shareholders, in attempting to achieve the maximum value of S(0) will desire to obtain all their receipts from the corporation in the form of share repurchases rather than dividends. In order to limit this possibility (i.e., close the tax loop-hole), the authorities restrict the ability of firms to repurchase shares. In our model, we have assumed no repurchase is possible.

Clearly, by the simplification that all debt is composed of one period bonds, debt accumulates according to

$$(10.1)\quad \dot{B} = b - B,\ B(0) \geq 0$$

while physical capital changes by the usual condition that

$$(10.2)\quad \dot{K} = I - \delta K,\ K(0) > 0.$$

The firm selects physical investment, new debt and share issues, which maximize the initial market value of the shares subject to the constraints given as $s \geq 0$, $D \geq 0$, $b \geq 0$, and $0 \leq E \leq E^r$. The solution to this optimal control problem is characterized by the following equations, with $\phi = 1 - 1/a < 0$, q_1, q_2, are the costate variables associated with K, and B; the λ's are the Lagrangean multipliers, with λ_1 to λ_5 associated with $s \geq 0$, $E \leq E^r$, $D \geq 0$, $b \geq 0$ and $E \geq 0$.

$$(11.1)\quad -(1+\lambda_3)(1-u_c)A' - p_I(1 + \lambda_2 + \lambda_3 - \lambda_5) + q_1 = 0$$

$$(11.2)\quad 1 + q_2 + \lambda_2 + \lambda_3 + \lambda_4 - \lambda_5 = 0$$

$$(11.3)\quad \phi + \lambda_1 + \lambda_2 + \lambda_3 - \lambda_5 = 0$$

$$(11.4)\quad \dot{q}_1 = (\rho + \delta)q_1 - (1 + \lambda_3)(1 - u_c)R' - (1 + \lambda_3)u_c \delta p_I$$

(11.5) $\dot{q}_2 = (\rho + 1)q_2 + (1 + \lambda_3)[(1 - u_c)r_b + 1]$.

There are also the transversality conditions and the relevant equations associated with the multipliers and K, B, as well as the Legendre-Clebsch conditions, which state that the matrix, comprised of the derivatives of (11.1) to (11.3) with respect to I, b and s, is negative definite. To understand equation set (11), let us look more closely at its components.

3. Corporate Financial Patterns

The determination of the capital structure centres around equations (11.2) and (11.3). Whether internal or external sources of funds (and which type of external sources) are used depends on their respective marginal costs.

By the definition of q_2, which is the shadow price of debt, we see that $-q_2 > 0$ (from (11.5) with $q_2 < 0$) is the per dollar marginal cost of debt. The per dollar cost of an additional dollar of internally generated funds is 1. Hence the difference between the per dollar marginal costs of debt and retention financing is $1 + q_2$. If $1 + q_2 > 0$ than at the margin debt is cheaper than internal funds as a financing source. We have the opposite result for $1 + q_2 < 0$ and with $1 + q_2 = 0$, the per dollar marginal costs are equal.

With respect to share financing the per dollar marginal cost is $(1 - u_g)/(1 - u_p)$. To see this consider a shareholder with an additional dollar to save after receiving a dividend payment. Since the shareholder has to pay $\$u_p$ of tax on the dividends, the amount distributed by the firm was $\$1/(1 - u_p)$. Suppose the shareholder does not receive the dividend but purchases $\$1/(1 - u_p)$ worth of shares. After paying the capital gains tax the individual has, in effect, purchased $\$1(1 - u_g)/(1 - u_p)$ worth of shares. Hence the per dollar marginal cost is $1 - \phi = (1 - u_g)/(1 - u_p)$. Thus the difference between the per dollar marginal cost of retentions and shares is $1 - (1 - \phi) = \phi = (u_g - u_p)/(1 - u_p) < 0$. Since the capital gains tax rate is less than the personal income tax rate, it is always cheaper for the firm to use retentions rather than issue shares.

3.1 Without Regulation

In the absence of regulation, $\lambda_2 = 0$. Let us begin with $1 + q_2 < 0$ so retentions are cheaper than bonds. Moreover, because retentions are cheaper than shares ($\phi < 0$) then all net internal funds will go to financing physical investment. If there is any residual then dividends will be paid. Therefore with $1 + q_2 < 0$, the maximum value of external financing is

$$b + s = \max(0, p_I I - F).$$

If $F > p_I I$ then $b=s=0$, $E=p_I I$ and $D=F-p_I I \geq 0$. Next if $F=p_I I$ then $b=s=0$, $E=p_I I$ and $D=0$. We have now established columns 4 and 5 in Table 1.

Suppose that net internal funds are insufficient to finance physical investment so $F < p_I I$. We must compare the marginal costs of bonds and

Table 1

Financing Characterization Without Regulation

	$1 + q_2 < 0$					$1 + q_2 \geq 0$
	$0 \leq F < p_I I$			$F > p_I I$	$F = p_I I$	
	$1 + q_2 \geq \phi$	$1 + q_2 \leq \phi$	$1 + q_2 = \phi$			
b	+	0	+	0	0	+
s	0	+	+	0	0	0
E	$F \geq 0$	$F \geq 0$	$F \geq 0$	$p_I I \geq 0$	$F > 0$	0
D	0	0	0	$F - p_I I > 0$	0	$F \geq 0$
	(1)	(2)	(3)	(4)	(5)	(6)

shares. If $1 + q_2 > 0$ then bonds are cheaper. Hence all internal funds are devoted to retentions, distributions are zero and the remaining physical investment is financed by debt. This gives us column 1 in Table 1. Columns 2 and 3 are established in a similar fashion.

Let us proceed to $1 + q_2 \geq 0$. Here debt is not more expensive than retentions. Thus investment is completely debt financed with all net internal funds devoted to dividends. This result is found as column 6 in Table 1.

3.2 With Regulation

Binding regulation ($\lambda_2 > 0$) means that retentions have been effectively limited to $E = E^r > 0$. Hence $F = 0$ is not possible for the regulated firm and $F \geq E^r$. These results establish columns 1-3 in Table 2. For the fourth column with $1 + q_2 < 0$ and $F \geq p_I I$, all physical investment is financed by retentions with $E^r = p_I I$ and the remaining net flow of funds are used for dividends.

Table 2

Financing Characterization With Regulation

	$1 + q_2 < 0$			
	$0 < F < p_I I$			$F \geq p_I I$
	$1 + q_2 > \phi$	$1 + q_2 < \phi$	$1 + q_2 = \phi$	
b	+	0	+	0
s	0	+	+	0
E	$E^r > 0$	$E^r > 0$	$E^r > 0$	$E^r = p_I I > 0$
D	$F\text{-}E^r > 0$	$F\text{-}E^r > 0$	$F\text{-}E^r > 0$	$F\text{-}E^r > 0$
	(1)	(2)	(3)	(4)

There is another difference between the financing patterns for regulated and non-regulated firms. In the absence of regulation with $1 + q_2 \geq 0$ there are no retentions. However, with regulation $E = E^r > 0$. Consequently from (11.2) with $\lambda_2 > 0$, $\lambda_5 = 0$, it is impossible for $1 + q_2 \geq 0$. The intuition is straightforward. Under effective regulation, the firm desires to finance a larger portion of physical investment through retentions than it is allowed. This implies that regulation must be constraining the firm when retentions are relatively cheaper than debt. In essence, this means $1 + q_2 < 0$.

The financing patterns described in Tables 1 and 2 can emerge at any instant in time. They are consistent with alternative share values in any time period. However, we want to guarantee that the initial share value $S(0) > 0$. A sufficient condition for this to occur is if

$$(12) \quad \Pi = (1 - u_c)[R(K)-A(I)-r_b B-\delta p_I K] > 0.$$

In other words revenues must exceed adjustment costs, interest payments, depreciation charges and taxes. This condition is quite reasonable. In addition, it imposes restrictions on the nature of the long-run equilibrium or stationary state. In the stationary state $\dot{K}=0$ so $I=\delta K$, and $\dot{B}=0$ so $b=B$. Therefore from equations (3) and (12)

$$D = \Pi + s > 0$$

But from Table 1 when $D > 0$, $s = 0$. Hence in the long-run equilibrium, the nonregulated firm does not issue new shares, only retentions or debt are issued.[6]

The regulated firm pays dividends since $D = \Pi + s > 0$. We observe from Table 2 with $E = E^r > 0$, $D > 0$ that $F - E^r > 0$. More importantly, all cases described in Table 2 are potential stationary states. Hence w find that in the long-run, simultaneous dividend payment and share issuance is a possible equilibrium for regulated firms.

4. Short-Run Equilibrium

Given values for B, K, q_1 and q_2, we investigate the determinants of physical investment, share and bond issues.

4.1 Without Regulation

We will deal with cases 4 and 6 since these are the only possible stationary states.

In case 4, $b = s = 0$, $E > 0$, $D > 0$, and the equation determining investment (from (11.1)) is

$$-(1 - u_c)A' - p_I + q_1 = 0$$

Hence $I = I^{m4}(q_1)$ and $\frac{dI}{dq_1} = 1/(1 - u_c)A'' > 0$

Investment is an increasing function of its demand price (q_1).

Next for case 6 with $b > 0$ $s = 0$ $E = 0$ $D = F > 0$, we have $p_I I = b$ and investment is determined from,

$$-(1 - u_c)A' + p_I q_2 + q_1 = 0.$$

Thus $I = I^{m6}(q_1,q_2)$ with $\frac{\partial I}{\partial q_1} = 1/(1-u_c)A'' > 0$ and $\frac{\partial I}{\partial q_2} = p_I(1-u_c)A'' > 0$.

Investment is an increasing function of its demand price. Moreover, as the per dollar marginal cost of financing decreases ($-q_2$ decreases so q_2 increases) then investment increases.

$b = B^{m6}(q_1,q_2)$ and $\frac{\partial b}{\partial q_i} = p_I \frac{\partial I}{\partial q_i} > 0$ $i = 1, 2$.

Since new bond issues finance investment then changes in bond issues must be proportional to changes in investment.

The importance of these results is two-fold. First, they point out how the short-run investment demand function differs according to the financing patterns. Second, these results are needed to analyse the dynamics and the stationary state.

4.2 With Regulation

Case 1 in Table 2 is defined for $b > 0$, $s = 0$ $E = E^r$ and $D > 0$. Investment and new debt are determined from

$$-(1 - u_c)A' + p_I q_2 + q_1 = 0$$

$$p_I I - b - E^r = 0.$$

Therefore $I = I^{m6}(q_1,q_2)$ with $\frac{\partial I}{\partial q_1} = 1/(1-u_c)A''>0$ and $\frac{\partial I}{\partial q_2} = p_I/(1-u_c)A''>0$.

In this context the investment demand function is identical to that found for case 6, in the absence of regulation. This occurs because in both situations debt is the financing instrument and dividends are paid out to the shareholders. For debt, $b = B^{r1}(q_1,q_2,E^r)$ with $\frac{\delta b}{\delta q_1} = p_I \frac{\delta I}{\delta q_i} > 0$

$i = 1,2$. The ceiling on retentions causes the change in bonds to be equal to the value of the change in investment.

The second case where $b = 0$, $s > 0$ $E = E^r$ and $D > 0$ leads to

$$-(1 - u_c)A' + p_I(\phi - 1) + q_1 = 0$$

with $I = I^{m4}(q_1)$, $\frac{dI}{dq_1} = 1/(1-u_c)A'' > 0$ and $s = S^{r2}(q_1, E^r)$ $\frac{\delta s}{\delta q_1} = p_I \frac{\delta I}{\delta q_1} > 0$.

The investment function is identical to case 4 in the absence of regulation, because equity is the financing source. Case 3 is similar to case 2, with respect to investment, although we are only able to solve for the sum of new debt and share issues. That is, $b + s = S^{r2}(q_1,E^r)$.

In the last case, because retentions are E^r and there is no external financing, then $p_I I = E^r$. The investment is determined by the regulator and it is fixed for all $t > 0$.

5. Dynamics and the Stationary State

The dynamic path to be followed by the firm depends on whether or not regulation is binding and which financing pattern arises. The relevant differential equations describing the dynamics are

(13.1) $\dot{K} = I(q_1, q_2, K, B) - \delta K$

(13.2) $\dot{B} = B((q_1, q_2, K, B) - B$

(13.3) $\dot{q}_1 = (\rho + \delta)q_1 - (1 + \lambda_3)[(1 - u_c)R' + u_c \delta p_I]$

(13.4) $\dot{q}_2 = (\rho + 1)q_2 + (1 + \lambda_3)[(1 - u_c)r_b + 1]$.

We now proceed to analyse the specific forms of equation set (13) under the alternative feasible cases.[7]

5.1 Without Regulation

There are only two possible cases to consider. Cases 4 and 6 from Table 1. First, for case 4, equation set (13) with $b = 0 = s$, $E = p_I I > 0$, $D = F - p_I I > 0$ becomes

$$\dot{K} = I^{m4}(q_1) - \delta K$$

$$\dot{B} = -B$$

$$\dot{q}_1 = (\rho + \delta)q_1 - (1-u_c)R' - u_c \delta p_I$$

$$\dot{q}_2 = (\rho + 1)q_2 + (1-u_c)r_b + 1.$$

Now in the stationary state with $\dot{B} = 0$, since $b = 0$ then all debt is retired. In addition, at $\dot{q}_2 = 0$, $q_2^{m4} = -[(1-u_c)r_b + 1]/(\rho + 1)$ and since $1 + q_2^{m4} < 0$ then $\rho = [(1-u_p)/(1-u_g)]r < (1 - u_c)r_b$. Thus we see that the rate of return required by shareholders (ρ) is less than the rate of return to bondholders and consequently equity (in terms of retentions) is the financing source.

We depict the dynamic path and the stationary state in Figure 1. The $\dot{K} = 0 = \dot{q}_1$ curves and the movement off the curves are derived by noting,

$$\frac{\partial \dot{q}_1}{\partial q_1} = \rho + \delta > 0 \quad , \quad \frac{\partial \dot{q}_1}{\partial K} = -(1-u_c)R'' > 0.$$

Figure 1

Stationary State in Case 4 Without Regulation

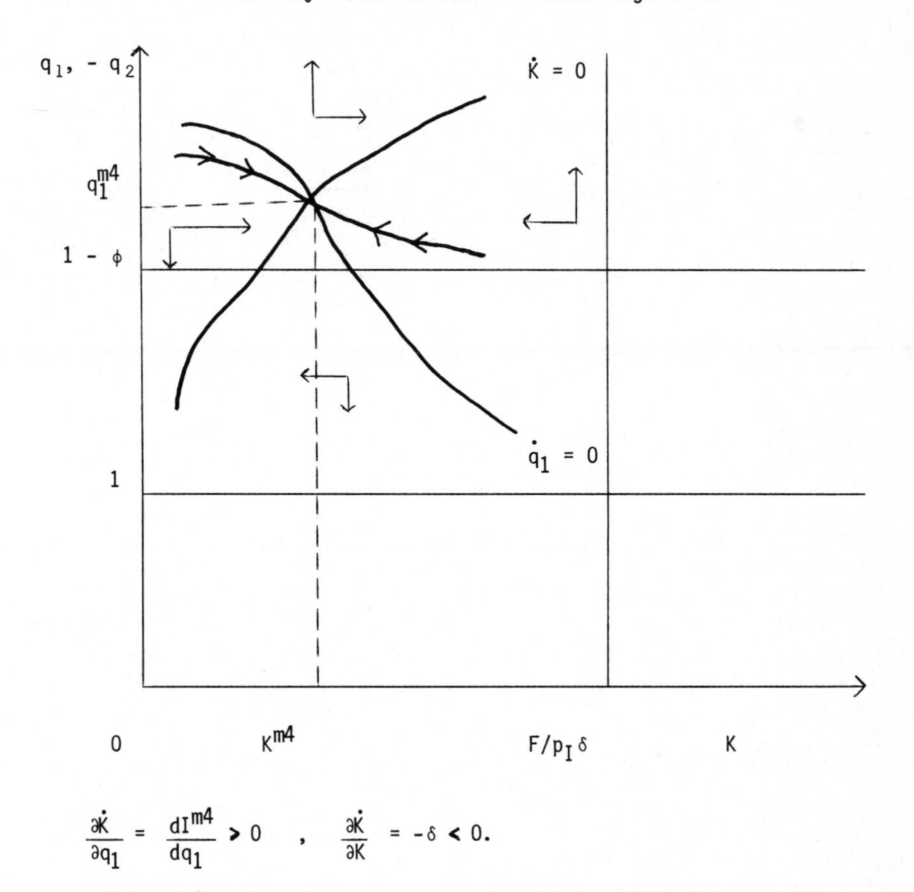

$$\frac{\partial \dot{K}}{\partial q_1} = \frac{dI^{m4}}{dq_1} > 0 \quad , \quad \frac{\partial \dot{K}}{\partial K} = -\delta < 0.$$

Thus the $\dot{K}=0$ locus is positively sloped with $\dot{K} < 0$ below the curve and $\dot{K} > 0$ above the curve. The $\dot{q}_1 = 0$ locus is negatively sloped with $\dot{q}_1 > 0$ above the curve and $\dot{q}_1 < 0$ below the curve. Therefore if the stationary state appears in case 4, it is unique and a saddle point.

In a similar fashion we can derive the same properties for case 6 where $1 + q_2 > 0$ and consequently $\rho > (1 - u_c)r_b$. Thus when there is only debt financing for a nonregulated firm the required rate of return to shareholders exceeds the rate of return the firm needs to offer bondholders.

The final aspect of the stationary state which needs to be discussed is the determination of the capital stock.[8] In the absence of regulation for case 4 the capital stock is determined from equations (11.1)-(11.3), (13.3) and (13.4),

$$(14)\quad R' = p_I[(\rho/1-u_c) + \delta] + (\rho + \delta)A'.$$

The right side of (14) is the user cost of capital. Clearly, we can observe that the personal and corporate tax rates affect the user cost, so that the tax system is not neutral. However $\rho = [(1-u_p)/(1-u_g)]r < r$ since $u_p > u_g$. Therefore the distortion created by the tax system is not as large as the situation when the personal income and capital gains tax rates are ignored.

In the case where only debt financing is used the capital stock is determined from,

$$(15)\quad R' = \frac{p_I}{(\rho+1)} \frac{[(\rho+\delta)r_b + \rho(1 - u_c\delta) + \delta]}{(1 - u_c)} + (\rho + \delta)A'$$

The user cost in case 6 is given by the right side of (15). Although this expression looks somewhat complicated, if we introduce the usual assumption for debt financed firms, that the accumulation of bonds matches capital stock changes, then the right side of (15) becomes,

$p_I\ (r_b + \delta) + (\rho + \delta)A'$. Here with the depreciation and bond retirement

rates the same, the corporate tax rate does not affect the user cost of capital.

5.2 With Regulation

There are four possible cases which can emerge as the stationary state under regulation. Table 2 characterizes the financing patterns for these different long-run equilibria. What remains to be discussed is the determination of the capital stock.[9]

Let us consider case 1 under regulation. Capital stock is determined from (with the bond retirement rate equal to the depreciation rate)

$$(16)\quad R' = p_I(r_b + \delta) + (\rho + \delta)A'.$$

Now we cannot compare the capital stock in this situation to that of the capital stock in the debt financing case in the absence of regulation. The reason is that under bond financing and no regulation $\rho > (1-u_c)r_b$, but with regulation we must always have $\rho < (1-u_c)r_b$. This means that we can compare the capital stock under retention financing for the non-regulated firm to that for the bond and retention-financed regulated firm. By inspection, we see that since $\rho < (1-u_c)r_b$ the right side of (16) exceeds the right side of (14). Thus the regulated firm in this case exhibits a smaller capital stock relative to the retention-financed non-regulated firm. This result is illustrated in Figure 2. In addition the regulated firm's capital structure contains a greater proportion of funds obtained from external sources.

Figure 2

Capital Stock Determination

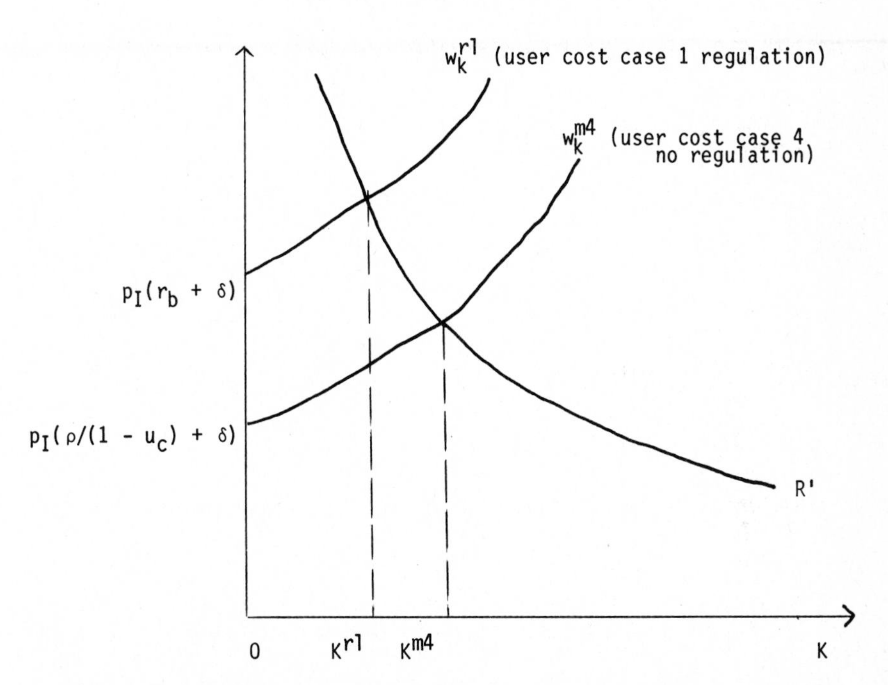

This conclusion arises for the following reason. A regulated firm faces a binding constraint on a financing source, namely internally generated funds. This constraint is effective when retentions are in fact the cheapest financing avenue available to the firm (i.e., $1 + q_2 < 0$). Thus the regulated firm is not able to use as much of its net flow of funds toward capital expansion as a nonregulated firm. The regulated firm must turn to the financial capital markets, which are a relatively more expensive financing source. Therefore the user cost of capital must be higher under regulation and consequently the capital stock smaller.[10]

6. Conclusion

In this paper we developed a model of investment and financing decisions for a flow of funds regulated firm. The regulatory constraint appeared as an upper limit on the amount of internally generated funds that can be devoted to capital accumulation.

We were able to characterize the differences between financing patterns for regulated and non-regulated firms. We found that when regulation is effective the return required by shareholders is less than the return to bondholders. This implies that regulation could only be a binding constraint when retentions are the cheapest sources of funds; otherwise the limit on internal funds would not affect the firm.

In the long-run, a non-regulated firm would either finance its capital stock by retentions or debt, but would not issue more shares. However, a regulated firm would use retentions in combination with either bond or share issues. Moreover, a comparison of the capital stocks between a retention-financed non-regulated firm and the regulated counterpart, illustrates that the latter's capital stock is smaller and it's capital structure exhibits a larger percentage of externally generated funds.

FOOTNOTES:

1. We drop the subscript t for notational convenience.

2. We assume away the problem of regulatory lag (see Klevorick [8]).

3. If we view regulation in terms of a limit on the rate of return to shareholders then the constraint is

   $$(1 - u_c)[R(K) - A(I) - r_b B] + u_c \delta p_I K \leq i p_s N_s$$

   where i is the allowed rate of return, p_s is the share price and N_s is the number of shares. Clearly, when regulation is effective (the equality holds), then, at any time period, when the stock variables are fixed (K, B, N_s), investment is not determined by the firm, but it is determined from the regulatory constraint. This implies that if the allowed rate of return is fixed for $t \geq 0$, the firm immediately adjusts to its long-run equilibrium capital stock. Since we do not observe such jumps in the capital stock, viewing regulation as a direct restriction on the rate of return to shareholders, rather than as a limit on the firm's flow of funds, may not be appropriate.

4. A dot over the variable signifies the time derivative.

5. Capital gains are taxed on an accrual basis rather than when they are realized, and the tax rates are independent of their respective

bases. This is the approach followed by Stiglitz [11] and King [7].

6. This conclusion is consistent with Stiglitz [11] and Auerback [2].

7. In order to guarantee that $0 \leq K \leq \infty$ in the stationary state, we assume that $R' > [(\rho + \delta)q_1/(1 - u_c)(1 + \lambda_3)] - u_c \delta p_I(1 - u_c)$ for $K = 0$ and $R' < [(\rho + \delta)q_1/(1 - u_c)(1 + \lambda_3)] - u_c \delta p_I(1 - u_c)$ for $K=\infty$.

8. To facilitate the discussion, we assume r, r_b, and p_I are constant over time.

9. In a similar fashion to the non-regulated firm we can develop phase diagrams for the regulated firm.

10. The same conclusion holds for case 2 under regulation where we find,

$$R' = p_I[\rho + \delta(1 - u_c) - \phi(\rho + \delta)]/(1 - u_c) + (\rho + \delta)A'.$$

Recall that $\phi < 0$ so that the user cost here exceeds that for the non-regulated retention-financed firm. Case 3 is the knife-edge case where $1 + q_2 = \phi$ and case 4 is trivial since $I = E^r/p_I$. This means that the regulator determines the capital stock, and after any initial jump, the stock is constant for all $t \geq 0$.

REFERENCES:

[1] Appelbaum, E. and Harris, R., Capital accumulation and investment in the regulated firm, 1978.

[2] Auerback, A.J. Wealth maximization and the cost of capital, Quarterly Journal of Economics (1979).

[3] Elton, E.J. and Gruber, M., Valuation and the cost of capital for regulated industries, Journal of Finance (1971).

[4] Elton, E.J. and Gruber, M., Valuation, optimum investment and financing for the firm subject to regulation, Journal of Finance (1975).

[5] Jaffe, J.F. and Mandelker, G., The value of the firm under regulation, Journal of Finance (1976).

[6] Joskow, P.L., Pricing decisions of regulated firms: A behavioral approach, Bell Journal of Economics (1973).

[7] King, M.A., Taxation and the cost of capital, Review of Economic Studies (1974).

[8] Klevorick, A.K., The behavior of a firm subject to stochastic regulatory review, Bell Journal of Economics (1973).

[9] Lucas, R.E., Adjustment costs and the theory of supply, Journal of

Political Economy (1967).

[10] Sherman, R., Financial aspects of the rate of return regulation, Southern Economic Journal (1977).

[11] Stiglitz, J.E., Taxation, corporate financial policy and the cost of capital, Journal of Public Economics (1973).

[12] Taggart, R.A. Jr., Rate of return regulation and utility capital structure decisions, Journal of Finance (1981).

[13] Treadway, A.B., Adjustment costs and variable inputs in the theory of the competitive firm, Journal of Economic Theory (1970).

Economic Analysis of Telecommunications:
Theory and Applications
L. Courville, A. de Fontenay and R. Dobell (eds.)

FINANCING AND INVESTMENT BEHAVIOUR OF THE REGULATED FIRM

M.K. Berkowitz and E.G. Cosgrove

University of Toronto

1. Introduction

Most studies of public utility regulation have concentrated on either the production decisions (e.g. Averch and Johnson (1962), Baumol and Klevorick (1970), etc.) or the financing decisions (e.g. Elton and Gruber (1971, 1977), Jaffe and Mandelker (1976), Arditti and Peles (1980), etc.). An exception to these myopic discussions is an article by Robert Meyer (1976) which attempts to integrate the "real" and financial aspects of the regulated firm. In doing so, Meyer merges the traditional weighted cost of capital concept and uncertainty with the basic Averch-Johnson model.

Unfortunately, Meyer's attempt is tainted by two fundamental misconceptions of the problem. First, it is not sufficient to simply replace the exogenously determined allowed rate of return by the weighted cost of capital in the regulatory constraint. While Meyers makes a distinction between the price of capital and the interest on funds used to finance that capital, clearly the rental price of capital includes the opportunity cost of funds so that the regulatory constraint is misspecified. Second, Meyers assumes that the correlation between the firm's returns and the returns on the market portfolio are unaffected by the firm's financing and investment decisions. It is generally accepted, however, that this systematic component of the firm's overall risk is related to the decisions undertaken by the firm. For example, Hamada (1969) and Rubinstein (1973) have demonstrated that the systematic risk is directly related to the firm's debt/equity choice. For both of these reasons, the results obtained by Meyers are subject to serious question.

In this chapter, we examine the integration of the financing and investment decisions for the regulated firm operating in an uncertain environment which includes the possibility of bankruptcy and the associated costs thereof. In doing so, our model corrects for the deficiencies in Meyer's formulation of the problem by more accurately reflecting the regulatory process and by being more consistent with generally accepted principles in finance theory. The firm must simultaneously decide on the level of capacity to employ, the method of financing, and the price to charge for its product. These decisions are compared for the regulated and unregulated firms and the impact of bankruptcy is examined within a regulatory framework in which the allowed rate of return is based upon the marginal costs of capital. These results are then compared to the decisions made under the current regulatory regime where the allowed return is based upon embedded costs. Finally, we discuss the problem of regulation within a principal-agent context in order to ascertain possible additional costs of regulation.

2. Development of the Model

Consider the formation of a new firm where K in capital, which is to be determined, is needed to finance its productive activities. For simplicity, we assume no depreciation and perfect second-hand markets so that after one period, the firm receives K upon liquidation[1]. In this situation, bankruptcy does not imply the termination of the firm's activities since the firm is already assumed to be liquidated after one period. Bankruptcy occurs if, at the end of the period when the settling-up with creditors takes place, the firm lacks the necessary funds to pay its bondholders both the principal and interest owed to them.

Throughout the subsequent discussion, we use the following notation:

- p : price per unit of output;
- $\tilde{D}(p)$: random demand for the firm's product which is assumed to be homoskedastic and of the form, $\bar{D}(p) + \tilde{\varepsilon}$, where $\frac{\partial \bar{D}}{\partial p} < 0$;
- c : constant operating cost per unit of output produced;
- i : before-tax interest rate on the firm's debt;
- ρ : riskless rate of interest;
- B : bookvalue of firm's debt;
- S : bookvalue of firm's equity;
- K : capacity employed by the firm, where the purchase price per unit of capacity is normalized to equal one;
- L : explicit costs associated with bankruptcy which are assumed to exceed the usual costs of liquidation at the end of the period;
- δ : debt-capital ratio employed by the firm, which equals B/K;
- k^* : after-tax allowed return to equity holders of the firm;
- τ : corporate tax rate;
- $\tilde{\Pi}$: before tax operating profit of the firm, which is equal to $(p-c)\,\tilde{D}(p)$.

As stated above, bankruptcy occurs if the firm is unable to meet its obligations to bondholders at the end of the period, i.e. the firm is bankrupt if

$$(1) \quad \tilde{\Pi} + K < (1+i)B$$

Associated with this bankruptcy condition is a probability distribution such that the cumulative probability of bankruptcy, $\text{Prob}[\tilde{\Pi}+K<(1+i)B]$, is expressed as G. The return to shareholders ($\tilde{Y}_S$) and bondholders ($\tilde{Y}_B$) during the period are therefore stochastic and dependent upon whether or not bankruptcy occurs[2]. Specifically,

(2)
$$\tilde{Y}_S = \begin{cases} (1-\tau)\ (\tilde{\Pi}-iB) & \text{if } \tilde{\Pi}+K>(1+i)B \\ 0 & \text{if } \tilde{\Pi}+K<(1+i)B \end{cases}$$

$$\tilde{Y}_B = \begin{cases} iB & \text{if } \tilde{\Pi}+k>(1-i)B \\ \tilde{\Pi}-L & \text{if } \tilde{\Pi}+K<(1+i)B \end{cases}$$

The combined returns to both shareholders and bondholders are then

$$\tilde{Y}^* = \begin{cases} \tilde{\Pi}(1-\tau) + i\tau B & \text{if } \tilde{\Pi}+K>(1=i)B \\ \tilde{\Pi}-L & \text{if } \tilde{\Pi}+K<(1+i)B \end{cases}$$

The value of the firm can be expressed under fairly general conditions by the valuation equation developed by Sharpe (1964), Lintner (1965), and Mossin (1966)[3].

(4)
$$V = -K + \frac{E(\tilde{Y}^*) + K - \lambda \text{cov}\ (\tilde{Y}^*,\tilde{Y}_M)}{1+\rho}$$

$$= \frac{E(\tilde{Y}) - \lambda \text{cov}\ (\tilde{Y},\tilde{Y}_M)}{1+\rho}$$

where $\tilde{Y}$ is defined as the difference betwen Y^* and ρK, $\tilde{Y}_M$ is the return on the market portfolio, and λ is the constant price per unit of risk which is equal to $[E(Y_M) - \rho V_M]/\sigma^2_M$.[4]

While actual returns may be above or below the allowed rate of return, it is assumed that on average the firm earns the allowed return. Suppose that the competitive regulatory measured weighted cost of capital in the absence of regulation is r[5], where

(5) $$r = \bar{k}\ (\frac{\bar{S}}{K}) + \bar{i}\ (\frac{\bar{B}}{K})$$

$\bar{k}$ and $\bar{i}$ represent the "true" opportunity cost of equity and debt funds, respectively, and $\bar{S}$ and $\bar{B}$ are the optimal competitively determined levels of equity and debt for any level of capacity (K) employed. Often in practice, $\bar{B}/K$ is referred to as the notional capital structure. In essence, r is the rental price of capital in a competitive market. In a world of certainty, the firm should be allowed to earn exactly r. However, lack of confidence in the estimated of r and the uncertainties that exist in the demand for the product suggest that a risk-averse strategy is often followed by the regulator. This results in the allowed rate being set above r in order to insure that investors earn the competitive return.

If an excess return is in fact realized, it is imputed to the shareholders so that the firm's allowed rate of return, s, can be expressed as

$$(6) \quad s = k^* \left(\frac{S}{K}\right) + i\left(\frac{B}{K}\right) ,$$

where k* is the allowed return to shareholders. The difference between s and r represents the upper bound of the excess returns per unit of capital available to firm. It should be recognized that as the firm alters its debt-equity mix from the (regulator's) perceived optimal level of δ, the regulator may also adjust the true opportunity cost of funds in determining r so as to reflect the actual capital structure used by the firm. In the initial formulation of the problem, regulation is assumed continuous so that any change in that changes the opportunity cost of funds is immediately reflected in a revised allowed return on equity so as to maintain a constant (or near constant) total excess return to shareholders.

To satisfy the condition that excess profit does not exceed (s-r)K, the firm has three degrees of freedom. It chooses its product price, level of capacity, and debt-equity mix so that the following inequality is satisfied.

$$(7) \quad [E(\Pi) - iB]\,(1-\tau) + iB - [E(\bar{\Pi}) - \bar{i}\bar{B}]\,(1-\tau) - \bar{i}\bar{B} \leq s-r)K,$$

where $E(\bar{\Pi})$ and $\bar{B}$ represent the optimal competitive expected operating profit and debt level respectively. Substituting the values of r and s in (5) and (6), denoting the debt-capital ratio by δ, and simplifying yields

$$(8) \quad E(\Pi) - E(\bar{\Pi}) \leq [\alpha^*(1-\delta) - \bar{\alpha}(1-\bar{\delta}) + i\delta - \bar{i}\bar{\delta}]K$$

where α^* and $\bar{\alpha}$ are $k^*/(1-\tau)$ and $\bar{k}/(1-\tau)$ respectively[6]. Implicit in the formulation of the above regulatory constraint is that any tax benefits associated with a debt-equity choice by the firm which exceeds the regulator's desired level (or notional capital structure) is regulated away through a mandated lower product price. To see this, let us look at the constraint. As long as α^* is greater than i, a marginal increase in δ above $\bar{\delta}$ results in a net reduction in the R.H.S. of the constraint equal to $(-\alpha^*+i)K$. The L.H.S. must, therefore, be similarly reduced and the firm is forced to lower the price of its product. Because, it is unclear whether or not such deviations from the perceived optimal capital structure do result in such reactive measures by the regulator, as well as the degree of the response, the regulatory constraint can more generally be expressed as

$$(9) \quad E(\Pi) - E(\bar{\Pi}) < [\alpha^*(1-\delta) - \bar{\alpha}(1-\bar{\delta}) + (i\delta - \bar{i}\bar{\delta})\phi]K ,$$

where $\phi = (1-\gamma\tau)/(1-\tau)$ and γ equals 1 when the regulator effectively regulates away all the excess tax benefits and γ equals $\underline{0}$ when all excess tax benefits are realized by the firm from having $\delta > \bar{\delta}$. Any value of δ between 0 and 1 represents the regulator's degreeof efficiency in performing this function.

Because the firm must set its price prior to knowing actual demand, the demand for its product may exceed capacity. The cost and problems of rationing the service in the event of a shortage suggest that firms might, ex ante, choose their prices and capacity to reflect their aversion to such an occurrence. One method for dealing with this problem is the inclusion of a chance-constraint of the form.

$$(10) \quad \text{Prob}[D(p) > K] \leq \S$$

The greater is the firm's aversion to unsatisfied demand, the smaller is equivalent form which enhances the tractability of solutions. The above inequality can be rewritten as

$$(11) \quad \bar{D}(p) + N\sigma_D \leq K \ ,$$

where N is the number of standard deviations above the mean necessary to reduce the area in the upper tail of the probability distribution to §.

If the firm's objective is to maximize the market value of shareholders' wealth, it will choose p, K and δ so as to maximize the valuation expression in (4) subject to the constraints outlined in (9) and (11). Denoting the Lagrangian multipliers associated with (9) and (11) by μ and v respectively, the Lagrangian expression can be formally written as:

$$(12) \quad \underset{p,k,\delta}{\text{Max}}\ L = \frac{1}{\rho'}[E(\tilde{Y}) - \lambda \text{cov}\,(\tilde{Y},\tilde{Y}_M)] + \mu[[\alpha^*(1-\delta) - \bar{\alpha}(1-\bar{\delta})$$

$$+ (i\delta - \bar{i}\bar{\delta})\phi]K - E(\tilde{\Pi}) + E(\bar{\Pi})] + v[K - D(p) - N\sigma_D]$$

The Kuhn-Tucker conditions for the problem outlined above are:

$$(13a) \quad \frac{\partial L}{\partial p} : E(MR) = c + \frac{1}{\psi}[v' + \tau(1-G)\frac{\partial p}{\partial D}E(\varepsilon/\varepsilon \geq A) + \frac{\lambda\sigma_{DM}}{\sigma^2_D}\frac{\partial p}{\partial D}$$

$$[\sigma^2_D - \tau(1-G)E(\varepsilon^2/\varepsilon > A)]] + \frac{1}{\psi}\frac{\partial G}{\partial p}\frac{\partial p}{\partial D}X$$

$$(13b) \quad \frac{\partial L}{\partial K}: v = \frac{\rho}{\rho'} - \frac{i\tau\delta(1-G)}{\rho'}\left[1 - \frac{\lambda\sigma_{DM}}{\sigma^2_D}E(\varepsilon/\varepsilon \geq A)\right] - \mu\left[\alpha^*(1-\delta)\right.$$

$$\left. - \bar{\alpha}(1-\bar{\delta}) + (i\delta - \bar{i}\bar{\delta})\phi\right] - \frac{\partial G}{\partial K}\frac{X}{\rho'}$$

$$(13c) \quad \frac{\partial L}{\partial \delta}: \frac{i\tau(1-G)(1+\eta)}{\rho'}\left[1 - \frac{\lambda\sigma_{DM}}{\sigma^2_D}E(\varepsilon/\varepsilon \geq A)\right] = \mu\left[\alpha^* - i\phi(i+\eta)\right] - \frac{\partial G}{\partial \delta}\frac{X}{K}$$

(13d) $p \geq 0,\ K \geq 0,\ \delta \geq 0,\ \mu \geq 0$, and $v \geq 0$

where $\rho' = 1 + \rho$;

$$\eta = \frac{\delta}{i}\frac{\partial i}{\partial \delta} > 0,$$

$$\psi = 1 - \tau(1-G)[1 - \frac{\lambda\sigma_{DM}}{\sigma_D^2} E(\varepsilon/\varepsilon \geq A)] - \mu\rho';$$

$$X = \tau p - c\,[\bar{D}(p) + E\,(\varepsilon/\varepsilon \geq A)] - i\tau B - L$$

$$+ \frac{\lambda\sigma_{DM}}{\sigma_D^2} E(\varepsilon/\varepsilon \geq A) - \left[(\frac{1-\tau G}{G})\, E(\Pi) - i\tau B - \frac{L(1-G)}{G}\right.$$

$$\left. - (p-c)\left[\frac{\sigma_D^2 - (1-\tau G)\, E(\varepsilon^2/\varepsilon > A)}{G}\right]\right];\ \text{and}$$

$$A = \frac{(1+i)B - K - (p-c)\bar{D}}{(p-c)}$$

Equation (13a) states that the firm should set the price of its product where the expected marginal revenue equals the expected marginal cost. Notice that the expected marginal cost includes the usual operating and capacity costs as well as an explicit adjustment for risk. The greater the covariance between the demand for the firm's product and the market, the lower the price of the product.

An interesting implication of this is that a firm which sells the same product to two or more customers can "legitimately" charge different prices if the demands of the different customer classes are correlated differently with the market. For the services offered by most public utilities, this is typically the case and would suggest that differentiated pricing practices must be evaluated on both cost and risk differences between customer classes.

Furthermore, because X appears to be positive for reasonable values of its arguments, the third expression on the R.H.S. of (13a) depends upon the change in the cumulative probability of bankruptcy as the price increases. If we assume that $\tilde{\varepsilon}$ is normally distributed, this change in the cumulative probability becomes:

(14)
$$\frac{\partial G}{\partial p} = -g(A)\left[\frac{(p-c)^2\frac{\partial \bar{D}}{\partial p} + (1+i)B-K}{(p-c)^2}\right]$$

where g(A) is the marginal probability of A. It should be recognized, however, that the theory suggests that the probability distribution is not continuous over the entire range, but instead, the probability of bankruptcy is discontinuous at some critical δ, say δ^*, at which point the probability of bankruptcy increases substantially for values of δ above δ^*. The above expression can therefore be thought of as an approximation of the actual distribution in order to explicitly show the variables affecting $\frac{\partial G}{\partial p}$. For values of δ below δ^*, it should be realized, however, that $\frac{\partial G}{\partial p}$ is negligible while for values of δ greater than or equal to δ^*, $\frac{\partial G}{\partial p}$ is indeed significant.

Looking to (14), we see that the sign of $\frac{\partial G}{\partial p}$ is related to the demand elasticity, interest and debt principal, and total capacity expenditure. The sign of these relevant variables suggests that $\frac{\partial G}{\partial p}$ is positive. Therefore, it follows that the third expression in (13a) is positive though small for $\delta < \delta^*$. This means that a higher price, which increases the risk of bankruptcy, reduces prices yet furhter than they would have been in the absence of the threat of bankruptcy.

Condition (13b) describes the firm's capacity decision. Capacity is added to the point where the expected marginal contribution of the last unit employed exactly equal its expected marginal cost. It is interesting that when looking at (13b), we see that the marginal capacity cost is the sum of the discounted riskless rate and an adjustment for risk which is related to the correlation of the demand with the market (as expressed in the second part of the second term). The effect of regulation, which is assumed binding ($\mu > 0$), is the not surprising result that more capacity is employed relative to an unregulated, but otherwise identical firm. Moreover, the fourth expression on the R.S.H. of (13b) is positive since it can be easily demonstrated that $\frac{\partial G}{\partial K}$ is negative,[7] though again quite small for $\delta < \delta^*$. As K increases and the probability of bankruptcy subsequently decreases, there appears to be a negative consequence of such behavior that can be seen by examining the components of X. While the firm loses the tax benefits from its debt and must incur the explicit costs of bankruptcy in the event of such a mishap, it also will not pay taxes on its ex ante expected income as it had planned to do when it made its capacity decision at the beginning of the period. Although this is unquestionably a small effect when δ is less than the critical value; this opportunity benefit is an inducement to a somewhat lower level of capacity, below that already distorted level induced by regulation.

What we have referred to above as an opportunity benefit of bankruptcy has typically been overlooked when the costs of bankruptcy have been examined in the literature. For example, Kim (1978) states that bankruptcy costs can be thought of as being comprised of three major components. First, there is the "short-fall" arising from liquidation or the "indirect" cost of reorganization, both of which are absent in a single-period model such as the one being analyzed in this paper. Second, various administrative expenses must be paid to third parties in the course of the bankruptcy proceedings, represented by L in our model. Third, firms lose tax credits which they would have received had they not gone bankrupt, as expressed by

$i\tau B$. In addition, however, there is the tax that was expected to be paid, but in the event of bankruptcy, will not be paid, and is represented as $\tau(p-c)[\bar{D}(p) + E(\varepsilon/\varepsilon \geq A)]$. This latter expression can be though of as a negative cost (or benefit) of bankruptcy.

The final condition (13c), aside from the non-negativity conditions summarized in (13d), describes the firm's debt-equity choice. This equation suggests that the optimal debt-equity mix is the one in which the discounted expected marginal benefits from an extra dollar of debt exactly equals the expected marginal cost associated with that dollar of debt. The expected marginal cost of debt, as expressed by the R.H.S. of (13c), includes two terms. The first denotes the opportunity cost of debt since the cost of equity exceeds the debt cost. Therefore, as debt is substituted for equity the allowed return is reduced. Notice that as the elasticity of the interest rate increases, the opportunity cost of having additional debt is reduced as long as the excess return to shareholders remains constant (i.e. regulation is continuous). The greater is the elasticity in this situation, the smaller is the financing inefficiency.

The degree of reduction in the opportunity cost depends as well on the extent to which the tax benefits from debt are regulated away through lower product prices. In the extreme case where γ is equal to one, implying all the tax benefits from excess debt are regulated away, ϕ equals one. At the other extreme, when γ equal zero, implying all the tax benefits from excess debt are realized by the firm, ϕ equals $(1/1-\tau)$. Because $(1/1-\tau)$ exceeds 1, the marginal cost of debt is lower when all tax benefits from debt are allowed to be realized, ceteris parabis, and the firm will choose a higher debt-equity mix than otherwise.

The second term on the R.H.S. of (13c) is negative since $\frac{\partial G}{\partial \delta}$ is positive[8]. That is, because an increase in the debt-equity ratio increases the probability of bankruptcy, there is a higher probability that the ex ante expected tax payment will not have to be made. Thus, the increased risk of bankruptcy has a small (for $\delta<\delta^*$), but positive effect on the firm's decision to employ more debt.

Before, leaving our discussion of the first-order conditions, we should mention that these results are quite general as well as robust. For example, suppose we assume that the regulatory constraint is not binding (i.e. $\mu=0$) and bankruptcy cannot occur. Then condition (13c) implies that in equilibrium the marginal benefit from an extra dollar of debt is zero which suggests 100 percent debt financing the well-known result arrived at by Modigliani and Miller (1963). Once the threat of bankruptcy is introduced, again assuming $\mu=0$, and the bankruptcy costs (L) are large enough to cause X to be negative, a finite level of debt is dictated by (13c) even in the presence of corporate taxes. That is, if the bankruptcy costs are large and the probability of bankruptcy takes a discontinuous jump at δ^*, the optimal mix will be marginally below δ^*. In a somewhat different manner, Stiglitz (1969) arrived at the same result in the presence of bankruptcy and taxes.

Finally, let us examine the debt-equity choice for the regulated firm relative to an otherwise identical but unregulated firm. If we assume for the moment that the second term on the R.H.S. of (13c) is constant, irrespective of the firm being regulated or not, then it appears that the unregulated firm ($\mu=0$) will employ more debt than its regulated ($\mu>0$)

counterpart. It is likely, however, that the second term is not constant. If we interpret the cost of bankruptcy, L, as including lost earnings in the event of bankruptcy[9], clearly the cost of bankruptcy for the unregulated firm exceeds that for the regulated firm. Therefore, the smaller the excess return allowed the regulated firm, the greater will be the relative loss in earnings for the unregulated firm. It is possible, moreover, that L is so large for the unregulated firm, as compared to the regulated firm, that X will reverse in sign to become negative so that the R.H.S. of (13c) for the unregulated firm exceeds the R.H.S. for the regulated firm. If this occured, the impact of regulation would be opposite to the earlier result, i.e. the unregulated firm would have a lower optimal debt-equity mix than it would if it were regulated. This conclusion was reached by Arditti and Peles (1980) who argued that the firm has less to lose when it is regulated and will therefore issue a greater amount of debt than the unregulated firm which has more to lose if bankruptcy occurs. While one is unable to unequivocably state the direction of the regulated and unregulated debt-equity levels in this model, we are able to discern the relevant factors and appreciate the complexity of this problem, something avoided in the simplified world envisioned by Arditti and Peles.

3. Regulation Based Upon Embedded Costs

When the regulatory constraint is formulated on an embedded cost basis, as is typically the case, the component costs of debt and equity are weighted averages of the outstanding and proposed issues. In this situation the firm enters the period with an accumulated stock of capital, $\hat{K}$, which has been financed by debt and equity of $\hat{B}$ and $\hat{S}$, respectively. Because we continue to maintain that excess returns accrue only to shareholders, and the excess return is constant, it is sufficient to compare the debt costs under the two regulatory regimes in order to evaluate the effect of marginal cost-based regulation.

The embedded cost of debt, i', can be represented as

$$(15) \qquad i' \quad \frac{\hat{i}\,\hat{B}}{\hat{B}+B} + \frac{i\,B}{\hat{B}+B}$$

The greater the increase in interest rates during the past years and the greater the proportion of the firm's outstanding debt that has been financed at the previous lower rates, the smaller that i' is relative to i. Once this embedded cost of debt is substituted into the regulatory constraint, the revised constraint becomes:

$$(16) \qquad E(\Pi) - E(\bar{\Pi}) < [\alpha^*(1-\delta) - \bar{\alpha}(1-\bar{\delta}) + (i'\delta - \bar{i}\bar{\delta})\phi]\,(\hat{K}+K)$$

Without formally presenting the first-order conditions associated with the amended problem, it is fairly straightforward to compare the results to those in (13a)-(13c). Because the embedded cost of debt is below the marginal cost, equation (13c) would dictate a greater substitution to equity from debt in order that the allowed return might be increased. The tradeoff between a higher regulated return and increased tax benefits is even more one-sided in favor of the former incentive than in the earlier problem. Furthermore, it is quite easy to show that upon solving (13c) for μ, the increase in allowed return, due to a lower interest rate on debt in the regulatory constraint, reduces the benefit, μ, from relaxing the constraint. In turn the lower μ and δ have the effect in (13b) of increasing the expected marginal capacity cost so that less capital is employed relative to marginal cost-based regulation. In essence, what embedded cost regulation has done compared to marginal cost regulation is to induce the firm to substitute greater financial inefficiency for less production inefficiency. Unfortunately the effect of this substitution upon the product price is directly related to the relative sizes of the inefficiencies and their respective effects can only be determined by assuming specific functional forms.

4. Agency Costs of Regulation

Throughout our discussion we have consistently referred to the firm's decisions where it has been implicitly assumed that the managers of the firm always act in the best interests of the shareholder-owners, or to be even more narrowly specified, the managers are the owners of the firm. The literature on regulation has followed a similar path. Though the divergence of interests between the managers and owners has been recognized, the resulting consumption of perquisites by managers has been overlooked for one reason or another. This apparent oversight has, however, been redressed in the general economics literature where the incentive problem has received a great deal of attention[10]. The agent-principal problem has also been discussed with respect to its impact on the firm's financing and investment decisions - most notably by Jensen and Meckling (1976).

In the Jensen-Meckling model, an agency problem arises from the fact that with a fixed money wage, a manager who owns less than 100 percent of the firm's stock, say κ, imputes to himself only the fraction κ of the lower value of the stock when he consumes more on the job. As a consequence, the manager consumes more shirking and perquisites the smaller in his fraction of ownership of the firm's common stock. It should be recognized that perquisites may take the form of (suboptimal) decisions which are not consistent with the interests of shareholders. Moreover, in the J-M model, monitoring costs are assumed to vary inversely with the fraction of the firm owned by the manager(s). The manager has, therefore, an even stronger incentive to shirk and consume perquisites the smaller in his ownership share because his consumption is more costly to monitor and control. As a result, the optimality of the investment package and method of financing the investment varies inversely with κ.

While the costs of inefficiency due to the regulation on both the production and financing sides have been examined, the effect of regulation on the size of the agency costs within the firm has thus far not been discussed in the literature. To appreciate the implications of regulation in this light, it is first necessary to examine the role of regulation in the principal-agent relationship.

In the absence of regulation, the firm can be described by the usual principal-agent relationship. The manager chooses a set of actions and then shares the consequences of the actions with the principal. In performing his managerial functions, the interests of the manager are not always consistent with those of the owners so that costs are incurred which are directly attributable to this relationship. To reduce these costs, the owners can institute monitoring at some cost. The greater the expenditure on monitoring, presumably the lower are the agency costs, i.e.

consumption of perquisites and shirking.

Because of the natural monopoly characteristics of these firms, regulation has most often been suggested as the remedy that allows the firm to operate at the scale of a monopolist yet not charge prices which reflect that degree of monopoly power. In this regard, the regulator is the watchdog of the public interest - i.e. consumers of the product and owners of the firm alike.

With the addition of regulation, the principal-agent relationship becomes more complex. The managers of the firm are no longer simply responsible to the owners, but must now satisfy the interest of consumers as well. The function of regulation in this respect is two-fold. First, it monitors the decisions of the manager so as to insure that the actions taken are in the best interests of the co-principals (consumers and owners) jointly. Because the agent now has a responsibility to two principals with separate interests, the actions taken by him are suboptimal from the standpoint of each principal. This has been witnessed in our earlier discussion of the induced inefficiencies ascribed to the regulatory process. The second function of the regulator is to provide an equitable distribution between the principals of the outcome of the manager's actions. This is the dynamic aspect of regulation. If the manager in the present period accepts a project and next period earns excess rents on that project, the regulator must decide the portion of those rents which should be imputed to consumers by way of lower prices and what portion should be imputed to shareholders.

Our concern is whether the managers of a regulated firm pursue actions so as to incur agency costs which are directly attributable to the process of regulation. While it is recognized that regulation itself may have its costs, both administrative and inefficiency, these are not agency costs since the manager was always implicitly assumed to operate in the owner's interest. Given the regulatory environment in which the manager operates, overcapitalization, for example, would be in the interest of shareholders.

While it is not the purpose of this discussion to identify all the possible costs of the agency relationship which are attributable to regulation, one example demonstrates that an evaluation of current regulatory practices should consider these additional costs. For our particular example, it is useful to examine the proceedings of a regulatory review. Typically, the managers, purportedly acting in the interest of the owners, argue that the allowed return be increased for one reason or another. Consumer-intervenors, on the other hand, vehemently argue that the allowed return should be lower than the rate suggested by the managers. Clearly, the managers are in a dubious position. They realize that part of their performance as managers is to achieve as high an allowed return as possible for the owner-share-holders. Yet, having been successful and being allowed a higher return, they must adopt investments of higher risk in order to realize the high return. Because managers are generally perceived as being risk averse and recognize that such an investment policy could jeopardize their position in the firm, they have a choice of actions. On the one hand, their arguments for an increased allowed return can be presented in a less than optimal manner so that only a marginal increase is allowed by the regulator. In this situation, they are able to make relatively less risky investments and achieve the allowed rate with little additional risk to themselves. On the other hand, they may argue and achieve a higher allowed return after which they adopt investments of a risk which assures that their realized return is below the allowed level. In both of these situations there are agency costs. The action taken by the manager is the one which is more difficult (or expensive) to monitor. It appears in this case that the manager will choose to act suboptimally during the regulatory review since the alternative action is quite easy to monitor.

As we stated above, we are not attempting here the task of identifying all the possible areas in which regulation might affect the costs of the agency relationship. Yet we should point out in closing that an avenue worthy of future research is the effect of regulation on the risk structure of the firm and the consequential actions by risk-averse managers in response to this change in risk due to regulation.

5. Summary and Conclusions

In this chapter we have examined the integration of the financing and investment decisions for the regulated and unregulated firm operating in an uncertain environment which includes the prospect of bankruptcy. Contrary to current regulatory practice which bases the allowed return on embedded costs, our model postulated a regulatory framework in which the allowed return was based on marginal costs and was therefore consistent with the investment decision. The model developed was shown to be both general and robust in that it is capable of demonstrating the early theories of capital structure much discussed in the finance literature. Furthermore, it was possible using this model to show that regulation induces both a production and financing inefficiency. Despite popular belief, it is likely that the regulated firm operates with too little debt. Moreover, when we compared the financing decisions for the regulated and unregulated firms, we were able to identify the relevant parameters and their respective magnitudes necessary for an unambiguous comparison.

The optimal decisions under marginal cost based regulation were then compared to those under embedded cost regulation. In doing so, it was shown that these different forms of regulation lead to different tradeoffs between the inefficiencies in investment and the method of financing the investment. The magnitude of the total cost of inefficiency associated with each method of regulation must be compared in order to identify the least inefficient practice.

Finally, the problem of regulation was characterized within the principal-agent relationship. Within this framework, we discussed the possibility of additional agency costs which can be directly attributable to the regulatory process. While regulation has consistently throughout the years been attacked for the inefficiencies which it causes, it may very well be that the additional costs of the agency relationship, which are attributable to regulation, are of sufficient order to further rebuke any benefits ascribed to the regulatory process. At the least, it is a subject worthy of further pursuit.

FOOTNOTES:

[1] If depreciation (obsolence) occured at some rate d throughout the period, the firm would realize (1-d)K at the end of the period. The

per unit cost of capacity in (4) would then be $(\rho+d)$ instead of ρ. This additional consideration would effect the bankruptcy condition in (1) and our results would change accordingly. Moreover, if the value of K was uncertain at the end of the period, the certainty equivalent end-of-period value of the capital would appear in (4) and the results again would be reinterpreted.

[2] We have simplified the model by assuming that in the event of bankruptcy the firm (bondholders) does not have to pay taxes. In a multi-period setting, the firm could suffer several period of losses or low profitability without being forced into bankruptcy as long as interest obligations (and any other bond covenants) were fulfilled. Since these losses are quite common in years preceeding the actual bankruptcy, the firm would probably be able to carry forward any previous losses and eliminate any tax liability that might arise should the firm show a "taxable profit" in its final period.

[3] Although the Capital Asset Pricing Model (CAPM) is necessary to provide a simple, but practical method for examining the problem of optimal capital structure, one does not need to assume CAPM. Instead, a more general theoretical model such as the state preference approach could be adopted, in which case it is likely that the additional complexity of the model will greatly inhibit implementation. See, for example, Kraus and Litzenberger (1973) who suggest within a state preference framework that a stochastic dynamic programming approach should be used to search for an optimal capital structure.

[4] After simplifying the expression,

$$E(\tilde{Y}) = (p-c)D[1-\tau(1-G)] - \rho K + i\tau B(1-G) - LG - \tau(p-c)(1-G)E(\varepsilon/\varepsilon \geq A)$$

and

$$Cov(\tilde{Y},\tilde{Y}_M) = \frac{\sigma_{DM}}{\sigma_D^2} [[(1-G)E(\varepsilon/\varepsilon \leq A)(i-\tau) + GE(\varepsilon/\varepsilon \geq A)]E(\tilde{\pi}) + (1-G()E(\varepsilon/\varepsilon \geq A)i\tau B - GLE(\varepsilon/\varepsilon \leq A) + (p-c) [(1-\tau)(1-G)E(\varepsilon^2/\varepsilon \geq A) + GE(\varepsilon^2/\varepsilon \leq A)]],$$

where $A = \frac{(1+i)B-K(p-c)\bar{D}}{p-c}$

The term $E(\varepsilon/\varepsilon \geq A)$ represents the expected value of conditional upon bankruptcy not occuring, i.e. $\varepsilon \geq A$. Similarly, $E(\varepsilon^2/\varepsilon \geq A)$ is the variance of given bankruptcy does not occur.

[5] The practice in regulatory proceedings is to measure the overall return, and hence the cost of funds in the absence of transaction costs, by weighting the sum of the after-tax return on equity and before-tax return on debt. We refer to this sum as the regulatory measured weighted cost of capital, as distinguished from the generally accepted cost of capital which is a weighted sum of the after-tax costs of both equity and debt.

[6] Though equity costs are measured after-tax and debt costs before-tax

in order to reflect actual regulatory behavior, the inequality in (8) is fully consistent since both sides are presented on a before-tax basis.

[7] Again, assuming $\tilde{\varepsilon}$ is normally distributed,

$$\frac{\partial G}{\partial K} = g(A)\left[\frac{(1+i)\delta - 1}{(p-c)}\right]$$

which is greater than, equal to, or less than zero as $(1+i)\delta \lesseqgtr 1$. For reasonable values of δ, it follows that $\frac{\partial G}{\partial K} < 0$.

[8] It follows from our assumption of $\tilde{\varepsilon}$ being normally distributed that

$$\frac{\partial G}{\partial \delta} = g(A)\left[\frac{(1+i)K}{(p-c)}\right] > 0$$

[9] It is possible, furthermore, to assume that L is stochastic within the model, where $\tilde{L} = 0$ if bankruptcy does not occur and $\tilde{L} = L(\tilde{\pi})$ if bankruptcy occurs. The specification was adopted by Kim (1978).

[10] One of the earliest reasons given for the incentive problem that exists between the principal and the agent was a difference in risk attitudes held by the two parties. Within this context, Arrow (1971) and Wilson (1968) examined the optimal sharing of purely exogenous risk. Later, Wilson (1969) and Ross (1973) considered situations in which risk could be affected by the actions of the agent. In contrast, Spence and Zeckhauser (1971) analyzed the problem of divergency in incentives as a result of differential information, which was subsequently extended by Harris and Raviv (1979). Following this, Shavell (1979) and Holmstrom (1979) have examined the problems associated with imperfect monitoring.

Economic Analysis of Telecommunications:
Theory and Applications
L. Courville, A. de Fontenay and R. Dobell (eds.)

THE VALUE OF THE FIRM UNDER REGULATION AND THE THEORY OF THE FIRM UNDER UNCERTAINTY: AN INTEGRATED APPROACH

Stylianos Perrakis[1]

University of Ottawa

1. Introduction

While the existence of uncertainty in economic decision-making is central to financial theory, the incorporation of such uncertainty in the microeconomic theory of the firm is a comparatively recent phenomenon. A related tendency, which again has only recently been abandoned, was that of separate consideration of production and financial decisions. This paper examines the interaction of finance and the microeconomics of uncertainty in the theory of the regulated firm, where both disciplines have traditionally played a major role.

The theory of the firm has paid special attention to regulation ever since the Averch-Johnson (AJ) study alleging input distortions induced by the rate-of-return regulatory constraint. Financial theory, on the other hand, has had a built-in role to play in regulation because of the requirement in the Hope Natural Gas decision that a regulated firm's product price provide a "fair" return, interpreted as being "commensurated with returns on investment in other entreprises having corresponding risk" and sufficient to "attract capital" Hence, financial theory must provide rules to evaluate the random income streams that the regulated firm generates, to determine the impact of regulation upon the value of the firm[2]. These random income streams are determined by the production decisions of the firm, which take place under a regulatory constraint, determined by using inputs from the capital markets' evaluation of the income streams of the firm. Hence, financial and production theory have to be examined in conjunction during the study of regulation.

Surprisingly, this did not happen in most studies[3], and for the most part the microeconomic discussions of the regulated firm took place in the absence of financial consideration, and vice-versa. The impact of such a separation was, in our opinion, more serious on the financial side, since most studies in the financial literature seem to have ignored the refinements that have appeared in the theory of the regulated firm. In particular, these studies have bypassed major controversies concerning the nature of the regulatory process itself (which will be shown to have an impact upon the value of the regulated firm).

In the financial literature many important contributions came as a result of the Miller-Modigliani (MM) empirical study [23] of the cost of capital in the electric utility industry. Although this study was not directly concerned with regulation, the fact that it dealt with a regulated industry provided several discussions and controversy ([6], [7], [10], [11], [13], [14], [16]) for a decade. The microeconomic model that determines the earnings stream of the regulated firm in that study was due to Gordon and his associated ([7], [13], [14]), and it was basically a reinterpretation of the AJ model of regulation under certainty, with expectations replacing the deterministic stream of profit. This reinterpretation was examined recently by Meyer [22] and was shown to parallel fairly closely the deterministic AJ results. Nonetheless, even within the context of the AJ-Gordon model several problems still remain, related to the nature of the uncertainty and the resulting consequences upon the value of the firm. These consequences are, in turn, dependent upon the model of valuation used, such as the Capital Asset Pricing Model (CAPM [2], [16], [22]), the simultaneous production and financial model based on the spanning property of the earnings stream ([9], [20]) and the MM cost of capital model ([7], [10], [11], [14], [16]).

The AJ-Gordon model of regulation is "forward-looking" in essence, given that what is constrained within that model is the by definition unobservable) mathematical expectation of earnings. Hence, it requires perfectly shared information and expectations between firm and regulator, as well as a test rule that is not based on observed past performance. Otherwise, if regulation of future performance is based on observed past firm behavior, the firm has an incentive to tailor its performance to fit regulatory expectations. Such "backward-looking" regulatory models have appeared frequently in the microeconomic literature of uncertainty[4]. Their justification can be found in detailed studies of the regulatory process such as those of Joskow ([17], [18]). The implications of backward-looking regulation for the value of the firm have not been explored until now; they form the main result of this paper.

In the next two sections a one-period model of the firm under forward-looking regulation is presented, and some of the earlier valuation results are re-examined in the context of new developments in the theory of the firm under uncertainty. In section IV backward-looking regulation is examined, given that a forward-looking valuation model of the firm has been established. It is shown that the value of the regulated firm under backward-looking regulation can be derived by combining financial instruments of this same firm under forward-looking regulation with simple options (call or puts) on these instruments.

Finally, in the last sections of this paper it is attempted to extend the model of backward-looking regulation beyond the single-period framework under simplified assumptions.

2. The General Model

We let p be the product price and Q (p, u) the random demand, where u is a random factor. The firm's capital is denoted by K, s is the allowed rate of return and $r < s$ is the riskless rate of interest. t is the tax rate and D the amount of debt in the case of levered firm. Subscript L and no subscript denote levered and unlevered firms respectively.

The firm selects the size of its assets K, the output price p, and the other production inputs by maximizing profits while keeping the rate of return on assets at or below s. Similarly, the regulatory rule is such that the profit stream is different for levered than for unlevered firms. Hence, the choices of the regulated firm are going to depend on leverage.

Let x_i, w_i, $i = 1,\dots n$ denote the variable production inputs and their prices. The non-levered firm under certainty chooses its inputs and its output price by solving the following problem

$$(1) \quad \text{Max}\ \{(1-t)\ [pQ(p,u) - rK - \sum_{i=1}^{n} w_i x_i]\}$$

subject to the production function and rate-of-return constraints $Q < F(x_1,..,x_n,K)$ and $(1-t)\ [pQ(p,u) - \sum_{i=1}^{n} w_i x_i] \leq sK$ respectively. It can be shown that, if $C(Q,W,K)$ denotes the variable cost function, where the vector $W = [w_1,...,w_n]$, problem (1) becomes

$$(2) \quad \underset{K,p}{\text{Max}}\ \{(1-t)\ [pQ(p,u) - rK - C(Q(p,u), W,K)]\}$$

subject to

$$(3) \quad (1-t)\ [pQ(p,u) - C(Q(p,u), W,K)] \leq sK$$

The solution is easier to visualize if we define

$$(4) \quad N(u, W, K) = \underset{p}{\text{Max}}\ \{pQ(p,u) - C(Q(p,u), W, K)\}$$

the quasi-rents function of the firm. Under certain common assumptions about the firm's revenue and production functions the function N is concave and increasing in K. Hence, under certainty the optimal K is the positive solution of the equation in K $N(u,W,K) = \frac{sK}{1-t}$, independent of r as long as $r < s$.

For the levered firm the maximand in (1) is augmented by the term trD, where the debt D is assumed exogenous[5]. This does not affect the analysis until the last step, in which the total assets K_L are determined from $(1-t)N(u,W,K_L) + trD = sK_L$, with $K_L > K$. We can compare p_L and p if we adopt the putty-clay-hypothesis [12] with homothetic production structure, according to which, although input substitution may be feasible ex ante (before K is selected), no such substitution is allowed ex post. The production function becomes of the fixed coefficients type, implying that $C(Q,W,K) = [\sum_{i=1}^{n} w_i b_i(K)]\ h(Q)$. In financial analysis it is also assumed ([10], [11], [16]) that the production function exhibits constant returns to scale in the variable inputs, i.e. that $h(Q) = Q$, (constant average variable cost). Then we have $p > p_L$, since $\sum_{i=1}^{n} w_i b_i(K_L) < \sum_{i=1}^{n} w_i b_i(K)$ for $K_L > K$ and $C(Q,W,K)$ decreasing in K.

3. The Value of the Firm Under Forward-looking Regulation

In the model under certainty we assume that there is a single time interval, at the end of which (time 1) the firm is dissolved. All uncertainty is revealed at the end of the period, at which point the variable inputs x_i, i=1,..., n are selected and the output produced instantaneously. The capital stock K is selected at the beginning of the interval and its price is unity. Uncertainty appears in u and in the input price vector W, unknown at the time K is selected. As before, debt D and the rate of return s are exogenous.

With this specification the first step of the certainty analysis (leading from (1) to (2)) remains unchanged in all models. Similarly, under forward-looking regulation the regulatory constraint (3) is satisfied with equality only when the expectation is taken in the LHS. A number of alternative objectives of the regulated firm in this uncertain world have appeared in the literature, explicitly or implicitly in the context of different problems. They will be surveyed briefly below.

a) Expected profit maximization: These models have been used extensively in the MM cost-of-capital for regulated firms controversy ([7], [10], [11], [14], [16]). There are two limiting versions of them depending on the regulatory lag. If the lag is "small" relative to the length of the static period then the output price selection takes place after uncertainty has been resolved, while capital K is selected initially from the expected rate of return constraint. If the lag is "large" then output price is selected simultaneously with K ([30], [31]). Both versions yield similar cost of capital conclusions, although the simultaneous price-capital determination is more complex.

The cost-of-capital controversy as expressed in [10], [11], [13], [14], and [16], refers to the validity for regulated firms of the well-known MM formula $\rho_L = \rho + (1-t)(\rho-r)\frac{D}{S}$, where ρ and ρ_L represent the cost of equity capital without and with leverage, and S is the value of the levered firm's equity. It was asserted by Gordon and disputed by Elton and Gruber (EG) ([10], [11], [13]) that the MM formula does not hold in regulated industries under most common specifications of uncertainty. In [16] the arguments were re-examined and it was concluded that the MM-EG arguments under regulation are valid under special (but commonly assumed) circumstances, such as multiplicative demand uncertainty; otherwise the relation between ρ_L and ρ is a nonlinear function dependent on leverage. However, under the more general formulation followed here the MM formula does not hold even in the cases examined in [11] and [16]. This occurs because of two features ignored in these previous studies: the random nature of the firm's cost function and the fact that K_L is > K, combined with the nonlinearity of earnings with respect to the capital stock[6].

To demonstrate this we denote by V and V_L = S+D the values of the unlevered and levered firms respectively and following [33], we denote by < > the value of the random cash flow in the brackets. It was shown in [33] that within the context of the theory of arbitrage valuation, the valuation operator < > is linear, and that the MM theory is a special case of arbitrage valuation. With this notation we obviously have

(5a) $V = (1-t) < N(u,W,K) >$, $V_L = (1-t) < N(u,W.K_L)> + tD$,

when price is determined ex post, or

(6a) $V = (1-t) [< pQ(p,u) > - < C(Q,W,K) >]$,

(6b) $V_L = (1-t)\,[\langle p_L Q(p_L,u)\rangle - \langle C(Q_L,W,K_L)\rangle] + tD$,

for price determined simultaneously with capital stock, and for $Q_L = Q(p_L,u)$. In (5) K and K_L are determined form $(1-t)E[N] = (1-t)\bar{N} = sK$ or $(1-t)E[N_L] = (1-t)\bar{N}_L = sK_L - trD$, while in (6a, b) expected profit maximization determines p and p_L by the equality of marginal revenue and marginal cost. In both cases it will be shown by a simple (but realistic) counter-example that the MM formula does not hold.

From the definitions of ρ_L and ρ it can be shown that the MM relation $\rho_L = \rho + (1-t)(\rho - r)\frac{D}{S}$ holds iff $\langle\frac{N}{\bar{N}}\rangle = \langle\frac{N_L}{\bar{N}_L}\rangle$ ([16], p. 707). Suppose, however, that $Q = Bup^{-2}$, and that $C(Q,W,K) = Q[w_1b_1(K) + w_2b_2(K)]$, with w_2 fixed and w_1 random. Then $N(u,W,K) = \frac{Bu}{4}[w_1b_1(K) + w_2b_2(K)]^{-1}$. The MM relation holds

$$\text{iff}\quad \frac{\langle u[w_1+w_2\frac{b_2(K)}{b_1(K)}]^{-1}\rangle}{E[u[w_1+w_2\frac{b_2(K)}{b_1(K)}]^{-1}]} = \frac{\langle u[w_1+w_2\frac{b_2(K_L)}{b_1(K_L)}]^{-1}\rangle}{E[u[w_1+w_2\frac{b_2(K_L)}{b_1(K_L)}]^{-1}]}$$

In general $\frac{b_2(K)}{b_1(K)}$ is not constant with respect to K. For instance, if inputs 1 and 2 are fuel and labor respectively it may be expected that the ex ante substitutability of capital is greater for fuel than for labor, implying that $\frac{b_2(K_L)}{b_1(K_L)} > \frac{b_2(K)}{b_1(K)}$ if $K_L > K$. Similarly, maximizing (6a), and for $\overline{uw}_1 = E(uw_1)$, we get $p = 2[\overline{uw}_1b_1(K) + w_2b_2(K)]$. As before, we must compare the value of the before tax earnings normalized by their expected value for levered and unlevered firms. It can be shown that the MM relation holds

$$\text{iff}\quad \frac{\langle u(w_1\ w_2\frac{b_2(K)}{b_1(K)})\rangle}{\overline{uw}_1 + w_2\frac{b_2(K)}{b_1(K)}} = \frac{\langle u(w_1\ w_2\frac{b_2(K_L)}{b_1(K_L)})\rangle}{\overline{uw}_1 + w_2\frac{b_2(K_L)}{b_1(K_L)}}\text{, which is not true}$$

in general for the reasons explained above.

The conclusion, therefore, is that the validity of the MM cost-of-capital theory for regulated industries is extremely restricted. The randomness in input prices makes the average variable cost uncertain, while regulation, by forcing the choice of a larger capital stock for a levered firm, brings systematic differences in the probability distributions of the streams of earnings between levered and unlevered firms.

b) Value maximization in a mean-variance world: Let R_m denote the return to the market portfolio, and λ the market price of risk. For p determined ex post the capital stock is again determined by the rate-of-return constraint. The before-tax value of the unlevered firm then becomes

$$(7) \quad <N(u,W,K)> = \frac{\bar{N} - \lambda Cov(N, R_m)}{1 + r}$$

and $V = (1-t) < N >$ for this unlevered firm. The effect of leverage is identical to that of the previous case, the key element in the comparaison of the distributions of $\frac{N}{\bar{N}}$ and $\frac{N_L}{\bar{N}_L}$ being their covariance with the market index return R_m. These covariances are, in general, unequal, as it can be easily seen in the counterexamples presented above.

If p and K are determined simultaneously then the analysis becomes more complex, and additional assumptions are needed in order to preserve the certainty AJ model. We denote by $R = R(p,u) = pQ(p,u)$ and $C = C(Q,W,K)$ the revenue and cost functions respectively, and by a bar represent the expectation. Then the before-tax value of the all-equity firm becomes

$$(8) \quad < R-C > = \frac{\bar{R} - \bar{C} - \lambda Cov(R-C, R_m)}{1 + r}$$

with $V = (1-t) < R-C >$, as before. Value maximization introduces a number of difficulties, which may be avoided by adopting the competitivity assumption of [2], under which λ is assumed constant to reach the Pareto-optimal solution. With this assumption the first-order conditions are:

$$(9a) \quad \frac{\partial V}{\partial p} + \mu(1-t)\left(-\frac{\partial \bar{R}}{\partial p} + \frac{\partial \bar{C}}{\partial p}\right) = 0$$

$$(9b) \quad \frac{\partial V}{\partial K} - 1 + \mu\left[s + (1-t)\frac{\partial \bar{C}}{\partial K}\right] = 0$$

$$(9c) \quad \mu[s K-(1-t)(\bar{R} - \bar{C})] = 0$$

where $\mu \geq 0$ is a Kuhn-Tucker multiplier and the price of capital was normalized and set equal to 1. The unregulated firm's choices are given by (9a, b) for $\mu = 0$. This model preserves the A-J certainty results (that regulation forces the choice of a lower price and a higher capital stock) under certain assumptions about the sign of the second derivatives of the value as a function of p and K. The effect of leverage is the same as above.

c) Other simultaneous production and financial equilibrium models: Such models are based on extensions of the Arrow-Debreu general equilibrium models to an economy with incomplete markets for state-contingent claims, by restricting the shape of the earnings function of the firm. They will be treated very briefly, since the resulting valuation models (other than the CAMP) are very restrictive in the case of regulated firms and have had very little impact on applied financial research.[7]

We consider for simplicity the case of insignificant regulatory lag, in which the value of the unlevered firm is equal to $(1-t) < N(u,W,K) >$. For the existence of a unanimously preferred simultaneous production equilibrium the earnings function N(u,W,K) must satisfy the so-called spanning condition ([2], [9], [20]): let J be the number of risky firms in the economy, and the subscript j denote the monopolist. Then the spanning condition implied that there exists a set of J coefficients a_h^J for the jth firm independent of (u,W) (but not necessarily of the capital stocks of the J firms) such that[8]

$$(10)\quad \frac{\partial N_j(u,W,K_j)}{\partial K_j} = \sum_{h=1}^{J} a_h^j N_h(u,W,K_h) + H_j(K_j) ,$$

where $H_j(K_j)$ is a known function, constant in (u,W).

If this condition holds then the value of the firm becomes a function of the entire matrix of coefficients $(a^m{}_h)$, m, h = 1,...,J. To prove this it is easiest to adopt the formulation of [9]. Let k = 1,.., K denote the set of possible values of (u, W) and Ω the J x J matrix $(a^m{}_h)$ of known coefficients. Following [2] and [30] we also denote by ω^i the K-vector of normalized marginal utilities of the ith consumer weighted by the probabilities of occurrence of each state, evaluated at the joint production - financial equilibrium. Then, we have, for any m = i,.., J

$$(11a)\quad (1-t)[\nabla N_{mk}]\ \omega^i = [(1-t)]$$

$$(11b)\quad [V_m](1+r) = (1-t)\ [N_{mk}]\ \omega^i$$

$$(11c)\quad [\nabla N_{mk}] = \Omega\ [N_{mk}] + H[1]$$

where H is a J x J diagonal matrix with $H_m(K_m)$ its mth element, $\nabla N_{mk} = (\frac{\partial N_{mk}}{\partial K_m})$, the brackets indicates an appropriately dimensioned matrix of the elements within the brackets, and (11c) is a matrix version of (10). Then,

$$(12)\quad V_j = \Omega^{-1}[1] - \frac{(1-t)}{1+r}\ \Omega^{-1}H[1].$$

In practice (10) has been applied under the following more restrictive version

$$(13)\quad N_j(u,W,K_j) = n_j(u,W)g_j(K_j) + h_j(K_j),$$

which implied (10) for the following values of Ω and H:

$$a_h^j = 0,\ h \neq j,\ a_j^j = \frac{g_j'(K_j)}{g_j(K_j)},\ H_j(K_j) = h_j'(K_j) - \frac{g_j'(K_j)}{g_j(K_j)}\ h_j(K_j).$$

Then the value V_j of the firm is equal to $\frac{g_j(K_j)}{g_j'(K_j)} - \frac{(1-t)}{1+r}$ $[\frac{h_j'(K_j)g_j(K_j)}{g_j'(K_j)} - h_j(K_j)].$

The analysis is almost identical when p and K are determined simultaneously, with the difference that the ex ante decision variable K_j becomes now a two-dimensional vector (p_j, K_j).

Equation (12) establishes the value of the firm in the absence of a regulatory constraint. As pointed out in [2] (p. 214), with free entry and perfect mobility of resources the RHS of (12) will be equal to K_j and the firm will not be able to generate rents for its owners. The key to this model and to all its variants ([2], [3], [9], [19], [20]) is the fact that the vector ω^1 is unaffected by a change in the capital stock K_j.

The analysis of such a model under a rate-of-return constraint has been examined in detail in [3] and will not be repeated here. It is sufficient to state that under "naive" rate-of-return regulation the AJ certainty results hold. If by contrast regulation takes place in a "sophisticated" manner so that the firm does not realize the link between pricing decisions and its own choice of capital then no technical inefficiency need result. Unfortunately, the practical difference between naive and sophisticated regulation is unclear.

The most serious drawback of these models is that the separability condition (13) is not satisfied in general with most commonly used production and demand specifications when both u and W are random ([30]). Consequently, and barring very respective assumptions, these capital market theories do not provide as yet valuation models for regulated firms.

As a general conclusion for the forward-looking regulatory models we note that they all share a number of characteristics: they are single-period models, whose multiperiod extension presents a number of difficulties. They do not preserve the MM cost-of-capital results except under very restrictive separability conditions. Finally, they do preserve most of the AJ certainty results under "acceptable" assumptions. In the next section it will be shown that under backward-looking regulation this last property disappears even though precise relations may be established between the values of the firm under backward and forward-looking regulation.

4. Backward-looking Regulation and the Value of the Firm

This type of regulation was described in detail in articles by Joskow ([17], [18]). The fundamental difference between the Joskow and the AJ - Gordon views of the regulatory process is that regulation in the former is based on observed past performance, rather than expected future performance as in the latter. The regulated firm, knowing this, adjusts its own performance in anticipation of the regulatory action.

In the Joskow model there are two regulatory constraints, a lower, as well as an upper, rate of return limit. Firms whose realized earnings approach the lower limit know that they have good chances of getting a rate increase. Firms whose earnings are in excess of the upper limit run the risk of a regulatory hearing initiated by the regulatory commission under the prodding of consumer interests. This hearing, in addition to being costly and time-consuming, will force the firm to earn below the upper limit, probably on the lower rate-of-return boundary. Between the two limits the firm will operate under a stable product price without any rate-of-return consideration ([17], pp. 133-134).

In such a world it would make sense for the firm to try to voluntarily limit its earnings in order not to exceed the upper limit. In Joskow's model the firm does this by voluntarily decreasing its output prices. An equally plausible way of reducing earnings is by operating above the minimum cost level. Although the efficiency implications of these two methods are very different, the results from the point of view of observed earnings and value of the regulated firm are similar.

Two special cases of the general Joskow model have also appeared in the literature. The first ([27], [28], [29]) assumes only an upper limit and the second [8] only a lower limit on the realized rate of return. Both can be treated within the context of the model developed in this section[9]. The general Joskow model of backward regulation is applicable only when p and K are simultaneously determined in the forward case, because when the earnings are equal to $N(u,W,K)$ there are no instruments available to raise them to the level defined by the lower limit on the rate of return. Hence, we shall assume that the before-tax earnings under forward-looking regulation are equal to $pQ-C(Q,W,K)$, where p and K are fixed.

Let $V_f(p,k) = (1-t) < pQ-C(Q,W,K) >$, where the appropriate valuation model of the previous section may be used. Likewise, denote by V_b the corresponding value of the backward-looking regulated firm with the same capital stock, initial output price, demand and technology. Instead of (3) we now have the following mechanism. Let s_1 and $s_2 > s_1$ denote the two limits on the realized rate of return. If the firm's earnings at minimum cost fall below s_1K then the output price is automatically adjusted to bring them to s_1K. If, on the other hand, these earnings are above s_2K then the firm reduces output price or does "cost-padding" to bring them to s_2K and avoid a rate hearing. Otherwise the price remains unchanged and production occurs at minimum cost. All adjustments are assumed "frictionless" (no lags).

We shall solve the valuation problem in detail for a single period first, and then we shall extend it recursively to more than one periods. The extension is difficult, principally because of the sequential pricing mechanism. A "period" is defined here as a time interval, during which at most one price change can take place. Hence, our single-period formulation is also valid for "long" time intervals of product price stability, such as the late fifties and early sixties.

The random variable $\tilde{X} = \lfloor p\,Q - C(Q,W,K)\rfloor\,(1-t)$ takes values in an interval on the real line, ,the distribution being assumed discrete (for simplicity and without loss of generality), i.e. $\tilde{X} \in [X_0, X_0 + m\Delta X]$, where $X_0 = \inf_{(u,W)} \lfloor p,Q - C(Q,W,K)\rfloor\,(1-t)$, and ΔX is the step size of the discrete distribution. Under the assumed regulatory mechanism the value V_b may be derived from $V_f(p,K)$ by using the theory of contingent claims prices.

We define the integers k_1 and k_2 so that $x_0 + k_i\,\Delta X = s_iK$, $i = 1, 2$, $k_1 < k_2$. The random returns accruing to the stockholders of the regulated firm are equal to s_1K for $\tilde{X} \varepsilon\, [X_0, X_0 + (k_1-1)\,\Delta X]$, to $\tilde{X}$ for $\tilde{X}\, \varepsilon [X, X_0 + k_1\,\Delta X$,

$X_0 + k_2\, \Delta X]$, and to s_2K for $\tilde{X} \in [X_0 + (k_2 + 1)\,\Delta X, X_0 + m\Delta X]$. They may be replicated by the following portfolio: a risk-free investment yielding s_1K after one period, a long position in European call options on the value $V_f(p,K)$, with an exercise price of s_1K, and a short position in European call options on $V_f(p,K)$ but with exercise price s_2K. These options respectively pay 0 when $\tilde{X} \leq s_1K$, and $\tilde{X} - s_1K$ otherwise, and 0 when $\tilde{X} \in [X_0, X_0 + k_2\Delta X]$ and $\tilde{X} - s_2K$ otherwise. By the arbitrage theory the value of the backward-looking regulated firm is equal to the value of the portfolio. If $O(S,Y,n)$ denotes the value of a European call option on equity whose value is S, with exercise price Y and which expires in n periods then we have

$$(14) \quad V_b = \langle s_1K \rangle + O(V_f, s_1K, 1) - O(V_f, s_2K, 1)$$
$$= \frac{s_1K}{1+r} + O(V_f, s_1K, 1) - O(V_f, s_2K, 1)$$

where $V_f = V_f\,(p,K)$ and r is the one-period riskless rate of interest.

Equation (14) is valid both before and after taxes, with suitable reinterpretation of the symbols. Since the valuation operator is linear, and the call price $O(S,Y,n)$ is linear homogeneous in (S,Y), it suffices to reinterpret V_f, s_1 and s_2 as before-tax value and limits on rate-of-return respectively.

Suppose now that the lower limit s_1K disapears and that only the upper limit s_2K exists as in [27], [28] and [29]. Then the earnings of the firm are $\tilde{X}$ for $\tilde{X} \in [X_0, X_0 + k_2\Delta X]$, s_2K otherwise. The value $\hat{V}_b$ becomes

$$(15) \quad \hat{V}_b = V_f - O(V_f, s_2K, 1).$$

With the same reasoning, if only the lower bound s_1K is kept as in [8] the value of the corresponding firm $\bar{V}_b$ is

$$(16) \quad \bar{V}_b = \frac{s_1k}{1+r} + O(V_f, s_1K, 1)$$

The well-known inequality ([21], p. 144) $O(S,Y,1) \geq \text{Max}\left[0, S - \frac{Y}{1+r}\right]$, when applied to (14) - (16) establishes clearly that $\bar{V}_b \geq V_b \geq \hat{V}_b$.

Suppose now that we wish to ascertain whether the AJ certainty results hold for backward-looking regulated firms given that they hold for forward-looking firms. Assume that we are in a value-maximizing world, and that p and K have both been determined by unconstrained maximization of $V_f - K$. Then, if p is kept constant at that level but K is varied in order to satisfy an expected profit rate-of-return constraint $E[\tilde{X}] \leq s_2K$ that is binding at the unconstrained optimum, it is easy to see that we get a larger capital stock for the forward-looking regulated firm than that derived by unconstrained value maximization. However, this excess capitalization does not hold for the backward-looking regulated firm that maximizes $V_b - K$ at the same output price. To see this we compute the quantities

$$\frac{\partial V_b}{\partial K} - 1 \text{ or } \frac{\partial \hat{V}_b}{\partial K} - 1 \text{ at the value of K, at which } \frac{\partial V_f}{\partial K} - 1 = 0.$$

As a specific example, consider $\frac{\partial \hat{V}_b}{\partial K}$ when the function $O(S,Y,n)$ is given by the well-known Black-Scholes [5] option pricing model. Then, if r is redefined to indicate the instantaneous rate of interest in continuous time, N_m ( ) is the cumulative standard normal distribution, and σ is the standard deviation of $\tilde{X}$, we have ([34], p. 24).

$$(17) \quad \frac{\partial \hat{V}_b}{\partial K} - 1 = s_2 e^{-r} N_m \left[\frac{\ln V_f - \ln(s_2 K) + r - \sigma^2/2}{\sigma} \right]$$

$$- N_m \left[\frac{\ln V_f - \ln(s_2 K + r + \sigma^2/2}{\sigma} \right]$$

Here s_2 is > e^r and the sign of the expression in (17) depends on the relative size of s_2 and σ. In fact, for any $s_2 > e^r$ it is easy to see that the "riskier" the firm becomes (the larger the σ) the more (17) declines and the larger the size of the backward-looking value-maximizing capital stock relative to that of the forward-looking firm. Hence, the certainty AJ results do not necessarily hold. A similar conclusion can be derived if the value of the firm is V_b, in combination with Black-Scholes option pricing.

As a final application of the one-period valuation theory for backward-looking regulation we examine the effect of capital structure upon the value of the regulated firm. Let V_{bL} denote this value for a levered firm with an exogenously given amount of debt D. The previous contingent claims analysis can be applied here as well, by redefining the k_i's as $k_i = \frac{s_i K - t(1+r)D - X_o}{\Delta X}$, i = 1,2, and noting that the value $< \tilde{X} >$ is given by the same expression as for the unlevered firm, but with the new k_i's replacing the old ones. Hence, we have

$$(18) \quad V_{bL} = tD + \frac{s_1 K}{1 r} + O(V_f,\ s_1 K - t(1+r)D, 1) - O(V_f,\ s_2 K - t(1+r)D, 1)$$

The interesting thing in (18) is that the positive effect of leverage on the value of the regulated firm exceeds the term tD even if we neglect the effect of leverage upon the choice of capital stock. While in forward regulation it is always true that $V_{f1}(p,K) = V_f(p,K) + tD$, by contrast it can be shown that $V_{bL}(p,K)$, is always > $V_b(p,K) + tD$ by showing that $O((S,Y_1 - \Delta,1) - O(S,Y_2 - \Delta,) > O(S,Y_1,1) - O(S,Y_2,1)$ for any pair (Y_1,Y_2) with $Y_1 < Y_2$ and any $\Delta > 0$.

5. Multiperiod Extension

As in the single-period case, we assume that the initial output price p_1 is given from historical or other factors, while the capital stock K is determined optimally at the beginning of the first period. The rate of return limits s_1 and s_2 are assumed fixed, and let (u_T, W_T) denote the value of the random factors during period T. Define, for

$Q_T = Q(p_T,u_T)$, $(1-t)\,[p_TQ_T - C(Q_T,W_T,K)] = X_T$

(19) $\omega_1 = \{(u_T,W_T) \mid X_T \geq s_1K\}$, $\bar{\omega}_1 = \{u_T, W_T) \mid (u_T,W_T) \notin \omega_1$

Then the following output pricing rule under negligible regulatory lag is adopter.

(20) $P_{T+1} = p_T$ if $(u_T,W_T)\ \varepsilon\bar{\omega}_1$, $P_{T+1} = \{p_{T+1} \mid p_{T+1}\,Q_{T+1} - C(Q_{T+1}, W_T, K) = s_1K\}$, for $(u_T, W_T)\ \varepsilon\omega_1$, where in (20) $Q_{T+1} = Q(p_{T+1}, u_T)$.

We shall also adopt two additional simplifying assumptions over and above those of the single period valuation. The first is that the firm operates in the region of positive marginal (with respect to price) earnings, which means that $p_{T+1} > p_T$ if $(u_T, W_T)\ \varepsilon\bar{\omega}_1$. The second is that the earnings follow a Markovian structure given the pricing rule (20), i.e. that p_{T+1} depends only on the revealed value of X_T and not on the individual values of u_T and W_T.

The valuation theory will be first developed in a two- period framework and then extended recursively to any number of periods. Let V_{bT}, T=1,2 denote the value of the backward-looking regulated firm at the beginning of period T. At the end of period 1 there are two possibilities: either X_1 is $\geq s_1K$, in which case $p_2=p_1$, or $X_1 < s_1K$, $p_2 > p_1$, and according to the Markovian assumption $p_2 = p_2(X_1)$, since the solution of the equation $(1-t)\,[p_2Q(P_2,u_1) - C(Q(p_2,u_1), W_1,K)] = s_1K$ depends only on X_1 for all $(u_1,W_1)\ \varepsilon\bar{\omega}_1$. According to the notation adopted in the previous section we define $V_{fT} = \langle X_T \rangle T$, T = 1,2, the value of the non-truncated earnings. However, while V_{f1} is unchanged, we have by contrast different values V_{f2} depending on whether $u_1\ W_1)\ \varepsilon\omega_1$ or $\bar{\omega}_1$. Let V^i_{f2}, i = 1,2 denote these two values with

(21) $V^i_{f2} = [\langle X_2 \rangle {}^2{}_1 \mid X_1 > s_1K]$, $V^2_{f2} = [\langle X_2 \rangle {}^2{}_1 \mid X_1 < s_1K]$. By the Markovian assumption $V^i_{f2}(X_1)$, i=1,2 are known functions of X_1 for given distributions of (u_2,W_2); such distributions at the beginning of period 1 are also assumed to depend only on X_1.

In this framework it is easy to develop expressions for V_{b1} based on contingent claims on X_1. For instance, it can be shown that

$$(22)\quad V_{b1} = \int_{X_0}^{s_1K} 0_{yy}\,[s_1K + V^2{}_{b2}(y)]\,dy + \int_{s_1K}^{s_2K} 0_{yy}\,[y + V^1{}_{b2}(y)]dy + \int_{s_2K}^{X_0+M} 0_{yy}\,[s_2K + V^1{}_{b2}(y)]dy$$

where we can substitute for $V^i_{b2}(y)$ from (14).

The above formula is applicable without reformulation to any number of periods, by replacing 1 and 2 by T and T+1 respectively in every subscript of V_b.

The analytical complexity of (22) stems from the conditional structure of the second-period rteturns, depending on the event of the occurrence of an output price change. Conversely, if such an event is not contemplated in valuation we have a straightforward and simple extension of (15a). Assume that we have an N-period horizon, and that the random factors (u,W) are intertemporally independent. Let also $V_f = V_f(p,K) = \sum_{T=1}^{N} < X_T \overset{T}{\underset{T-1}{>}}$, the value of the stream of cash flows if the self-imposed upper limit did not exist. If (15a) is valid in every single-period then under backward-looking regulation each period T contributes $< X_T \overset{T}{\underset{T-1}{>}} - O(V_{fT}, s_2K, T)$, where $V_{fT} = < X_T \overset{T}{\underset{T-1}{>}}$ and the sort option becomes a T-period European call. Hence, we have

$$(23) \quad \hat{V}_{bT} = V_f - \sum_{T=1}^{N} O(V_{fT}, s_2K, T)$$

This simple generalization of (15a) may have been applicable to the "long" periods of regulatory price stability of the fifties and early sixties.

6. Discussion and Conclusions

This paper has examined the impact of Joskow-type regulatory behavior or backward-looking regulation upon the value of the firm. This impact is qualitatively different from forward-looking regulation, that has been considered almost exclusively in the financial literature until now. The difference lies in the fact that, while forward-looking regulation affects the stream of earnings of the firm only through the decision parameters, backward-looking regulation also changes the distribution of the stream of earnings for given decision parameters. This change is tied-in directly into the regulatory process itself and depends upon exogenous or self-imposed upper and lower limits.

The expressions that were developed did not depend upon any particular valuation model. Section 3 contained a brief survey of the valuation models that have appeared in the financial literature in the context of forward-looking regulation. It was concluded that most of the certainty results were preserved in these forward-looking regulatory models.

Given now any single-period basic valuation model, the theory of arbitrage pricing was used to derive the single-period value of the backward-looking regulated firm. This value was then used in examining a number of well-established results. It was found that backward-looking regulation changes these results in significant ways. Thus, when used in connection with value maximization, it invalidates the AJ certainty results, and it introduces a systematic bias into the MM value of the levered firm theory. Hence, it is quite possible that the large and growing empirical evidence against the AJ certainty results reflects the existence of backward-looking regulation rather than the reasons cited in these studies ([4, note 25], [31]).

In section 5 the value of the firm under backward-looking regulation was derived for more than one periods. Most of the resulting expressions became quite complex, and some additional assumptions were necessary for the derivation. These difficulties are not peculiar to regulation, since the problem of multiperiod valuation of random income streams has been solved only under very restrictive assmuptions, as in [26] under the CAPM. Had similar assumptions been also adopted here it would have been possible to achieve considerable simplifications.

As a final remark it should be noted that backward-looking regulation has a number of disturbing efficiency and equity implications. Thus, the optimality of "cost-padding", already noted in [28], becomes evident. Similarly, the size of the regulator-controlled parameters s_1 and s_2 determines whether regulation confers capital gains or losses upon the firm's stockholders. Such questions were not examined in this paper, but they should form the object of further studies.

FOOTNOTES:

[1] Professor, Faculty of Administration, University of Ottawa. This research was partially supported under contract OSU80-00157 with the Canadian Department of Communications. An extended version was presented at the Conference on Telecommunications in Canada, Montreal, March 1981.

[2] This dual role raises a well-known problem of circularity, since the value of the rate base depends on the revenue that it generates, while these revenues, in turn, depend on the allowed rate of return and the rate base. See [32].

[3] Exceptions are in [3], [4] and [19], which will be discussed in detail.

[4] The earliest such model is probably the simple example presented by Myers [25]. Systematic studies of backward-looking regulation were in [27]. [28], [29] and (in a different context) [8].

[5] Otherwise, if D is selected endogenously we reach the MM corner solution of an all-debt firm as in [24].

[6] EG mention ([10], note 10) that K may be $\neq K_L$, but they assume linearity of earnings or treat the two capital stocks as equal (see also [11], p. 1153), while their analysis is accepted uncritically in [16], p. 807.

[7] For an exception see [4].

[8] This formulation of the spanning condition differs slightly from [9] because it follows the normalization procedure of [2] and [30].

[9] It should be noted, however, that in [8] it is clearly stated that their model is only an approximation to Joskow's. In [27], [28] and [29], on the other hand, the upper limit is interpreted in the spirit of the AJ certainty model. As Baron and Taggart pointed out in [3] this interpretation requires ex post lump-sum transfers of wealth from the firm to the consumers that are not observed in real life. However, these models make a lot of sense if the constraint is reinterpreted as a self-imposed upper limit in order to avoid a potentially more damaging regulatory intervention.

REFERENCES:

[1] Averch, H. and Johnson, L., Behavior of the Firm Under Regulatory Constraint, American Economic Review 52,5 (December 1962) 1053-1069.

[2] Baron, D.P., Investment Policy, Optimality and the Mean-Variance Model, Journal of Finance 34,1 (March 1979) 207-232.

[3] Baron, D.P. and Taggart, R.A., Regulation and the Investor-Owned Firm Under Uncertainty, Working Paper, Northwestern University (November 1976).

[4] Baron, D.P. and Taggart, R.A., A Model of Regulation Under Uncertainty and a Test of Regulatory Bias, The Bell Journal of Economics 8,1 (Spring 1977) 151-167.

[5] Black, F. and Scholes, M., The Pricing of Options and Corporate Liabilities, Journal of Political Economy 81,3 (May-June 1973) 637-654.

[6] Brennan, M., Valuation and the Cost of Capital for Regulated Utilities: Comment, Journal of Finance, 27,5 (December 1972) 1147-1149.

[7] Brigham, E.F. and Gordon, M.J., Leverage, Dividend Policy, and the Cost of Capital, Journal of Finance 23,1 (March 1968) 85-103.

[8] Burness, H.S., Montgomery, W.D. and Quirk, J.P., Capital Contracting and the Regulated Firm, American Economic Review 70,3 (June 1980), 342-354.

[9] Ekern, S. and Wilson, R., On the Theory of the Firm is an Economy with Incomplete Markets, The Bell Journal of Economics and Management Science 5,1 (Spring 1974) 171-180.

[10] Elton, E.J. and Gruber, M., Valuation and the Cost of Capital for Regulated Industries, Journal of Finance 26,3 (June 1971) 661-670.

[11] Elton, E.J. and Gruber, M., Valuation and the Cost of Capital for Regulated Industries: Reply, Journal of Finance 27,5 (December 1972) 1150-1155.

[12] Fuss, M., The Structure of Technology Over Time: a Model for Testing the Putty-Clay Hypothesis, Econometrica 45,8 (November 1977) 1797-1821.

[13] Gordon, M.J., Some Estimates of the Cost of Capital to the Electric Utility Industry, 1954-1957: A Comment, American Economic Review 56,5 (December 1967) 1267-1277.

[14] Gordon, M.J. and McCallum, J., Valuation and the Cost of Capital for Regulated Utilities: Comment, Journal of Finance 27,5 (December 1972) 1141-1146.

[15] Greenwald, B.C., Admissible Rate Bases, Fair Rates of Return and the

Structure of Regulation, Journal of Finance 35,2 (May 1980) 359-368.

[16] Jaffe, J.F. and Mandelker, G., The Value of the Firm Under Regulation, Journal of Finance 31,2 (May 1976) 801-813.

[17] Joskow, P.L., Pricing Decisions of Regulated Firms: a Behavioral Approach, The Bell Journal of Economics and Management Science 4,1 (Spring 1973) 118-140.

[18] Joskow, P.L., Inflation and Environmental Concern: Structural Change in the Process of Public Utility Regulation, Journal of Law and Economics 17, (October 1974) 291-327.

[19] Leland, H., Regulation of Natural Monopolies and the Fair Rate of Return, The Bell Journal of Economics and Management Science 5,1 (Spring 1974) 3-15.

[20] Leland, H., Production Theory and the Stockmarket, The Bell Journal of Economics and Management Science, 5,1 (Spring 1974) 125-144.

[21] Merton, R.C., The Theory of Rational Option Pricing, The Bell Journal of Economics and Management Science 4,1 (Spring 1973) 141-183.

[22] Meyer, R.A., Regulated Monopoly Under Uncertainty, Southern Economic Journal 45,4 (April 1979) 1121-1129.

[23] Miller, M. and Modigliani, F., Some Estimates of the Cost of Capital to the Electric Utility Industry, 1954-57, American Economic Review 56,3 (June 1966) 333-391.

[24] Modigliani, F. and Miller, M., Corporate Income Taxes and the Cost of Capital - A Correction, American Economic Review, 53,3 (June 1963) 433-443.

[25] Myers, S.C., A Simple Model of Firm Behavior Under Regulation and Uncertainty, The Bell Journal of Economics and Management Science, 4,1 (Spring 1973) 304-315.

[26] Myers, S.C. and Turbull, S., Capital Budgeting and the Capital Asset Pricing Model - Good News and Bad News, Journal of Finance 32,2 (May 1977) 321-332.

[27] Peles, Y. and Stein, J.L., The Effect of Rate-of-Return Regulation is Highly Sensitive to the Nature of Uncertainty, American Economic Review 66,3 (June 1976) 278-289.

[28] Perrakis, S., Rate-of-Return Regulation of a Monopoly Firm with Random Demand, International Economic Review 17,1 (February 1976) 149-162.

[29] Perrakis, S., On the Regulated Price-Setting Monopoly Firm with a Random Demand Curve, American Economic Review 66,3 (June 1976) 410-416.

[30] Perrakis, S., On the Technological Implications of the Spanning

Theorem, Canadian Journal of Economics 12,3 (August 1979) 501-511.

[31] Perrakis, S. and Zerbinis, J., An Empirical Analysis of Monopoly Regulation Under Uncertainty, Applied Economics 13,1 (March 1981) 109-125.

[32] Robichek, A., Regulation and Modern Finance Theory, Journal of Finance 33,3 (June 1978) 693-705.

[33] Ross, S.A., A Simple Approach to the Valuation of Risky Streams, Journal of Business 51,3 (July 1978) 453-476.

[34] Smith Jr., C.W., Option Pricing: A Review, Journal of Financial Economics 3,1/2 (January - March 1976) 3-52.